AF361757

3D/4D Geological Modeling for Mineral Exploration

3D/4D Geological Modeling for Mineral Exploration

Editors

Gongwen Wang
Lizhen Cheng
Nan Li
Weisheng Hou

MDPI • Basel • Beijing • Wuhan • Barcelona • Belgrade • Manchester • Tokyo • Cluj • Tianjin

Editors

Gongwen Wang
School of Earth Sciences and
Resources
China University of
Geosciences
Beijing
China

Lizhen Cheng
Research Institute in Mines
and the Environment (IRME)
University of Québec in
Abitibi-Témiscamingue
Rouyn-Noranda
Canada

Nan Li
Big Data Center of Mineral
Resources
Institute of Mineral
Resources, Chinese Academy
of Geological Sciences
Beijing
China

Weisheng Hou
School of Earth Sciences and
Engineering
Sun Yat-sen University
Guangzhou
China

Editorial Office
MDPI
St. Alban-Anlage 66
4052 Basel, Switzerland

This is a reprint of articles from the Special Issue published online in the open access journal *Minerals* (ISSN 2075-163X) (available at: www.mdpi.com/journal/minerals/special_issues/GMME).

For citation purposes, cite each article independently as indicated on the article page online and as indicated below:

LastName, A.A.; LastName, B.B.; LastName, C.C. Article Title. *Journal Name* **Year**, *Volume Number*, Page Range.

ISBN 978-3-0365-7436-3 (Hbk)
ISBN 978-3-0365-7437-0 (PDF)

Contents

Editorial

Editorial for Special Issue "3D/4D Geological Modeling for Mineral Exploration"

Gongwen Wang

School of Earth Sciences and Resources, China University of Geosciences Beijing, Beijing 100083, China; gwwang@cugb.edu.cn

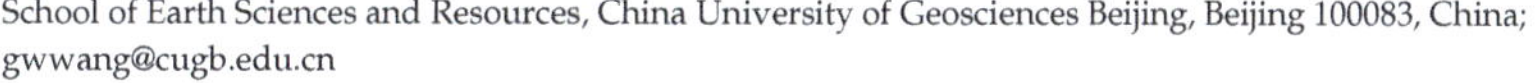

Citation: Wang, G. Editorial for Special Issue "3D/4D Geological Modeling for Mineral Exploration". *Minerals* **2023**, *13*, 198. https://doi.org/10.3390/min13020198

Received: 12 January 2023
Accepted: 18 January 2023
Published: 30 January 2023

With the development of high-precision geological observation technology, in situ mineral microanalysis technology, isotope geochemical analysis technology, deep geophysical exploration technology, deep drilling, real-time mining, remote sensing high-resolution hyperspectral image technology, and supercomputer and industrial intelligence, geoscience has entered an era of big data and artificial intelligence in the 21st century. Three-dimensional/four-dimensional (3D/4D) geoscience modeling with the multi-disciplinary intersection of geosciences has been used as the basis for mineral exploration and the extraction of geosciences information for mineral resources assessment. The European Geosciences Union launched the second phase of the OneGeology plan using 3D/4D modeling, artificial intelligence technology, and a big data methodology at the Resources for Future Generations conference (RFG2018) [1].

Three-dimensional/four-dimensional geological modeling is a key technology and methodology for geologists to understand geological events and quantitatively analyze multiscale metallogenic models for mineral exploration. The Special Issue aims to improve decision-making processes using 3D/4D geological modeling for mineral exploration and multiple innovative methodologies and technologies (e.g., conventional explicit and implicit modeling, real-time mining and 5G+ information technology, artificial intelligence decision-making, 3D/4D simulation, and digital twin). The geological concept model can be quantitatively analyzed by typical deposit research with an exploration engineer; thus, 3D/4D models can be built, simulated, and integrated via multisource geosciences datasets or big data from the field survey and the analysis of geosciences methods. Constructing 3D/4D certainty models for mineral exploration using multiscale and multisource datasets can be challenging [1,2]; mineral resource assessment and environment protection are associated with regional mining development and strategic planning.

Lv et al. [2] used GIS technology to integrate geological information, geophysical, geochemical, and remote sensing data and then applied RBFLN model in-depth learning to carry out two-dimensional metallogenic prediction and finally successfully delineated favorable 3D target zones. It can be seen that the deep metallogenic prediction method based on machine learning provides a scientific basis for the exploration and deployment of mineral resources. Mineral resource prediction and evaluation methods are developing toward the direction of model integration and intelligent information analysis. In addition to 3D geological modeling and metallogenic prediction of solid minerals, Zhang et al. [3] applied 3D Structural Modeling (3D SM) and Joint Geophysical Characterization to hydrocarbon reservoir characterization. The results show that this method is helpful in better understanding the structural and stratigraphic characteristics of the reservoir, the spatial distribution of associated facies and the petrophysical properties to conduct reliable reservoir characterization. Liu et al. [4] discussed the mineralization and oxidation transformation process of Shangfanggou molybdenite in the supergene stage by using three-dimensional (3D) multi-parameter geological modeling and microanalysis. Meanwhile, from macro to micro, the temporal–spatial–genetic correlation and exploration constraints are also established through the 3D geological modeling of industrial Mo orebodies and Mo oxide orebodies. This Special Issue also includes a study on Micro-Mechanisms and Implications of Continental Red Beds by He et al. [5].

The method for deep targeting based on a stepped metallogenic model and the method for predicting the deep mineral resource potential of gold deposits based on the shallow resources of ore-controlling faults, Song et al. [6] predicted five deep prospecting target areas of the Jiaojia fault and Sanshandao fault, and the total gold resources of three ore-controlling faults, Sanshandao, Jiaojia, and Zhaoping, are about 3377–6490 tons at −5000 to −2000 m. Wu et al. [7] proposed the bi-Coons interpolation surface modeling method for an ore body model based on a set of cross-contour polylines (geological polylines interpreted from the raw geological sampling data) in his article, which can be used to effectively recover the complex ore body model from a set of cross contours polylines, providing a new technology for the establishment of the ore body model. With the development of deep prospecting, Short-Wave Infrared-spectrum 3D alteration mapping has been applied. With the rise of artificial intelligence, the combination of machine learning and geological big data has become a hot issue in the field of 3D Mineral prospectivity modeling (3DMPM) [8]. On the basis of quantitative extraction of metallogenic characteristics, Meng et al. [8], Kong et al. [9], and Yu et al. [10] constructed the deep 3D geological model of the Xuancheng-Mashan area and the deep geochemical model of the Zaozigou gold deposit, and then adopted the machine learning method to predict the deep quantitative mineral resources, and evaluated their effects. The results indicate that the machine learning method has good performance in the quantitative prediction of deep mineral resources, which is worthy of popularization and application in 3D quantitative metallogenic prediction. Li et al. [11] combined 3D spectral modeling and 3D geological modeling techniques to establish the altered mineral model and the multi-parameter model of the ore body of the Zhaoxian gold deposit in the northwestern Jiaodong Peninsula and then extracted and synthesized the metallogenic information to analyze the exploration targeting.

Abbassi et al. [12] developed a 3D statistical tool to extract geological features from inverted physical property models based on the synergy between independent component analysis and continuous wavelet transform. Multiple 3D geophysical images are also automatically interpreted by a hybrid Spectral Feature Subset Selection (SFSS) algorithm based on a generalized supervised neural network algorithm to reconstruct limited geological targets from 3D geophysical maps. Therefore, 3D geological and geophysical modeling are key exploration criteria for mineral resources [1,12].

Generally, 4D numerical simulations based on 3D geological models are used to extract the 3D exploration criteria for 3D targeting and mineral resources assessment [1]. In addition, machine learning and high-grade geostatistics can be used to extract 3D exploration criteria. It is hoped that this Special Issue is a valuable learning and research resource for anyone interested in the study of 3D/4D geological modeling research for mineral exploration and that it will serve as the basis for further research, including real-time mining and wisdom mine.

Funding: National Key R&D Program: 2022YFC2903604; No.51 of Shandong Provincial Department of Natural Resources. Provincial geological exploration project in 2022.

Conflicts of Interest: The author declares no conflict of interest.

References

1. Wang, G.; Zhang, Z.; Li, R.; Li, J.; Sha, D.; Zeng, Q.; Pang, Z.; Li, D.; Huang, L. Resource prediction and assessment based on 3D/4D big data modeling and deep integration in key ore districts of North China. *Sci. China Earth Sci.* **2021**, *51*, 1594–1610. [CrossRef]
2. Lv, X.; Yang, W.; Liu, X.; Wang, G. Applications of Radial Basis Functional Link Networks in the Exploration for Lala Copper Deposits in Sichuan Province, China. *Minerals* **2022**, *12*, 352. [CrossRef]
3. Zhang, B.; Tong, Y.; Du, J.; Hussain, S.; Jiang, Z.; Ali, S.; Ali, I.; Khan, M.; Khan, U. Three-dimensional structural modeling (3D SM) and joint geophysical characterization (JGC) of hydrocarbon reservoir. *Minerals* **2022**, *12*, 363. [CrossRef]

4. Liu, Z.; Zuo, L.; Xu, S.; He, Y.; Wang, C.; Wang, L.; Yang, T.; Wang, G.; Zeng, L.; Mou, N. 3D Multi-Parameter Geological Modeling and Knowledge Findings for Mo Oxide Orebodies in the Shangfanggou Porphyry–Skarn Mo (–Fe) Deposit, Henan Province, China. *Minerals* **2022**, *12*, 769. [CrossRef]
5. He, W.; Yang, Z.; Du, H.; Hu, J.; Zhang, K.; Hou, W.; Li, H. Micro-Mechanisms and Implications of Continental Red Beds. *Minerals* **2022**, *12*, 934. [CrossRef]
6. Song, M.; Li, S.; Zheng, J.; Wang, B.; Fan, J.; Yang, Z.; Wen, G.; Liu, H.; He, C.; Zhang, L. A 3D Predictive Method for Deep-Seated Gold Deposits in the Northwest Jiaodong Peninsula and Predicted Results of Main Metallogenic Belts. *Minerals* **2022**, *12*, 935. [CrossRef]
7. Wu, Z.; Bi, L.; Zhong, D.; Zhang, J.; Tang, Q.; Jia, M. Orebody Modeling Method Based on the Coons Surface Interpolation. *Minerals* **2022**, *12*, 997. [CrossRef]
8. Meng, F.; Li, X.; Chen, Y.; Ye, R.; Yuan, F. Three-Dimensional Mineral Prospectivity Modeling for Delineation of Deep-Seated Skarn-Type Mineralization in Xuancheng–Magushan Area, China. *Minerals* **2022**, *12*, 1174. [CrossRef]
9. Kong, Y.; Chen, G.; Liu, B.; Xie, M.; Yu, Z.; Li, C.; Wu, Y.; Gao, Y.; Zha, S.; Zhang, H. 3D Mineral Prospectivity Mapping of Zaozigou Gold Deposit, West Qinling, China: Machine Learning-Based Mineral Prediction. *Minerals* **2022**, *12*, 1361. [CrossRef]
10. Yu, Z.; Liu, B.; Xie, M.; Wu, X.; Kong, Y.; Li, C.; Chen, G.; Gao, Y.; Zha, S.; Zhang, H. 3D Mineral Prospectivity Mapping of Zaozigou Gold Deposit, West Qinling, China: Deep Learning-Based Mineral Prediction. *Minerals* **2022**, *12*, 1382. [CrossRef]
11. Li, B.; Peng, Y.; Zhao, X.; Liu, X.; Wang, G.; Jiang, H.; Wang, H.; Yang, Z. Combining 3D Geological Modeling and 3D Spectral Modeling for Deep Mineral Exploration in the Zhaoxian Gold Deposit, Shandong Province, China. *Minerals* **2022**, *12*, 1272. [CrossRef]
12. Abbassi, B.; Cheng, L.Z.; Jébrak, M.; Lemire, D. 3D Geophysical Predictive Modeling by Spectral Feature Subset Selection in Mineral Exploration. *Minerals* **2022**, *12*, 1296. [CrossRef]

 *minerals*

Article

Applications of Radial Basis Functional Link Networks in the Exploration for Lala Copper Deposits in Sichuan Province, China

Xiumei Lv *, Wangdong Yang, Xiaoning Liu and Gongwen Wang

School of Earth Sciences and Resources, China University of Geosciences Beijing, Beijing 100083, China; yangwdcugb@163.com (W.Y.); 15605219798@163.com (X.L.); gwwang@cugb.edu.cn (G.W.)
* Correspondence: lvxm163@163.com; Tel.: +86-18810616595

Abstract: The Lala copper area in Huili County, Sichuan Province, China, is favored by superior regional metallogenic geological conditions due to its location in an extremely important copper–iron metallogenic belt in southwest China, and it has witnessed the formation of a series of unique iron–copper deposits following the superposition of multiple tectonic events. In recent years, major mineral exploration breakthroughs have been achieved in the deep and peripheral zones of this area. Using the Lala copper mining area in Sichuan as an example, this paper describes metallogenic prediction research carried out based on multivariate geoscience information (geological information, geophysics, geochemistry, and remote sensing data) and the application of geographic information system (GIS) technology and the radial basis function neural network (RBFLN) model. The five specific aspects covered in this paper are as follows: (1) we collected geology–geophysics–geochemistry remote sensing data and other information, adopted GIS technology to extract multivariate geoscience ore-forming anomaly information, and established a geoscience prospecting information database; (2) we applied the RBFLN algorithm for information on integrated analysis of ore-forming anomalies in the study area; (3) we applied a statistical method to divide the threshold value to delineate favorable ore-prospecting target areas; (4) we applied three-dimensional (3D) visualization technology, through which sample assistance was verified, to evaluate the performance of the RBFLN model; and (5) the results revealed that the RBFLN model can integrate multivariate and multi-type geoscience information and effectively predict metallogenic prospective areas and delineate favorable target areas. The metallogenic prediction method based on RBFLN technology provides a scientific basis for the exploration and deployment of minerals in the study area. It is obvious that the methods to predict and evaluate mineral resources are developing towards model integration and information intelligent analysis.

Keywords: GIS; multivariate geoscience datasets; RBFLN; metallogenic prospective area

Citation: Lv, X.; Yang, W.; Liu, X.; Wang, G. Applications of Radial Basis Functional Link Networks in the Exploration for Lala Copper Deposits in Sichuan Province, China. *Minerals* **2022**, *12*, 352. https://doi.org/10.3390/min12030352

Academic Editor: Amin Beiranvand Pour

Received: 20 January 2022
Accepted: 10 March 2022
Published: 15 March 2022

Publisher's Note: MDPI stays neutral with regard to jurisdictional claims in published maps and institutional affiliations.

1. Introduction

Mineral resources play a significant role in China's economic development. At present, China is relatively limited in terms of mineral resource storage, and will especially be reliant on the importation for copper for a period in the future. This has affected China's industrial development to a large extent. Therefore, scientific predictions and evaluations of potential mineral resources are essential and serve as a key guarantee for the sustainable development of China's national economy. Currently, resource prediction and evaluation has become a research focus in the field of mineral exploration [1].

Geostatistics is a branch of statistics established by French statistician G. Matheorn. Based on the theory of regionalized variables and with variation functions as a tool, it is a science that explores natural phenomena which occur both at random and with a certain structure in spatial distribution [2]. It initially targeted the single spatial parameter statistical model as a focus, which gave rise to linear geostatistics [2–4]. With the development of spatial parameters that tend to be more complex and diverse, nonlinear geostatistics [5,6] have emerged when linear geostatistics failed to provide solutions.

Since the 1970s, the development of methodological systems of regional metallogenic prediction has accelerated. It mainly covers three directions, two of which are regional metallogenic predictions based on the research of geological metallogenic theory and one which is based on the research of metallogenic dynamics theory [7]. The regional metallogenic prediction method adopted in this study relies on comprehensive geological research as the foundation, deposit models or geological anomalies as the bases, the computer as a means of research, and comprehensive quantitative analysis of multivariate information, such as geology, geophysics, geochemistry, and remote sensing data, as key approaches [8–10].

GIS technology, a computer system which integrates collection, storage, management, analysis, display, and application, is a general technology for analyzing and processing multivariate geographic data [11]. In metallogenic prediction, its application has advantages such as the integrated management of multivariate geoscience data, data space simulation analysis, and data quantification [12]. In recent years, the methodology system of metallogenic predictions based on GIS technology has been continuously deepened and developed. In the late 1980s, the Geological Survey of Canada used GIS technology for the potential mapping of mineral resources, mineral exploration, and mineral resource evaluation [13]. In 1988, Wyborn et al., from Australia, adopted GIS technology to evaluate mineral resource potentials [14]. In 2012, Silva et al. applied the ArcGIS-SDM fuzzy logic method to generate mineral exploration maps and evaluate potential exploration areas [15,16]. Additionally, Li et al. successfully developed the GeoCube software by integrating mathematical modeling methods [17] and quantitatively extracted and integrated 3D geoscience prospecting information of the Luanchuan mining area in Henan Province with the help of the software in question [18]. Combining GIS resource evaluation technology and 3D modeling technology, Wang et al. extracted and integrated comprehensive metallogenic information from the zones at different depths in the mining area [19].

Artificial neural networks (ANNs) have been widely applied in regional metallogenic prediction as a nonlinear classification method [20]. It classifies and recognizes data by imitating human biological neurons with the capacity to process complex nonlinear spatial datasets. To date, a wide range of ANNs has been developed, such as RBFLN, generalized regression neural networks (GRNNs), and probabilistic neural networks (PNNs) [21–23]. Previous studies have revealed that RBFLN is superior in metallogenic prediction.

Based on the metallogenic geological conditions and the metallogenic model of the Lala copper concentration area in Sichuan (Figure 1), this paper describes how multivariate geoscience ore-forming information was integrated with the help of GIS technology and the use of the RBFLN model. Predictive prospective outcomes and superimposed verification results of ore body models indicated that this method is capable of rendering theoretical bases and practical suggestions for mineral prospecting effort at deep zones in this area.

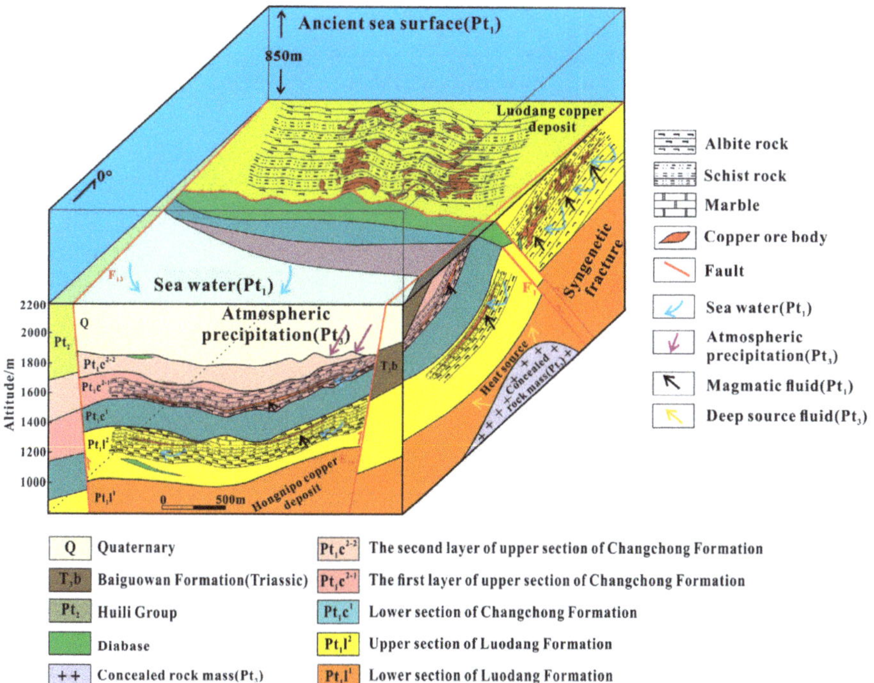

Figure 1. Geological model for prospecting predictions of the Lala copper deposit (modified from [24]).

2. Geological Setting

2.1. Regional Geology

The Lala region is located on the eastern edge of the midsection of the Kangding–Yunnan Axis of the Yangtze Platform which extends from north to south, and north of the Huili–Eastern Sichuan Aulacogen, which stretches from east to west (Figure 2). Spreading across the border between China and Vietnam, the Kangding–Yunnan copper belt lies on the western fringe of the Yangtze Plate, with a length of about 300 km. A famous iron-oxide copper gold (IOCG) metallogenic area [25,26], the province has more than 50 IOCG deposits, the most representative of which are the Luodang Copper Mine and Hongnipo Copper Mine (Figure 3).

The strata in this region developed from the Early Proterozoic Erathem to the Cenozoic Erathem. Except for the Ordovician–Carboniferous strata, all other strata can be found here, with those of the Proterozoic Erathem and the Mesozoic Erathem most developed. Stratigraphic combinations in this region mainly cover the Pre-Sinian System, the Sinian to Silurian Systems, the Permian System, the Triassic to Cretaceous Systems, and the Cenozoic Erathem [24]. The Hekou Group and the Huili Group are the major ore-bearing strata, whereas the industrial ore bodies are mainly hosted in the middle and lower parts of the Luodang Subgroup of the Hekou Group.

This region is typical of frequent and intense magmatic activities, which feature multiple cycles and multiple stages—the major stages are Jinning, Chengjiang, Variscan, and Indosinian Periods. Due to its unique tectonic–magmatic conditions and relatively developed fluid activities, this region is abundant in metal minerals [24].

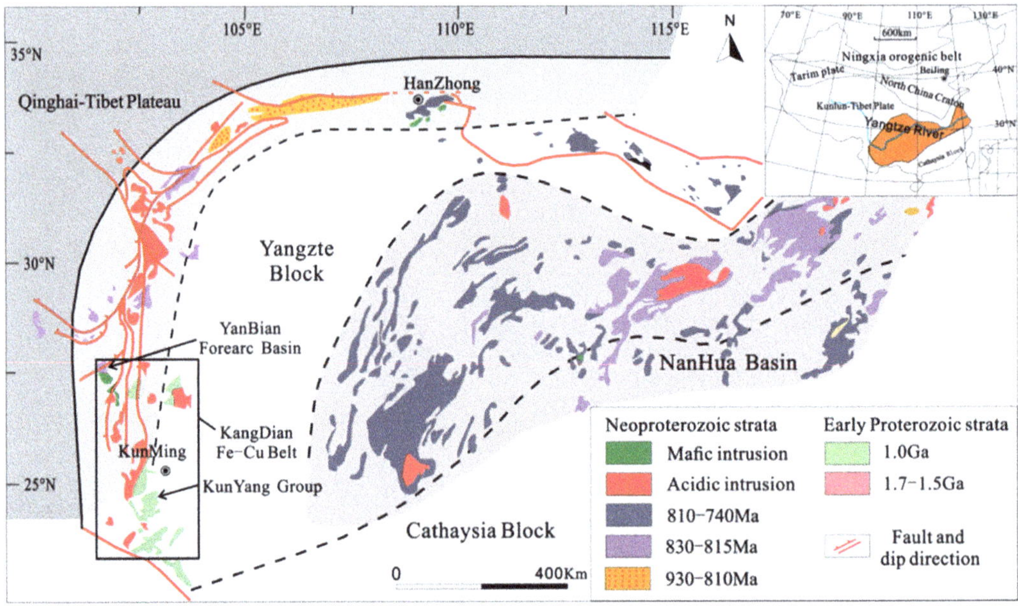

Figure 2. Geotectonic location of the Lala area (modified from [24]).

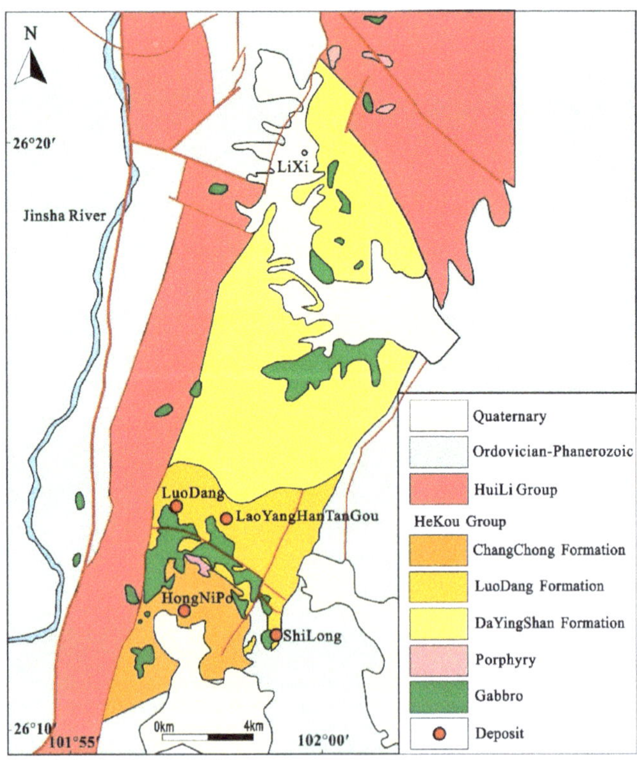

Figure 3. Distribution map of copper deposits in the Lala area (modified from [24,27]).

2.2. Deposit Geology

2.2.1. Luodang Copper Deposit

Comprising 32 ore bodies, the Luodang copper deposit is 1960 miles long from east to west and 900 miles wide from north to south, covering an area of 1.76 km^2. The upper section of the volcanic sedimentary cycle—Luodang formation complex, in the middle of the Hekou Group—is the major ore-bearing stratum. The ore body is stratiform, stratiform-like, and lentoid, and is generated in an overlapping-imbricate shape [28]. The attitude of the ore body, basically resembling that of the surrounding rocks, is controlled by lithology and stratum position and fluctuates with the folds of the surrounding rocks. The ore body presents an overall trend from east to west and tilts southward with an angle of 15°–40° [29].

2.2.2. Hongnipo Copper Deposit

The Hongnipo deposit sits about 5 km south of the Luodang deposit, with the ore body hosted in the Hekou Group strata in the Early Proterozoic metamorphic–volcanic sedimentary sequence and controlled by stratum position and lithology to a large extent [30]. The Tianshengba Group is the primary ore-bearing stratum, followed by the Luodang Group. The ore body is stratiform, stratiform-like, and podiform, and is generated in a bedding mode. The ore-bearing surrounding rocks mainly comprise carbonaceous quartz albitite, dolomitic quartz albitite, and quartz albitite at the bottom of the lower section of the Tianshengba Group.

3. Radial Basis Functional Link Networks

RBFLN, proposed by Broomhead in 1988 [31], is a kind of ANN that can process complex nonlinear spatial datasets. RBFLN is of great importance in dealing with the nonlinear relationship between known deposits and evidence factor layers. Taking the known deposits and non-deposits as training samples [32–34], RBFLN identifies the RBF network relationship between these samples and evidence factors as a nonlinear neural network algorithm (NNA), comprising a three-layer feedforward structure: input layer, hidden layer, and output layer [35,36] (Figure 4). The RBFLN structure includes an input layer composed of N nodes which receive the input feature vector x and a hidden layer comprising M neurons, each of which is represented by a Gaussian RBF. Each neuron in the hidden layer receives the input feature vector x and outputs the value y. If x^q is inputted into the mth neuron, the output value y_m^q can be expressed as [37]:

$$y_m^q = e^{[-\|x^q - v^m\|^2 / 2\sigma_m^2]}, \ 0 < y \leq 1 \tag{1}$$

where the value q ranges from 1 to N (N is the number of training samples), and the value M ranges from 1 to M (M is the number of neurons in the hidden layer). x^q represents the nth input feature vector, v^m refers to the center of the mth RBF in the hidden layer, which corresponds to the maximum likelihood point of the RBF, and σ^m represents the width or spread function of the mth neuron [34,38].

The overall output feature vector z contributes to the linear combination of output weights. If the abovementioned y_m^q is output to the jth output neuron, the formula can be expressed as [37]:

$$z_j^q = \left[\frac{1}{M+N}\right] \left\{ \left[\sum u_{mj} \times y_m^q\right] + \sum [w_{nj} \times x + b_j] \leftarrow t \right\} \tag{2}$$

where u_{mj} represents the synaptic weight between the hidden layer and the input layer, whereas w_{nj} represents the synaptic weight between the input layer and the output layer. Additionally, the constant b_j is added to the formula. The two synaptic weights were repeatedly modified through iteration processes until the output layer Z approached the target T.

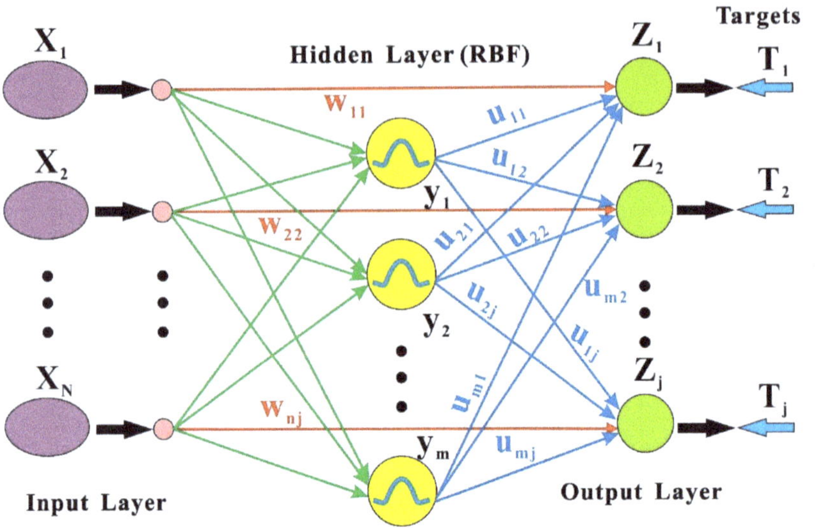

Figure 4. Architecture diagram of RBFLN (modified from [34]).

4. Extraction and Integration of Geoscience Datasets

GIS technology was used to extract and process information on ore-forming anomalies based on the collected data of mineral geology (1:50,000), aeromagnetic anomalies (1:50,000), copper geochemistry in the Huili–Huidong area, comprehensive anomalies of chemical elements, and the remote sensing images presented in this paper.

The Beijing 54 projection coordinate system (PCS) was adopted in datasets, and the storage format was a grid (unit: m).

The information extraction of multivariate geoscience datasets in the study area is shown below:

(1)   Geological Information Extraction and Interpretation

Using the mineral geological map (1:50,000) as a reference, we used MorPAS and ArcGIS software to extract four kinds of information on ore-forming anomalies: strata configuration entropy (Figure 5a), tectonic fracture isodensity (Figure 5b), tectonic fractal dimension (Figure 5c), and rock buffering (Figure 5d). Specifically, entropy can reflect the complexity of the structure. For instance, strata configuration entropy reveals the complexity of strata. There is a positive correlation between strata complexity and strata configuration entropy, and the latter changes with changes in the former [39]. From the perspective of space, tectonic fracture isodensity can be used to construct the development degree of information on anomalies in certain areas, which represents the cumulative sum of the tectonic length in a specific unit grid. Tectonic fractal dimension, an indicator to evaluate the complexity of fault structure, is positively correlated with the metallogenetic probability and has advantages over tectonic fracture isodensity. Magma intrusion is intense in this study area, and some magmatic rocks are occurrence ores. In addition, ores can be produced through magmatic rocks' contact metasomatism with the surrounding rocks; thus, the peripheral parts of magmatic rocks are also favorable metallogenic sites.

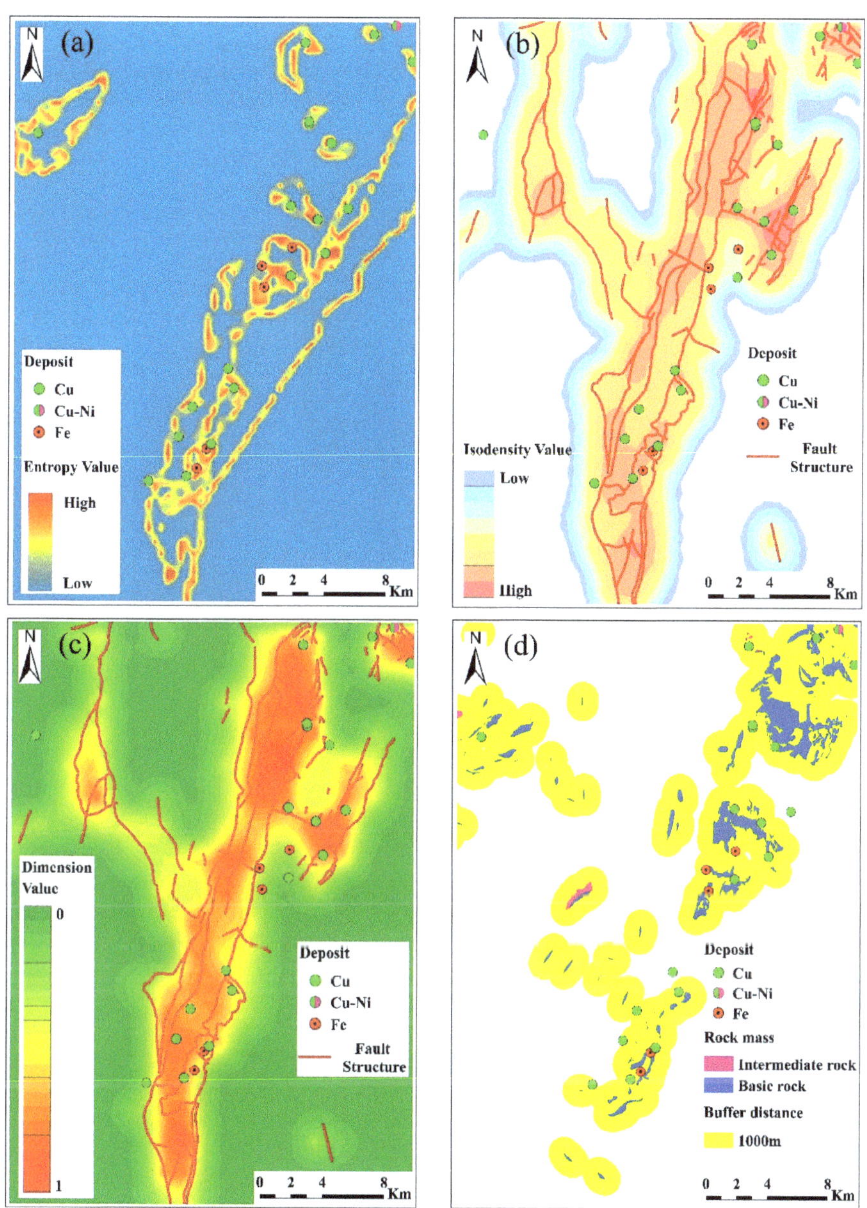

Figure 5. (**a**) Formation combination entropy; (**b**) isodensity map of structural faults; (**c**) dimension value map of structural faults; (**d**) rock mass buffer distribution map.

(2)   Geophysical Information Extraction and Interpretation

According to the aeromagnetic anomaly data (1:50,000), information on the aeromagnetic ore-forming anomalies of the study (Figure 6) area was extracted. In our research, we collected physical property data measured by our predecessors in this study [26], providing an important reference for interpreting magnetic anomalies. From the aeromagnetic isoline map, most deposits were located in the zero-line area at the junction of positive and negative anomalies. In terms of the magnetic field characteristics, we found that the gentle negative magnetic fields in the study area were a comprehensive reflection of a weak magnetic basement and sedimentary cover; the strong magnetic arched, beaded, and disordered magnetic anomalies were a reflection of basic volcanic rocks; and some relatively regular strong magnetic anomalies were a reflection of basic and intermediate-basic intrusive rocks.

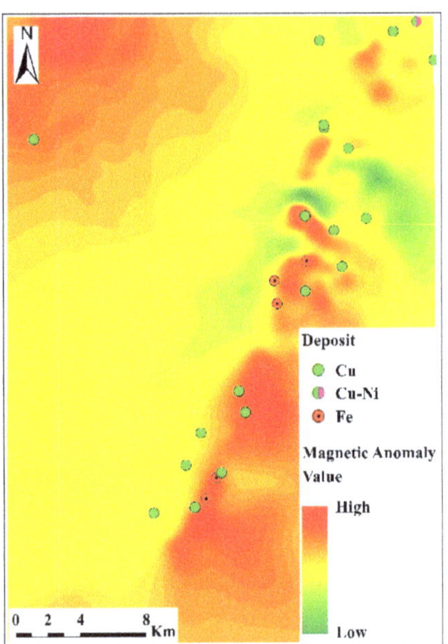

Figure 6. Aeromagnetic anomaly map.

(3)   Geochemical Information Extraction and Interpretation

The Cu element anomaly map (Figure 7a) and the comprehensive anomaly map (Figure 7b) of Cu–Co–Mo elements in this study area were provided by the Development and Research Center of the National Geological Archives of China. The Cu anomaly directly indicated the existence of Cu deposits or mineralization, and the comprehensive Cu–Co–Mo anomaly indirectly indicated the existence of Cu deposits or mineralized zones.

(4)   Remote Sensing Information Extraction and Interpretation

Different surface features of the earth display various spectral characteristics, and the altered surrounding rocks characterizing the existence of deposits and mineralized zones are no exception. Remote sensing technology can detect the altered surrounding rocks (if any) exposed on the earth's surface. The abnormal data of remote sensing alteration information are also an important basis for indicating mineralization. The distribution map of iron-stained alterations (Figure 8a) and remote sensing tectonic information in this study area were provided by the Development and Research Center of the National Geological Archives of China. In ArcGIS, we first divided the study area into grids, in accordance with a certain grid size. After that, we counted the total degree of linear fracture in each grid, and we assigned this value to the center point of the grid. Finally, we spatially interpolated the point data, and we obtained the isodensity map of the line–ring structure (Figure 8b).

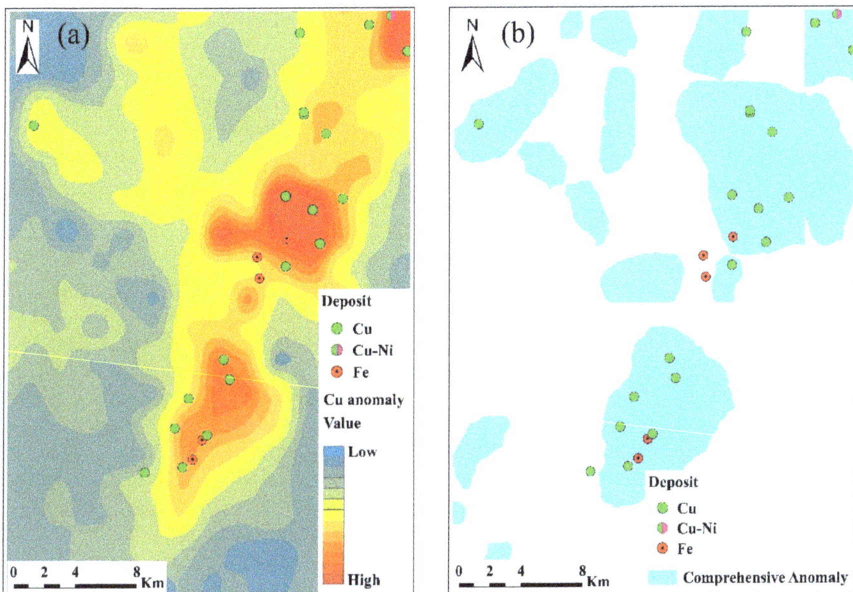

Figure 7. (**a**) Map of Cu anomalies; (**b**) map of comprehensive anomalies.

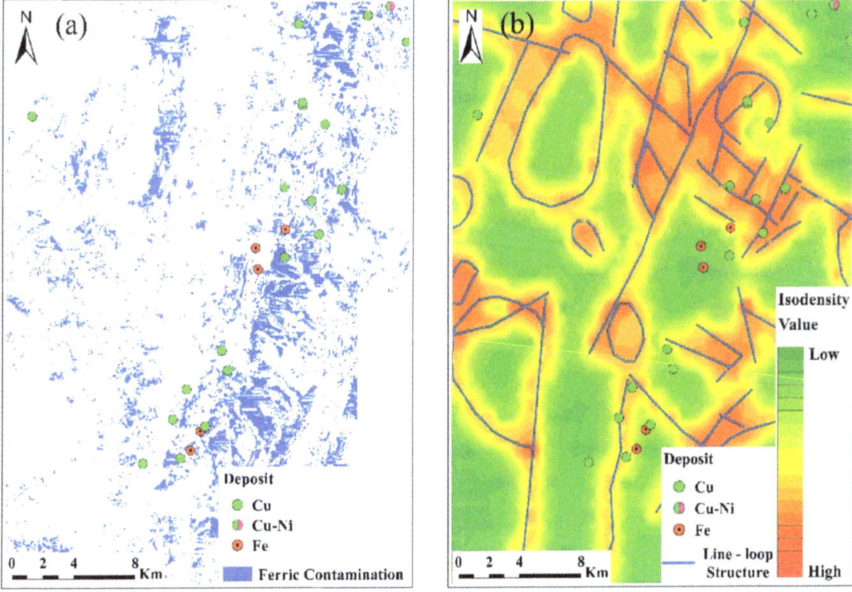

Figure 8. (**a**) Distribution map of iron-stained alteration information; (**b**) iso-density map of line–loop structure.

5. Application of RBFLN in the Lala Copper Deposit

5.1. Training and Classification

According to the aforementioned geology–geophysics–geochemistry–remote sensing ore-forming information, we used GeoXplore 5.1 software, the SDM module of ArcGIS 10.5

software, which were both developed by Environmental Systems Research Institute (ESRI), and the RBFLN model to perform metallogenic prospect predictions in the study area.

Twenty-five deposit sites and twenty-four non-deposit sites were involved in this research. A common way to evaluate the trained RBFLN was to use one-third of the training sites for verification and use the rest of the training sites to evaluate the network performance. Therefore, seven deposits and eight non-deposits were selected for verification, and eighteen deposits and sixteen non-deposits were chosen for training (Figure 9).

The nine abovementioned geoscience layers were combined by rasterization and reclassification, thus forming a unique grid. We divided the study area into a 100 m × 100 m grid scale, and the cumulative number of grids was 33,864. Each pixel in the grid data comprised feature vectors X (X_1, X_2, X_3, X_4, X_5, X_6, X_7, X_8, and X_9) in the nine-dimensional space, which served as the input layer of the RBFLN model.

In the training stage, the number of RBFs and the number of iterations were changed to obtain the minimum sum of squared error (SSE) between the output result and the target vector [34]. SSE represented the mismatch degree between the output result and the target vector, and a smaller value contributed to a higher matching degree. With the gradual increase in the number of iterations, SSE showed a downward trend and tended to be stable (Figure 10). When the number of iterations increased by more than 160, the network seemed to be affected by over-learning, which led to overfitting [32,34].

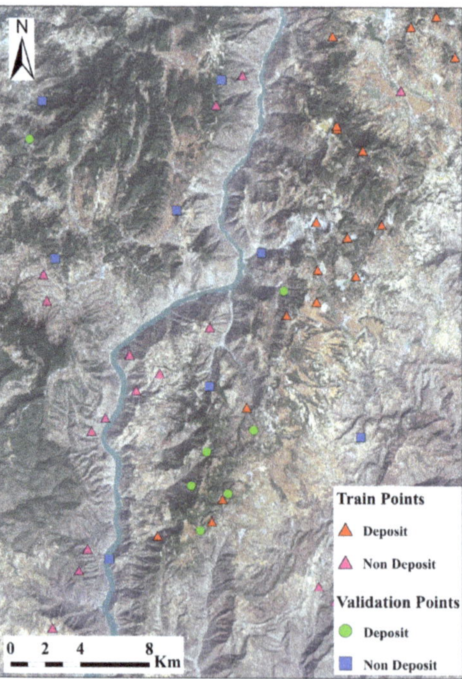

Figure 9. Location map of deposits and non-deposit training and validation sites used for training the neural networks.

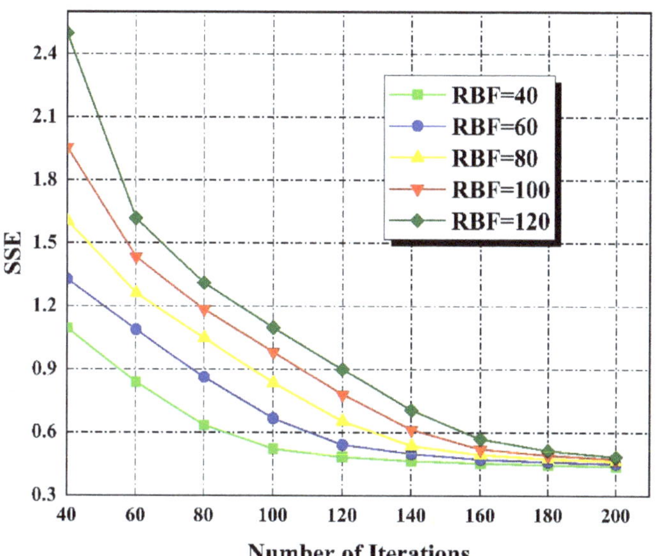

Figure 10. Number of iterations versus SSE for selected number of RBFs.

5.2. Validation

Accordingly, the results obtained through the RBFLN model with 40 RBFs and 120 iterations were selected to generate a probability output table with a value range of 0–1.577, which was normalized between 0 and 1. A relation graph of probability values and cumulative percentages of the study area was drawn (Figure 11), and the inflection points on the graph were calibrated to define the category intervals. Specifically, the posterior probability interval of the higher metallogenic target area was $p \geq 0.65$, whereas that of the medium metallogenic target area was $0.49 < p < 0.65$. Based on the features of the abovementioned intervals and the geological characteristics of the Lala concentration area, eight ore-prospecting prospective areas were delineated in this paper (Figure 12).

The validation data in the previous section were used to evaluate the performance of the RBFLN. The results demonstrated that five of the seven verification deposits were delineated in high prospective areas, indicating that the classification accuracy of unknown features was 71%. Additionally, one verification deposit and all non-verification deposits were delineated in low prospective areas, and one verification deposit was delineated in a medium prospective area. The comprehensive verification manifested that 86% of the verification deposits were in high and medium prospective areas (Figure 12).

Moreover, 3D visualization technology can be used to verify the prediction results and the existing ore bodies in the Luodang mining area, further proving the performance of the RBFLN model (Figure 13). The results showed that the distribution range of ore bodies was consistent with the prediction results of the RBFLN model.

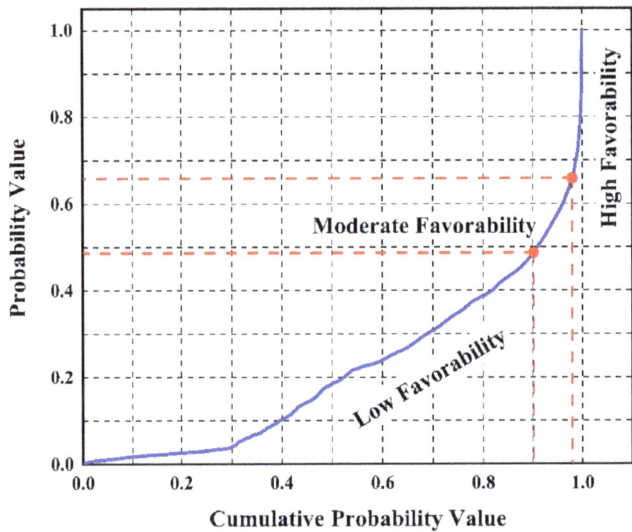

Figure 11. The probability values and the cumulative percentages for the study area.

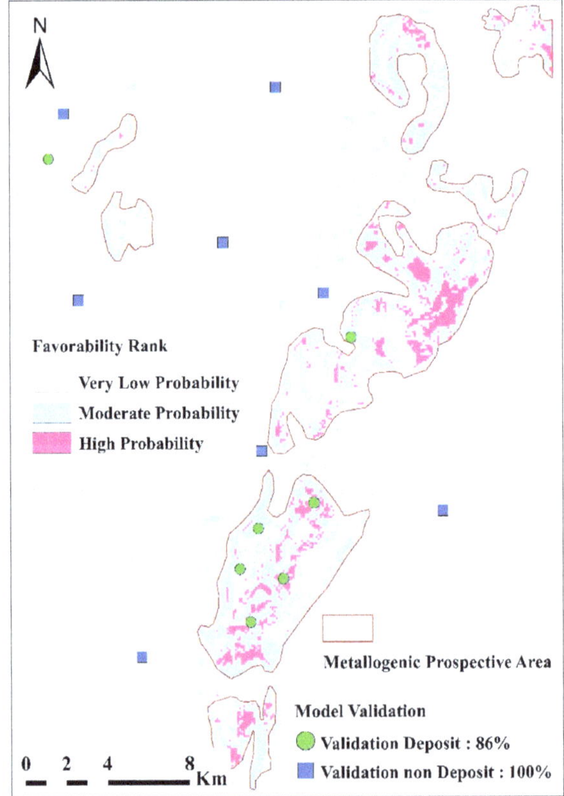

Figure 12. Prospective map of mineralization.

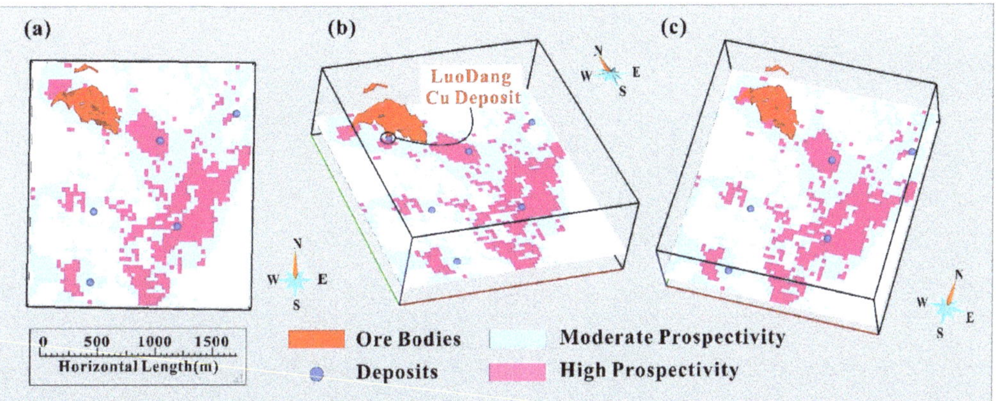

Figure 13. Superposition map of the ore body model and metallogenic prospect area. (**a**) Top View; (**b**) Left view; (**c**) Right view.

6. Conclusions

(1) We established a multivariate comprehensive ore-prospecting information model of geology–geophysics–geochemistry–remote sensing in the study area based on GIS technology, and we determined the existence of several ore-prospecting indicators (strata configuration entropy, tectonic fracture isodensity, tectonic fractal dimension, rock buffering 1 km, Cu geochemical anomalies, Cu–Co–Mo composite anomalies, aeromagnetic anomalies, alteration information, etc.). We adopted the RBFLN algorithm to conduct the metallogenic prediction of the Lala ore concentration area in Sichuan Province. Consequently, we successfully divided the secondary intervals of the higher metallogenic target area and medium metallogenic target area, and we determined eight ore-prospecting prospective areas by taking the above variables as predictors;

(2) The distribution of high and middle prospective areas in the metallogenic prospective map revealed that the metallogenic conditions would be more favorable and the possibility of deposits would be higher if the strata were more complex, the fracture intersection was denser, and the aeromagnetic anomaly was higher;

(3) In contrast to traditional metallogenic prediction methods, the RBFLN algorithm can solve complex nonlinear classification problems effectively and quickly. This algorithm features a simple calculating process, a rapid training speed, and an optimum approximation ability, which is conducive to processing a large amount of geological data and carrying out metallogenic prediction research;

(4) The cross-validation of multivariate geoscience information in favorable ore-prospecting areas was made possible by 3D visualization technology, which provides a better reference for ore prospecting in deep areas. However, the disadvantage of this research lies in the absence of on-site verification. No on-the-spot investigations have been conducted because of the impacts of the COVID-19 pandemic. Therefore, the reliability of the RBFLN model results should be further verified through field exploration.

Author Contributions: Conceptualization, X.L. (Xiumei Lv); methodology, X.L. (Xiumei Lv); software, X.L. (Xiumei Lv) and X.L. (Xiaoning Liu); validation, W.Y. and G.W.; writing—original draft preparation, X.L. (Xiumei Lv); writing—review and editing, G.W. All authors have read and agreed to the published version of the manuscript.

Funding: This research received no external funding.

Data Availability Statement: Data sharing is not applicable.

Conflicts of Interest: The authors declare no conflict of interest.

References

1. Xiao, K.; Xiang, J.; Fan, M.; Xu, Y. 3D Mineral Prospectivity Mapping Based on Deep Metallogenic Prediction Theory: A Case Study of the Lala Copper Mine, Sichuan, China. *J. Earth Sci.* **2021**, *32*, 348–357. [CrossRef]
2. Journel, A.G.; Huijbregts, C.J. *Mining Geostatistics*; The Blackburn Press: New York, NY, USA, 1991; pp. 77–218.
3. Matheron, G. Principles of geostatistics. *Econ. Geol.* **1963**, *58*, 1246–1266. [CrossRef]
4. Matheron, G. The theory of regionalized variables and its appplication. *Les Cah. Cent. Morphol. Mathématigue Fontainebleau.* **1971**, *5*, 212.
5. Verly, G.; David, M.; Journel, A.G.; Marechal, A. *Geostatistics for Natural Resources Characterization*; D. Reidel Publishing Company: Dordrecht, The Netherlands, 1984.
6. Journel, A.G. Nonparametric estimation of spatial distributions. *Math. Geol.* **1983**, *15*, 445–468. [CrossRef]
7. Yu, C.; Jiang, Y. The dynamic mechanisms of primary metal-zoning of cassiterite-sulfide deposits in the gejiu ore district, yunnan province. *Acta Geol. Sin.* **1990**, *44*, 226–237. (In Chinese)
8. Xiao, K. The new development of minerogenetic regulation and prediction -the comprehensive information methods. *Adv. Earth Sci.* **1994**, *9*, 18–23. (In Chinese)
9. Leite, E.P.; Filho, C.R.D.S. Probabilistic neural networks applied to mineral potential mapping for platinum group elements in the Serra Leste region, Carajás Mineral Province, Brazil. *Comput. Geosci.* **2009**, *35*, 675–687. [CrossRef]
10. Wang, G.; Zhang, S.; Yan, C. 3D geological modeling with multi-source data integration in polymetallic region: A case study of Luanchuan, Henan Province, China. *Earth Sci. Front.* **2009**, *16*, 166–167.
11. Yousefi, M.; Kreuzer, O.P.; Nykänen, V.; Hronsky, J.M. Exploration information systems—A proposal for the future use of GIS in mineral exploration targeting. *Ore Geol. Rev.* **2019**, *111*, 103005. [CrossRef]
12. Chi, S.; Zhao, P.; Li, J. Application of GIS to Geo-anomaly-Based Delineation of Mineral Resources. *J. China Univ. Geosci.* **2000**, *11*, 164–167.
13. Bonham-Carter, G.F.; Agterberg, F.P.; Wright, D.F. Integration of geological datasets for gold exploration in Nova Scotia. *Photogramm. Eng. Remote Sens.* **1988**, *54*, 1585–1592. [CrossRef]
14. Wyborn, L.; Gallagher, R.; Raymond, O. Using GIS for mineral potential evaluation in areas with few known mineral occurrences. Second National Forum on GIS in the Geosciences-Forum Proc., Australian Geol. *Surv. Organ. AGSO Record.* **1995**, *46*, 199–211.
15. Silva, E.C.E.; Silva, A.; Toledo, C.L.B.; Mol, A.G.; Otterman, D.W.; de Souza, S.R.C. Mineral Potential Mapping for Orogenic Gold Deposits in the Rio Maria Granite Greenstone Terrane, Southeastern Para State, Brazil. *Econ. Geol.* **2012**, *107*, 1387–1402. [CrossRef]
16. Zhang, Z.; Zhang, J.; Wang, G.; Carranza, E.J.M.; Pang, Z.; Wang, H. From 2D to 3D Modeling of Mineral Prospectivity Using Multi-source Geoscience Datasets, Wulong Gold District, China. *Nonrenew. Resour.* **2020**, *29*, 345–364. [CrossRef]
17. Li, R.; Wang, G.; Carranza, E.J. GeoCube: A 3D mineral resources quantitative prediction and assessment system. *Comput. Geosci.* **2016**, *89*, 161–173. [CrossRef]
18. Li, R.; Wang, G.; Zhang, S.; Qu, J.; Zhu, Y.; Huang, L.; Yan, C.; Song, Y.; Han, J.; Ma, Z. Three dimensional quantitative ex-traction and intergration for geosciences information: A case study of the Luanchuan Mo ore district. *Geol. Bull. China* **2014**, *33*, 883–893. (In Chinese)
19. Wang, G.; Zhang, S.; Yan, C.; Song, Y.; Sun, Y.; Li, D.; Xu, F. Mineral potential targeting and resource assessment based on 3D geological modeling in Luanchuan region, China. *Comput. Geosci.* **2011**, *37*, 1976–1988. [CrossRef]
20. Köhler, M.; Hanelli, D.; Schaefer, S.; Barth, A.; Knobloch, A.; Hielscher, P.; Cardoso-Fernandes, J.; Lima, A.; Teodoro, A.C. Lithium Potential Mapping Using Artificial Neural Networks: A Case Study from Central Portugal. *Minerals* **2021**, *11*, 1046. [CrossRef]
21. Skabar, A.A. Mapping Mineralization Probabilities using Multilayer Perceptrons. *Nonrenew. Resour.* **2005**, *14*, 109–123. [CrossRef]
22. Behnia, P. Application of Radial Basis Functional Link Networks to Exploration for Proterozoic Mineral Deposits in Central Iran. *Nonrenew. Resour.* **2007**, *16*, 147–155. [CrossRef]
23. Abedi, M.; Norouzi, G.-H.; Bahroudi, A. Support vector machine for multi-classification of mineral prospectivity areas. *Comput. Geosci.* **2012**, *46*, 272–283. [CrossRef]
24. Chen, H.; Lin, L.; Pang, Z.; Cheng, Z.; Xue, J.; Tao, W.; Gong, L.; Shen, H. Construction and demonstration of an ore prospecting model for the Lala copper deposit in Huili, Sichuan. *Earth Sci. Front.* **2021**, *28*, 309–327. (In Chinese)
25. Zhao, X.; Zhou, M. Fe-Cu deposits in the Kang dian region, SW China:a Proterozoic IOCG (iron-oxide-copper-gold) metallo-genic province. *Miner. Deposita.* **2011**, *46*, 731–747. [CrossRef]
26. Zhu, Z.; Sun, Y. Direct re-os dating of chalcopyrite from the lala iocg deposit in the kangdian copper belt, China. *Econ. Geol.* **2012**, *108*, 871–882.
27. Zhou, M.; Zhao, X.; Chen, W.; Li, X.; Wang, W.; Yan, D.; Qiu, H. Proterozoic Fe-Cu metallogeny and supercontinental cycles of the southwestern Yangtze Block, southern China and northern Vietnam. *Earth Sci. Rev.* **2014**, *139*, 59–82. [CrossRef]
28. Shen, H.; Gong, L.; Chen, H.; Lin, L.; Deng, Y.; Zeng, L. A disscussion on the genetic model of the copper deposit in Lala copper orefied, Sichuan Province. *Geol. Bull. China* **2020**, *39*, 1233–1246. (In Chinese)
29. Zhu, Z.; Tan, H.; Liu, Y.; Li, C. Multiple episodes of mineralization revealed by Re-Os molybdenite geochronology in the Lala Fe-Cu deposit, SW China. *Miner. Depos.* **2018**, *53*, 311–322. [CrossRef]

30. Lin, L.; Chen, R.; Pang, Z.; Chen, H.; Xue, J.; Jia, H. Sulfide Rb-Sr, Re-Os and In Situ S Isotopic Constraints on Two Mineralization Events at the Large Hongnipo Cu Deposit, SW China. *Minerals* **2020**, *10*, 414. [CrossRef]
31. Broomhead, D.S.; Lowe, D. Multivariable functional interpolation and adaptive networks. *Complex Syst.* **1988**, *2*, 321–355.
32. Porwal, A.; Carranza, E.J.M.; Hale, M. Artificial Neural Networks for Mineral-Potential Mapping: A Case Study from Aravalli Province, Western India. *Nonrenew. Resour.* **2003**, *12*, 155–171. [CrossRef]
33. Nykänen, V. Radial basis functional links nets used as a prospectivity mapping tool for orogenic gold deposits within the Central Lapland Greenstone Belt, Northern Fennoscandian Shield. *Nat. Resour. Res.* **2008**, *17*, 29–48. [CrossRef]
34. Tessema, A. Mineral Systems Analysis and Artificial Neural Network Modeling of Chromite Prospectivity in the Western Limb of the Bushveld Complex, South Africa. *Nonrenew. Resour.* **2017**, *26*, 465–488. [CrossRef]
35. Niros, A.D.; Tsekouras, G.E. A novel training algorithm for RBF neural network using a hybrid fuzzy clustering approach. *Fuzzy Sets Syst.* **2012**, *193*, 62–84. [CrossRef]
36. Ghezelbash, R.; Maghsoudi, A.; Carranza, E.J.M. Performance evaluation of RBF- and SVM-based machine learning algorithms for predictive mineral prospectivity modeling: Integration of S-A multifractal model and mineralization controls. *Earth Sci. Inform.* **2019**, *12*, 277–293. [CrossRef]
37. Looney, C.G. Radial basis functional link nets and fuzzy reasoning. *Neurocomputing* **2002**, *48*, 489–509. [CrossRef]
38. Maepa, F.; Smith, R.S.; Tessema, A. Support vector machine and artificial neural network modelling of orogenic gold prospectivity mapping in the Swayze greenstone belt, Ontario, Canada. *Ore Geol. Rev.* **2021**, *130*, 103968. [CrossRef]
39. Chi, S.; Zhao, P. Application of combined-entropy anomany of geological ormations to delineation of preferable ore-finding area. *Mod. Geol.* **2000**, *14*, 423–428. (In Chinese)

Article

Three-Dimensional Structural Modeling (3D SM) and Joint Geophysical Characterization (JGC) of Hydrocarbon Reservoir

Baoyi Zhang [1], Yongqiang Tong [1], Jiangfeng Du [2], Shafqat Hussain [3], Zhengwen Jiang [1], Shahzad Ali [3], Ikram Ali [3], Majid Khan [4] and Umair Khan [1,*]

[1] Key Laboratory of Metallogenic Prediction of Nonferrous Metals & Geological Environment Monitoring (Ministry of Education), School of Geosciences & Info-Physics, Central South University, Changsha 410083, China; zhangbaoyi@csu.edu.cn (B.Z.); tongyongqiang@csu.edu.cn (Y.T.); jiangzhengwen@csu.edu.cn (Z.J.)

[2] CNOOC Research Institute Co., Ltd., Beijing 100028, China; dujf@cnooc.com.cn

[3] Khyber Pakhtunkhwa Oil and Gas Co., Ltd. (KPOGCL), Peshawer 25000, Pakistan; shafqat.dpc@kpogcl.com.pk (S.H.); shehzadali.gng@kpogcl.com.pk (S.A.); ikram.dpc@kpogcl.com.pk (I.A.)

[4] School of Civil and Resource Engineering, University of Science and Technology Beijing, Beijing 100083, China; majid@ustb.edu.cn

* Correspondence: umair77@csu.edu.cn

Citation: Zhang, B.; Tong, Y.; Du, J.; Hussain, S.; Jiang, Z.; Ali, S.; Ali, I.; Khan, M.; Khan, U. Three-Dimensional Structural Modeling (3D SM) and Joint Geophysical Characterization (JGC) of Hydrocarbon Reservoir. *Minerals* **2022**, *12*, 363. https://doi.org/10.3390/min12030363

Academic Editor: Michael S. Zhdanov

Received: 2 February 2022
Accepted: 13 March 2022
Published: 16 March 2022

Abstract: A complex structural geology generally leads to significant consequences for hydrocarbon reservoir exploration. Despite many existing wells in the Kadanwari field, Middle Indus Basin (MIB), Pakistan, the depositional environment of the early Cretaceous stratigraphic sequence is still poorly understood, and this has implications for regional geology as well as economic significance. To improve our understanding of the depositional environment of complex heterogeneous reservoirs and their associated 3D stratigraphic architecture, the spatial distribution of facies and properties, and the hydrocarbon prospects, a new methodology of three-dimensional structural modeling (3D SM) and joint geophysical characterization (JGC) is introduced in this research using 3D seismic and well logs data. 3D SM reveals that the field in question experienced multiple stages of complex deformation dominated by an NW to SW normal fault system, high relief horsts, and half-graben and graben structures. Moreover, 3D SM and fault system models (FSMs) show that the middle part of the sequence underwent greater deformation compared to the areas surrounding the major faults, with predominant one oriented S30°–45° E and N25°–35° W; with the azimuth at 148°–170° and 318°–345°; and with the minimum (28°), mean (62°), and maximum (90°) dip angles. The applied variance edge attribute better portrays the inconsistencies in the seismic data associated with faulting, validating seismic interpretation. The high amplitude and loss of frequency anomalies of the sweetness and root mean square (RMS) attributes indicate gas-saturated sand. In contrast, the relatively low-amplitude and high-frequency anomalies indicate sandy shale, shale, and pro-delta facies. The petrophysical modeling results show that the E sand interval exhibits high effective porosity ($\varnothing_{eff}$) and hydrocarbon saturation (S_{hc}) compared to the G sand interval. The average petrophysical properties we identified, such as volume of shale (V_{shale}), average porosity ($\varnothing_{avg}$), $\varnothing_{eff}$, water saturation (S_W), and the S_{hc} of the E sand interval, were 30.5%, 17.4%, 12.2%, 33.2% and, 70.01%, respectively. The findings of this study can help better understand the reservoir's structural and stratigraphic characteristics, the spatial distribution of associated facies, and petrophysical properties for reliable reservoir characterization.

Keywords: 3D structural modeling (3D SM); 3D fault system models (FSMs); seismic attribute models; reservoir properties; facies; hydrocarbon-bearing zones

1. Introduction

The recent global increase in fuel demand has increased hydrocarbon production at established reservoirs [1,2]. According to the United States Energy Information Administration (USEIA), Pakistan may have over 9 billion barrels of oil and 105 trillion cubic feet of natural gas (including shale gas) reserves [3]. Pakistan's gas fields are only expected to last for about another 20 years at the most due to heavy industrial usage. Therefore, many recent works on the Lower Indus Basin (LIB) are focused on unconventional resources such as shale gas and coal seam gas [4–6]. However, focused characterization and re-evaluations of already-discovered petroleum systems are also required to evaluate reservoirs' abilities to meet the hydrocarbon requirements [7].

Reservoir characterization is a scheme that quantifies the physical and fluid properties of rock [8–12]. It also involves an understanding of reservoir structure, sedimentological heterogeneity, facies, and the quantity of hydrocarbon that exist in structural traps driven by tectonic movements, discontinuities such as faults, folds, anticlinal structures, horst and graben pop-up geometries, and duplex structures [13–18]. Advancements in 3D seismic data analysis and borehole geophysics have made it possible to characterize structural and stratigraphic features and their associated petrophysical properties with high reliability and precision, thereby reducing the risks associated with hydrocarbon exploration [19,20].

The development of three-dimensional structure models (3D SMs) is essential for reservoir description, e.g., true structural dip, fault system models (FSMs), up-dip hydrocarbon migration pathways, and quantitative geometric characterization in 3D space [21,22]. Three-dimensional structural modeling (3D SM) is divided into entity-based modeling and volume-based modeling (VBM) [23–28]. The former develops 3D SMs using a combination of four different geometric entities, i.e., point, line, surface, and body. It emphasizes the shape of geological structures and the relationship among geological bodies [29]. The latter subdivides the 3D space into discrete fields by regular or irregular voxels and emphasizes the spatial distribution of geophysical and geochemical properties. With the continuous development of reservoir geological modeling technology, the VBM, objective function, variation function, multipoint geostatistics, and static geological modeling with knowledge-driven methodology have been widely applied, significantly promoting the development and technology of 3D reservoirs geological modeling [30–36]. However, VBM is a step-change reservoir geological modeling technique that creates horizons based on depositional sequence instead of considering horizons as discrete surfaces [2,13,37].

Identifying reservoir facies and properties using seismic attributes and petrophysical analyses play an essential role in reservoir characterization. Reservoir facies classification and properties evaluation can be achieved via laboratory studies on core plugs, which are costly and time-consuming. However, seismic attributes can be used for stratigraphic-based basin depiction within a composite deposition-based structure and classify reservoir facies, thereby increasing the rate for adequate reservoir characterization. Seismic attributes, such as dip magnitude, edge enhancement, variance edge, sweetness, and root mean square (RMS) amplitude, are essential tools for delineating structural and stratigraphic characteristics, lithofacies changes, and hydrocarbon potential zones [19,38,39]. On the other hand, petrophysical analysis based on well logs offers practically continuous reservoir properties, e.g., lithology, the volume of shale (V_{shale}), average porosity ($\varnothing_{avg}$), effective porosity ($\varnothing_{eff}$), water saturation (S_W), and hydrocarbon saturation (S_{hc}). The proper analysis of these properties can significantly improve the ability to distinguish hydrocarbon-bearing zones [18,20,40,41].

The Middle Indus Basin (MIB) and Lower Indus Basin (LIB) are well known for hydrocarbon exploration in Pakistan [40]. The Kadanwari field in MIB is a significant hydrocarbon-producing field, with the early–late Cretaceous Lower Goru formation (LGF) acting as the potential reservoir. The previous studies conducted on the Kadanwari field were based on the assessment and development of the 2D fault system, porosity prediction [42], formation evaluation [40,43], and impact of diagenesis on reservoir quality prediction [44].

A systematic review of prospective observational studies found that the Kadanwari field in MIB has not yielded sufficient results for understanding the complex structural depositional environment [45]. Moreover, the distribution of key petrophysical properties and facies in the Kadanwari field are difficult to predict due to fluctuating deltaic conditions, mutable geological influences, varying hydrocarbon concentrations, regional tectonic settings, and changes in geometries. However, comprehensive research on the structural characteristics and associated tectonic extensional fault system models (FSMs) and the evaluation of reservoir characteristics and sweet spots for future drillings in the Kadanwari field is still missing.

This study, therefore, aims at evaluating the complex and heterogeneous depositional environment broadly comprised of structural and stratigraphic characteristics, distribution of associated facies, and petrophysical properties for reliable reservoir characterization, which was lacking in the previous studies. This was achieved by utilizing three-dimensional structural modeling (3D SM) and joint geophysical characterization (JGC) using seismic and well logs data. In this study, 3D SM and JGC have made novel contributions and facilitated a clearer representation of the complex and heterogeneous depositional environments, with the lateral and horizontal structural extent of reservoir horizons, FSMs (including fault geometry and orientation in 3D space), spatial facies, reservoir properties (e.g., lithology, V_{shale}, $\varnothing_{avg}$, $\varnothing_{eff}$, S_W, and S_{hc}), and direct hydrocarbon indicators (DHIs). In short, our presented study is crucial to characterize and evaluate reservoirs' geometrical characteristics, facies, and properties, in order to reduce uncertainties and improve the success rate of future exploration and development plans pertaining to hydrocarbons in the study area, as well as potentially in other regions around the world.

2. Background Geology

Pakistan is located at the triple junction of the Indian, Eurasian and Arabian plates (Figure 1a). During the middle/late Jurassic to early Cretaceous, the Indian plate rifted away from the Gondwana landmass, forming an island continent that drifted northwards into the Tethyan Ocean [46]. This tectonic event predominantly influenced the structures and sedimentation of the MIB and LIB, which could have resulted in NE—SW to N—S rift systems. The Kadanwari field is located in the District Khairpur of Sindh Province, southeast MIB, a prolific gas-prone basin in Pakistan (Figure 1b). The latitude of the study area ranges from 27°04′83″ N to 27°07′12″ N, and its longitude varies from 69°12′98″ E to 69°17′57″ E. Tectonically, the Kadanwari field lies between two extensive regional highs, i.e., the Mari-Kandhkot High and the Jacobabad-Khairpur High (Figure 1b). In the east, it is bounded by the Indian shield; in the north by the Sargodha high; in the west by the fold and thrust belt of the Kirthar and Sulaiman Ranges, and in the south by the Jacobabad-Khairpur High [7,45–47]. Three tectonic events were responsible for the structural configuration of the study area, i.e., the late Cretaceous uplift and erosion, the late Paleocene wrench faulting, and the late Tertiary to Quaternary uplift/inversion of Jacobabad High (Figure 1b) [46]. The Jacobabad-Khairpur High was a primary contributor to the study area's structural traps and surroundings [45,47]. The final tectonic event of the late Tertiary to Quaternary was an inversion of the Jacobabad-Khairpur High, which significantly affected the Kadanwari area [45,46]. In the Kadanwari field and its surroundings, the trapping mechanism is a complex combination of structural dip, sealing faults, and loss of reservoir quality to the north. The Kadanwari field consists of several low-relief faults, forming dip closures in the subsurface and providing a stratigraphic trapping component [7,45]. The fault dip closures and the wrench faults are particularly significant as they divide the Kadanwari field into reservoir compartments [40].

The lithology stack of MIB is depicted in Figure 2a, highlighting the basin fill sedimentary deposits. The lithostratigraphic columns show the rock units encountered in the Kadanwari-10 and Kadanwari-11 wells (Figure 2b). According to [48], the shales of the Sember Formation serve as the source rock for the regional petroleum systems of the MIB. However, the reservoir sections (e.g., G and E sands) in the Kadanwari field belong to the lower Goru sand (the Cretaceous age), while the sealing is provided by the upper Goru shaly sequence [43].

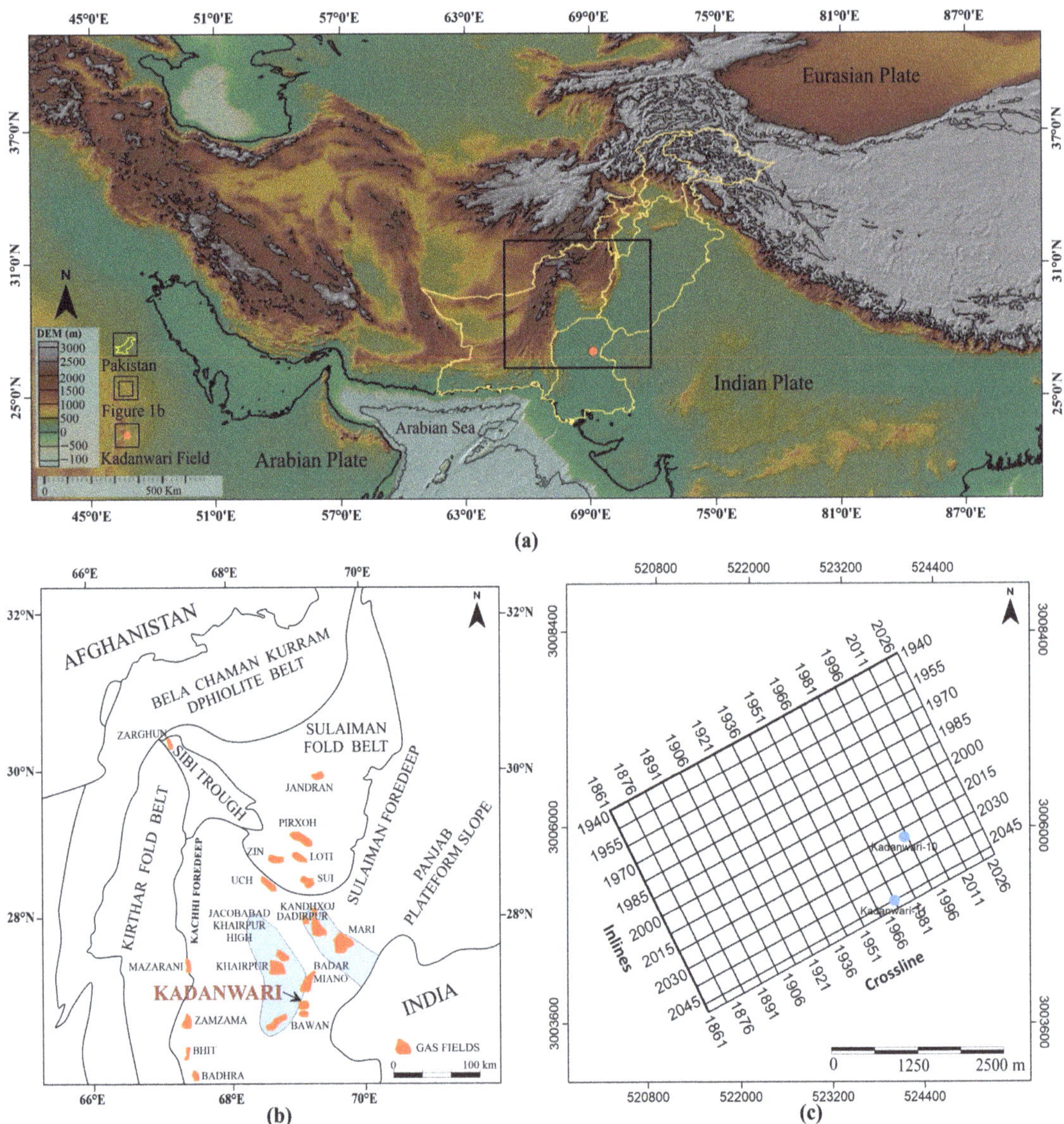

Figure 1. (**a**) Location of the study area; (**b**) generalized tectonic map with the location of major oil- and gas-producing fields in the study area, bounded by other gas fields, modified from [49]; (**c**) the base map shows the orientation and general information of the 3D seismic lines and wells in the Kadanwari field, MIB, Pakistan.

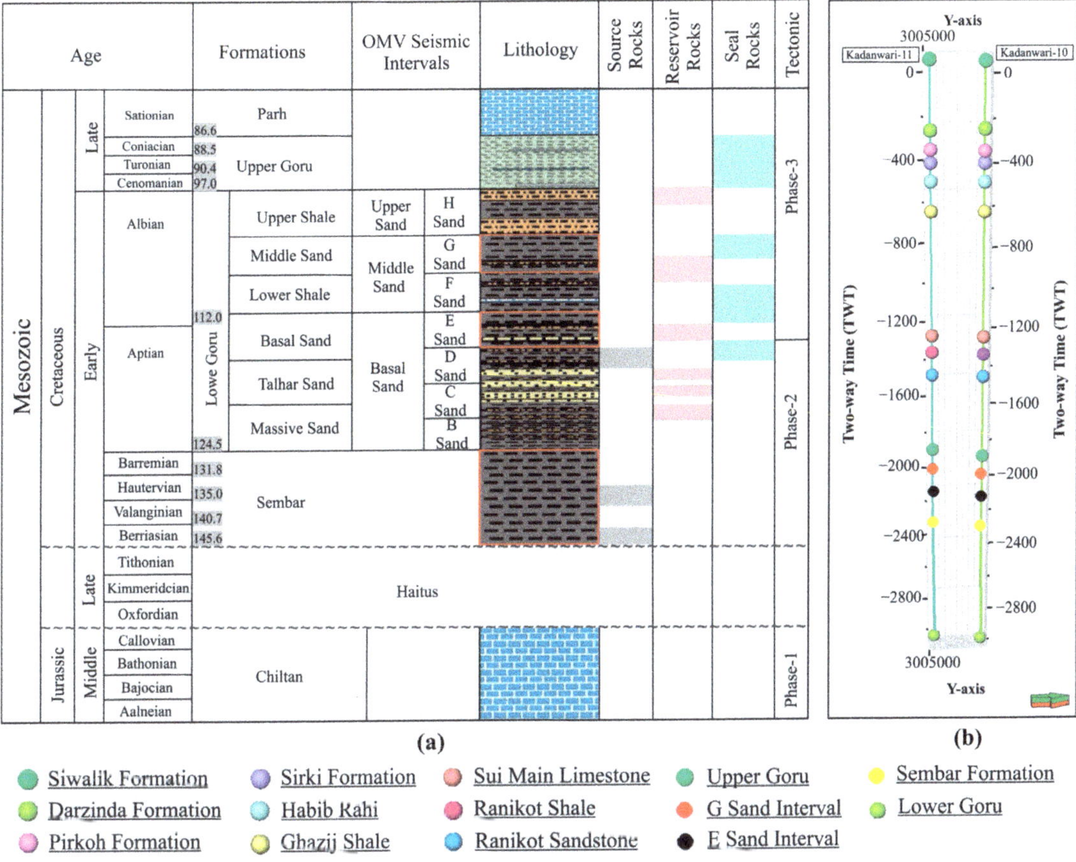

Figure 2. (a) Generalized stratigraphic column of the study area, modified from [48]; (b) lithostratigraphic columns showing the rock units encountered in Kadanwari-10 and Kadanwari-11 wells.

3. Material and Methods

3.1. Datasets Description and Processing

A vast volume of seismic reflection data was acquired from the MIB, Pakistan, to facilitate hydrocarbon exploration activities at different times, from the 1970s up to the modern-day [7]. The Pakistan branch of OMV (www.omv.com, Vienna, Austria) recently began a new exploration phase in the Kadanwari field by conducting a 3D seismic survey. The seismic data used in this study are 3D post-stack time migrated seismic reflection cubes stored in the SEG-Y format, wherein approximately 116 seismic inlines and approximately 181 cross-lines were used. The geometric information, e.g., total coverage area, inline interval, crossline interval, and time slice range of available 3D seismic data, were 12 km^2, 24.17, 25.17, and 1800–2600 (ms), respectively. The well logs data comprise lithology, resistivity, and porosity logs (e.g., GR, SP, LLD, LLS, MSFL, RHOB, NPHI, and DT) of the Kadanwari-10 and Kadanwari-11 wells (Table 1). The available 3D seismic data and well log data were collected from the Landmark Resources (LMKR) (www.lmkr.com, Calgary, Canada) upon the request of the Directorate General of Petroleum Concessions (DGPC) (www.mpnr.gov.pk, Islamabad, Pakistan); which are available to the public domain and can be utilized for scientific and research purposes. The dataset quality was first checked and harmonized in a clearly defined database. Accordingly, a base map of the 42-N Trans-

Mercator Macrocosm (UTM) zone was created using navigation and SEG-Y records to determine the orientation (dip or strike) and location of seismic lines and wells (Figure 1c).

Table 1. Metadata of the utilized well logs and their uses in this study.

Well Logs	Measured Property	Petrophysical Properties Estimated	Product
Caliper	Diameter	Borehole structure with depth	CALI
Gamma-ray	Radioactivity	Shale volume (V_{shale})	GR
Laterolog deep	Resistance to electric current	Uninvaded resistivity	LLD
Laterolog shallow	Resistance to electric current	Invaded zone resistivity	LLS
Micro-spherical focused log	Resistance to electric current	Mud cake resistivity	MSFL
Sonic	Velocity of sound waves	Porosity	DT
Spontaneous Potential	Electric potential	Formation water resistivity	SP
Neutron	Hydrogen concentration	Porosity	NPHI
Density	Bulk density	Porosity	RHOB

3.2. Methods

In this study, three-dimensional structural modeling (3D SM) and joint geophysical characterization (JGC) use an integrated 3D SM approach, involving structurally constrained geological models and seismic attribute implications and petrophysical properties that significantly enhance the understanding of the reservoir characteristics leading to reliable reservoir assessment. Figure 3 shows the complete workflow of the present case study. Firstly, seismic and well log data interpretation was carried out, which involves synthetic seismogram generation, and extracting and interpreting specific stratigraphic interfaces (geological period) and faults as geometric features in 2D cross-sectional (i.e., vertical) slices of a 3D seismic volume. Secondly, 3D FSMs and their attribute models such as dip angles, fault rose diagram and histogram were constructed to evaluate the fault mechanics and geometric distribution. Thirdly, 3D SMs of the early Cretaceous stratigraphic sequence were constructed using the VBM algorithm, incorporating all geometrical definitions (e.g., constraints from well tops, geologic horizons, and FSMs). Fourthly, several seismic attributes such as variance edge, sweetness, and RMS amplitude were incorporated into the 3D seismic data, which involves extracting the corresponding qualitative and quantitative geological features to validate the interpreted spatial forecasts of the geological structure, and to then evaluate the lithofacies distribution and direct hydrocarbon indicators (DHIs). Finally, petrophysical modeling based on various well logs (CALI, GR, SP, LLD, LLS, MSFL, DT, NPHI, and RHOB, explained in Table 1) was performed to determine the reservoir properties (e.g., lithology, V_{shale}, $\varnothing_{avg}$, $\varnothing_{eff}$, S_W, and S_{hc}).

3.2.1. Seismic Data Interpretation

Seismic interpretation in this study is focused on a particular area of interest within the two-way time (TWT) range of -1800 to -2600 (ms) (Figure 4). Depending on the seismic reflection discontinuities and terminations, manual horizon picking followed by seeded horizon auto-tracking was adopted to interpret the target horizons on 2D cross-sectional (i.e., vertical) slices of a 3D seismic volume (Figure 4a). In the middle parts of the seismic cross-sections, the reflections are mostly moderately chaotic and difficult to correlate due to the complexity of the geology and faults resulting from tectonic compression (Figure 4b). The modeling step of the seismic interpretation involves 3D TWT contour surfaces construction. Consequently, 3D TWT contour surfaces were constructed by marking the tops of each stratigraphic interface on the extended 3D seismic volume. The smooth function was then applied to the generated surfaces at three iteration levels to produce geologically reasonable stratigraphic surfaces. These 3D TWT contour surfaces were then used to interpret the prevailing structural trends in the study area via conventional methods.

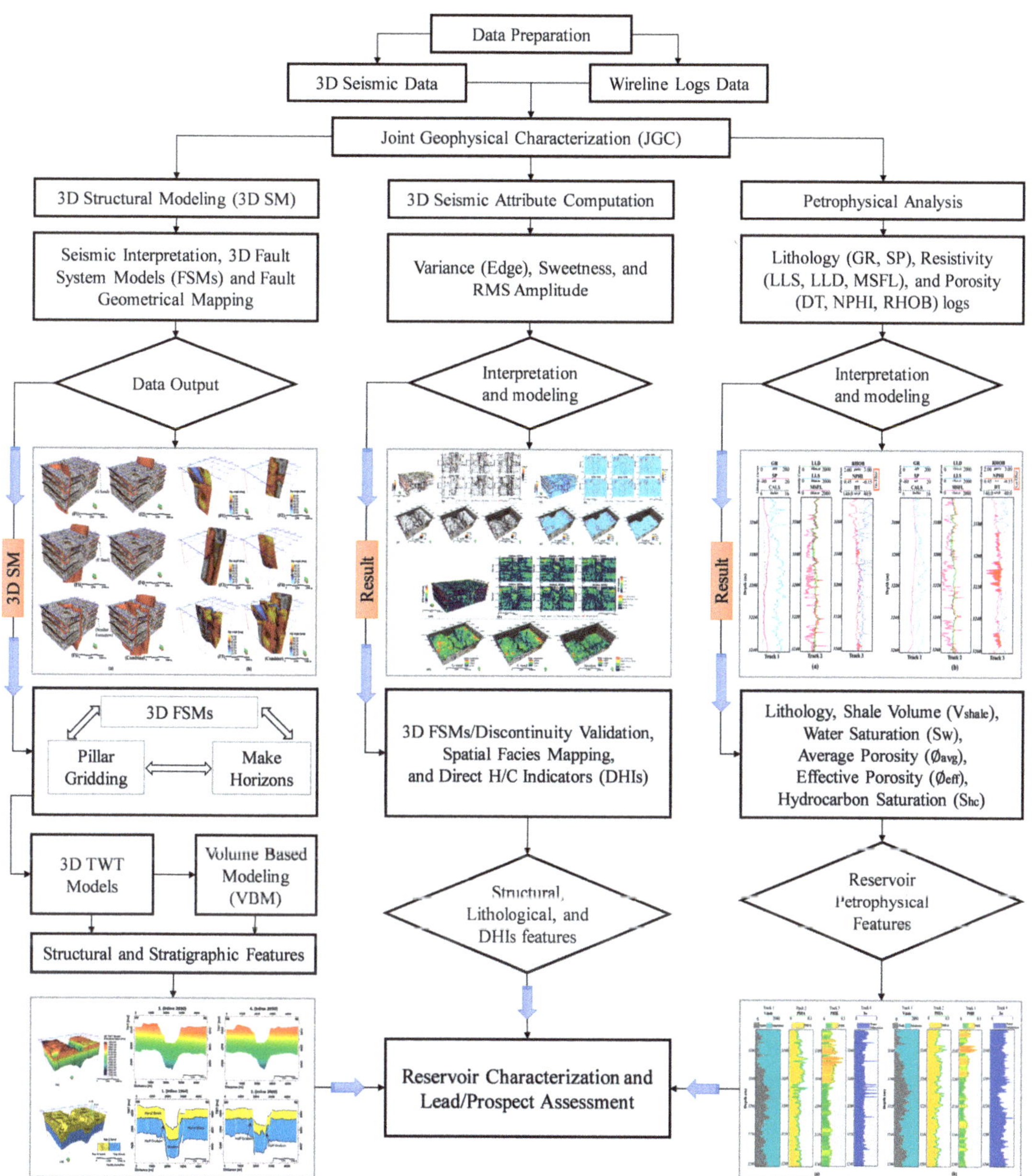

Figure 3. Workflow highlighting various steps and methods employed in this study for reservoir characterization and prospect assessment of the Kadanwari field, MIB, Pakistan.

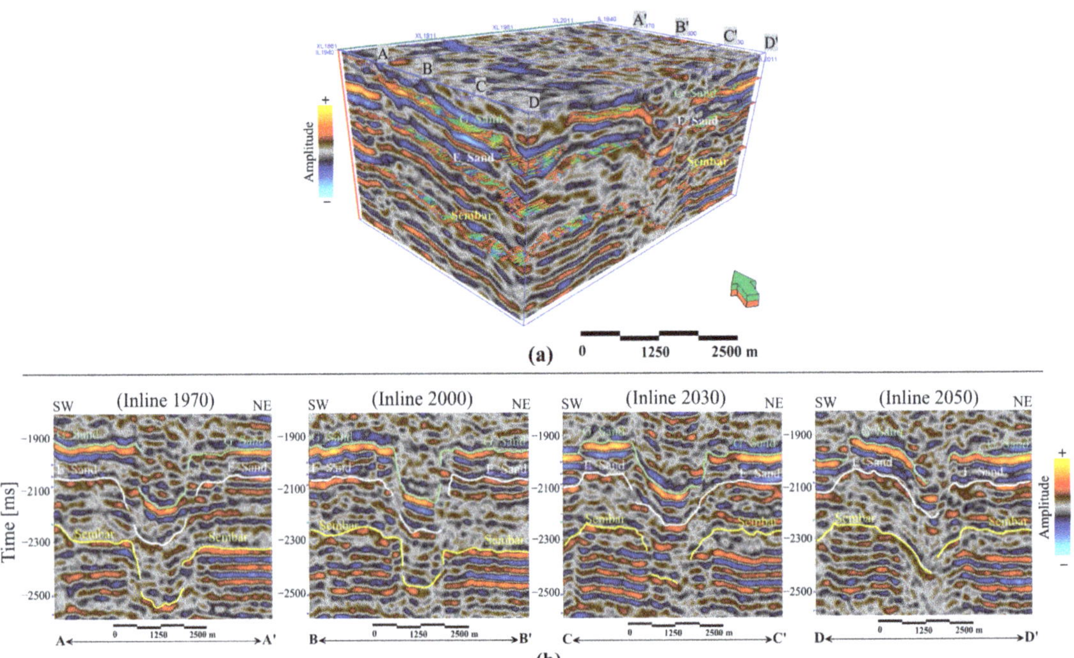

Figure 4. (**a**) 3D time-seeded horizon auto-tracking; (**b**) amplitude display of vertical time inline cross-sections (e.g., 1970, 2000, 2030, and 2050) in the SW to NE direction, showing horizon interpretation. Mint green, white, and yellow colors represent the lateral extent of the G sand interval, the E sand interval, and the Sembar Formation, respectively.

3.2.2. Three-Dimensional Structural Modeling (3D SM)

Employing 3D fault system modeling (FSM) in seismic analysis is a crucial step to constrain horizon interpretation [21,50]. One recent significant advance in FSM is the rendering available of 3D seismic data that provide detailed images of large volumes of rock and the often-complex 3D fault networks [51]. Meanwhile, most seismic datasets have signal disturbance zones, particularly in highly faulted areas. Furthermore, the discontinuities in seismic reflectors can be poorly resolved, resulting in the approximate localization or misinterpretation of faults. However, seismic discontinuities can be more clearly defined if the detection is based on multiple attributes and suitable filters [51,52]. Therefore, in this study, before proceeding to the 3D FSM, the precision of the geologic fault boundaries was assessed first for accuracy using dip magnitude and seismic edge enhancement attributes (Figure 5). Accordingly, the dominance of a normal fault system was inferred in the dip magnitude and edge enhancement cross-sections by aligning reflector discontinuities and degree of displacement. In the modeling phase, the fault surfaces were constructed by employing the fault polygon in Petrel^TM software for each type of fault with various geometrical structures. Finally, 3D FSMs and their attribute models (e.g., dip angle, fault rose diagram and histogram) were constructed to evaluate the fault system's geometric distribution and mechanics within the early Cretaceous stratigraphic sequence.

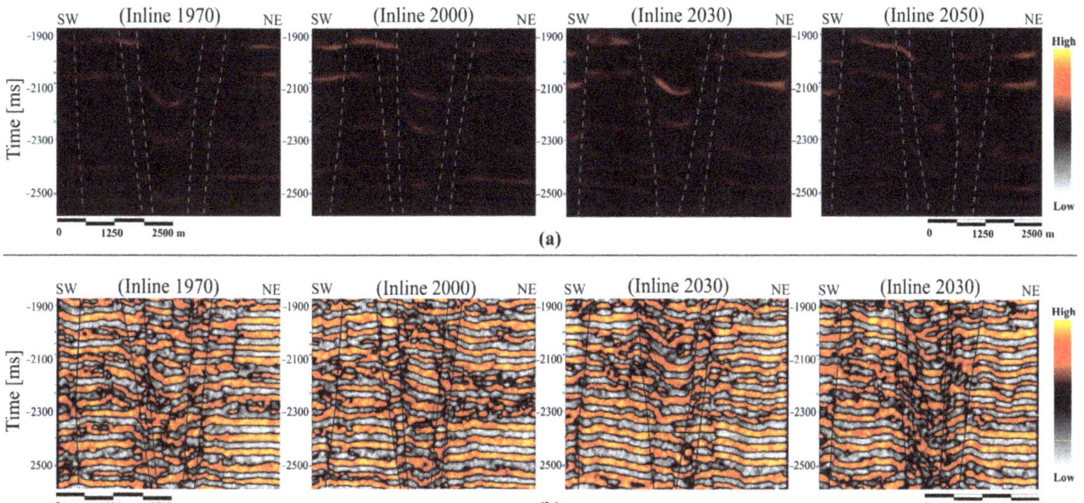

Figure 5. Fault system tracing and interpretation in the Kadanwari field using resemble brittle deformation features (seismic reflection discontinuities, terminations, and amount of displacement) on (**a**) dip magnitude and (**b**) seismic edge enhancement attributes cross-sections.

Three-dimensional SM aims to better image and understand complex reservoirs' geological structures. The local tectonics of the Kadanwari field in MIB have resulted in various structural deformations that produce many uncertainties when assessing the reservoirs' 3D structural framework [7]. Understanding such deformations can be better achieved via 3D SM using relevant algorithm models, e.g., the volume-based modeling (VBM) approach in the Petrel^{TM} software based on inferred seismic data integrated with borehole information [50,53,54]. Traditional modeling methods generally involve the oversimplification of geological settings; however, grid-generated VBM captures a realistic reservoir architecture [2]. This can be easily transferred into the dynamic realm, providing a better understanding of the reservoir for future field management and development activities [50]. In this study, 3D SM based on the VBM approach was performed in three main steps (e.g., fault modeling, pillar gridding, and horizon creation).

3.2.3. Three-Dimensional Seismic Attribute Analysis

The seismic attributes were computed via mathematical manipulation of the original seismic data to highlight specific geological, physical, or reservoir properties [50,55]. Variations in the amplitude, phase, frequency, and bandwidth of the seismic waves were subsequently used to validate the spatial forecasts of the geological structure and to evaluate the spatial distribution of the facies and the DHIs in the G sand interval, the E sand interval, and the Sember Formation time windows.

The variance edge attribute can visualize seismic amplitude discontinuities related to faulting or stratigraphy. It delineated the prominent faults and seismic amplitude discontinuity in both the horizon slices and the vertical seismic profiles, thereby validating the manual interpretation of faults. The sweetness attribute was applied to both horizon slices and vertical seismic profiles to identify sweet spot zones that are hydrocarbon-prone. It can be defined as the reflection strength (instantaneous amplitude) divided by the square root of the instantaneous frequency. The high sweetness anomalies highlight a seismic signal's high amplitudes and low-frequency contents and vice versa. Therefore, combining these two physical quantities helps distinguish sand bodies from shale and predicted gas-prone zones [56]. The RMS amplitude attribute was also applied to both horizon slices and vertical seismic profiles to measure amplitude anomalies to identify the spatial distribution of facies and DHIs at the G and E sand reservoir intervals. The RMS amplitude computes the square root of the sum of squared amplitudes divided by the number of samples within the window used [19].

3.2.4. Petrophysical Modeling

Petrophysical modeling is critical in a reservoir study because it represents a primary input data source for integrated reservoir characterization and resource evaluation [54,57]. This study performed petrophysical modeling based on well log data and internal geological reports to evaluate the G and E sand reservoir intervals' properties, e.g., lithology, V_{shale}, $\varnothing_{avg}$, $\varnothing_{eff}$, S_W, and S_{hc} (Table 1). Successful evaluation of these properties is necessary when determining the hydrocarbon potential of a reservoir system. The detailed plots of the well log curves and their depth ranges within the G and E sand reservoir intervals are shown in Figure 6. These log curves express the physical motifs of the stacked geological strata as a function of depth, which can help identify lithologies and porosities, and differentiate between porous and non-porous rocks and pay zones.

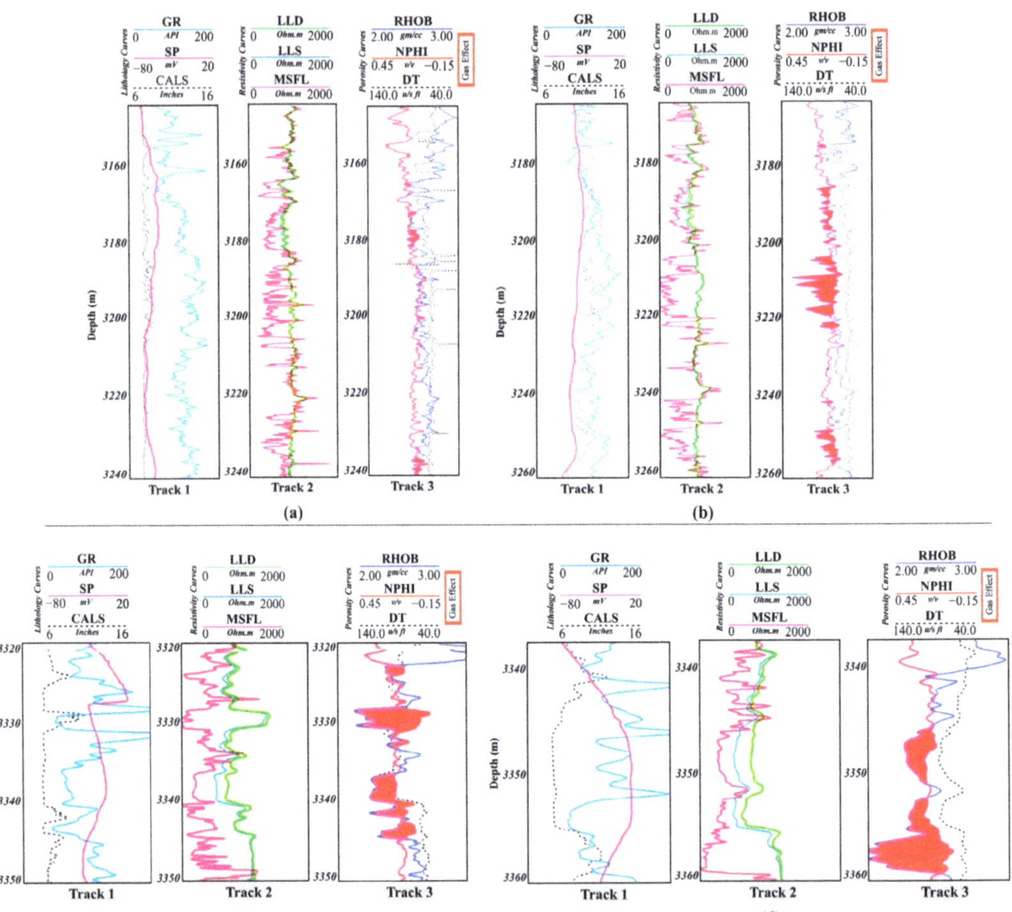

Figure 6. Input conventional log curves for the petrophysical modeling of the G sand interval in the (**a**) Kadanwari-10 (**b**) and Kadanwari-11 wells and for E sand interval in the (**c**) Kadanwari-10 (**d**) and Kadanwari-11 wells.

The following five key steps were followed to evaluate the reservoirs' fundamental properties.

(1) Volume of Shale (V_{shale})—The presence of shale in the productive zone severely impacts the petrophysical properties and can cause a reduction in the $\varnothing_{eff}$, $\varnothing_{avg}$, and permeability [58]. We used the GR log technique for V_{shale} estimation by firstly estimating the gamma ray index (I_{GR}). The I_{GR} was initially adopted using Equation (1) to estimate V_{shale}, utilizing the GR_{log} in track 1 of Figure 6. Secondly, to obtain a realistic V_{shale} estimation without overestimating the content of shale (first-order approximation: $V_{shale} = I_{GR}$), a non-linear relationship (Equation (2)) proposed by Dolan was employed [59].

$$I_{GR} = \frac{GR_{log} - GR_{min}}{GR_{max} - GR_{min}} \tag{1}$$

$$V_{shale} = 1.7 - [(3.38 - (I_{GR} + 0.7)^2)]^{\frac{1}{2}} \tag{2}$$

where I_{GR} stands for gamma ray index, GR_{log} shows the reading of the gamma ray log, GR_{min} and GR_{max} are the lower and upper limits of the GR_{log} value in shale, respectively.

(2) Average Porosity ($\varnothing_{avg}$)—Total porosity or $\varnothing_{avg}$ represents all the voids or pore spaces of the rock, including interconnected and isolated pores and pore spaces occupied by clay-bound water [2]. In this study, DT, RHOB, and NPHI logs that are sensitive to sedimentary micro-facies were selected to calculate $\varnothing_{avg}$, by which process the conventional logging responses of the G and E sand reservoir intervals can be summarized (Figure 6). The DT log measures the sound waves' traveling times in the rock unit. The sound waves in the rock unit depend on the shape, matrix material, and cementation (Equation (3)). Accordingly, the Sonic–Raymer (SR) porosity model was used to evaluate sonic porosity ($\varnothing_S$) (Equation (4)) [40].

$$t_{log} = t_m - [(1 - \varnothing_S) + t_f \, \varnothing_S] \tag{3}$$

$$\varnothing_S = \frac{\Delta t_{log} - \Delta t_m}{\Delta t_m - \Delta t_m} \tag{4}$$

where $\varnothing_S$ represents sonic porosity, t_{log} represents the log reading in $\mu sec/m$, t_m represents the matrix interval transient time, Δt_{log} represents the formation interval transient time in $\mu sec/m$, and Δt_m represents formation fluids' interval transient time in $\mu sec/m$. The density porosity (ϕ_D) was calculated using the RHOB log via Equation (5) [2].

$$\phi_D = [(\rho_{mat} - \rho_b)/(\rho_m - \rho_f)] \tag{5}$$

where ρ_{mat} represents the density of the matrix, ρ_f represents the fluid density, and ρ_b is the bulk density. The NPHI log measures the neutron porosity ($\varnothing_N$) by assuming that the pores are filled with fluid. Therefore, it measures the hydrogen concentration and energy loss. The $\varnothing_N$ can be expressed via Equation (6).

$$\phi_N = [aN + b] \tag{6}$$

where ϕ_N is the neutron-derived porosity, a and b are constants, and N is the neutron count in the formation intervals. In addition, the density–neutron cross-plot in track 4 in Figure 6 determines the cross-over gas effects. After identifying the porosities (e.g., $\varnothing_S$, $\varnothing_D$, and $\varnothing_N$) from the DT, RHOB, and NPHI logs, Equation (7) was used to calculate $\varnothing_{avg}$.

$$\varnothing_{avg} = \frac{\varnothing_S + \varnothing_D + \varnothing_N}{3} \tag{7}$$

(3) Effective Porosity ($\varnothing_{eff}$)—The $\varnothing_{eff}$ was calculated using Equation (8) [40,43].

$$\phi_{eff} = \left[(\phi_{avg}) \times (1 - V_{shale}) \right] \tag{8}$$

where $\varnothing_{avg}$ represents the average porosity, and V_{shale} is the shale content in volume units.

(4) Water Saturation (S_W)—The Poupon–Leveaux Indonesian (PLI) model is one of the best models for estimating S_W in a shaly sand reservoir [60]. In this study, the constraints of V_{shale} (Equation (2)) and $\varnothing_{eff}$ (Equation (8)), and the resistivity variation in V_{shale} and water formation, were subsequently integrated using the PLI model, as in Equation (9), to determine S_W.

$$S_w = \left\{ \left[\left(\frac{V_{shale}{}^{2-V_{shale}}}{R_{shale}} \right)^2 + \left(\frac{\varnothing_{eff}{}^2}{R_w} \right)^{\frac{1}{2}} \right]^2 R_t \right\}^2 \tag{9}$$

where R_t is the true resistivity of the formation obtained from the LLD log response, R_{shale} is the resistivity variation in the V_{shale}, and R_w is the resistivity of water formation.

(5) Hydrocarbon Saturation (S_{hc})—The S_{hc} was calculated by subtracting the percentage of pore volume occupied by S_w from 1; the remaining percentage pore volume gives the S_{hc} (Equation (10)).

$$S_{hc} = 1 - S_w, \tag{10}$$

4. Results

4.1. Stratigraphic Interfaces Interpretation

The interpretation of the stratigraphic interfaces of the early Cretaceous sequence is integrated into an internally consistent 3D workflow, which is easy to incorporate and use to form conclusions that exceed known points by constraining 3D structural models. The interpreted cross-sectional (i.e., vertical) slices of a 3D seismic volume show the displacement and deformation of the stratigraphic interface (Figure 4b). The structural variations in the stratigraphic interfaces are predicated primarily on the TWT contour surfaces because contour lines connect the same elevation; this is why they are essential tools for analyzing and interpreting seismic data. The time surfaces plot the TWT of seismic signals from the surface to the horizon and reflect the interpreted stratigraphic interface's distribution. In Figure 7, the TWT surfaces indicate the extension and propagation of the regional stratigraphic structure in the subsurface. The TWT values of the interpreted stratigraphic interfaces decrease towards the central part of the field, which gives rise to structural highs in the southwestern and southeastern portions. The southwestern and southeastern portions of the interpreted stratigraphic interfaces are structurally high; hence, this is a region of interest for hydrocarbon exploration. The minimum, mean and maximum TWT variations of the G sand interval, the E sand interval, and the Sembar Formation in the southwestern and southeastern transect are presented in Table 2.

Table 2. Minimum, mean, and maximum TWT variations of the stratigraphic interfaces in the study area.

Stratigraphic Interfaces	TWT Minimum	TWT Mean	TWT Maximum	Shallow Structure (TWT)	Deep Structure (TWT)
G sand interval	−1900 (ms)	−2037.5 (ms)	−2175 (ms)	−1900 to −2025 (ms)	−2026 to −2175 (ms)
E sand interval	−2025 (ms)	−2187.5 (ms)	−2350 (ms)	−2025 to −2100 (ms)	−2101 to −2350 (ms)
Sembar Formation	−2250 (ms)	−2400 (ms)	−2550 (ms)	−2250 to −2375 (ms)	−2376 to −2550 (ms)

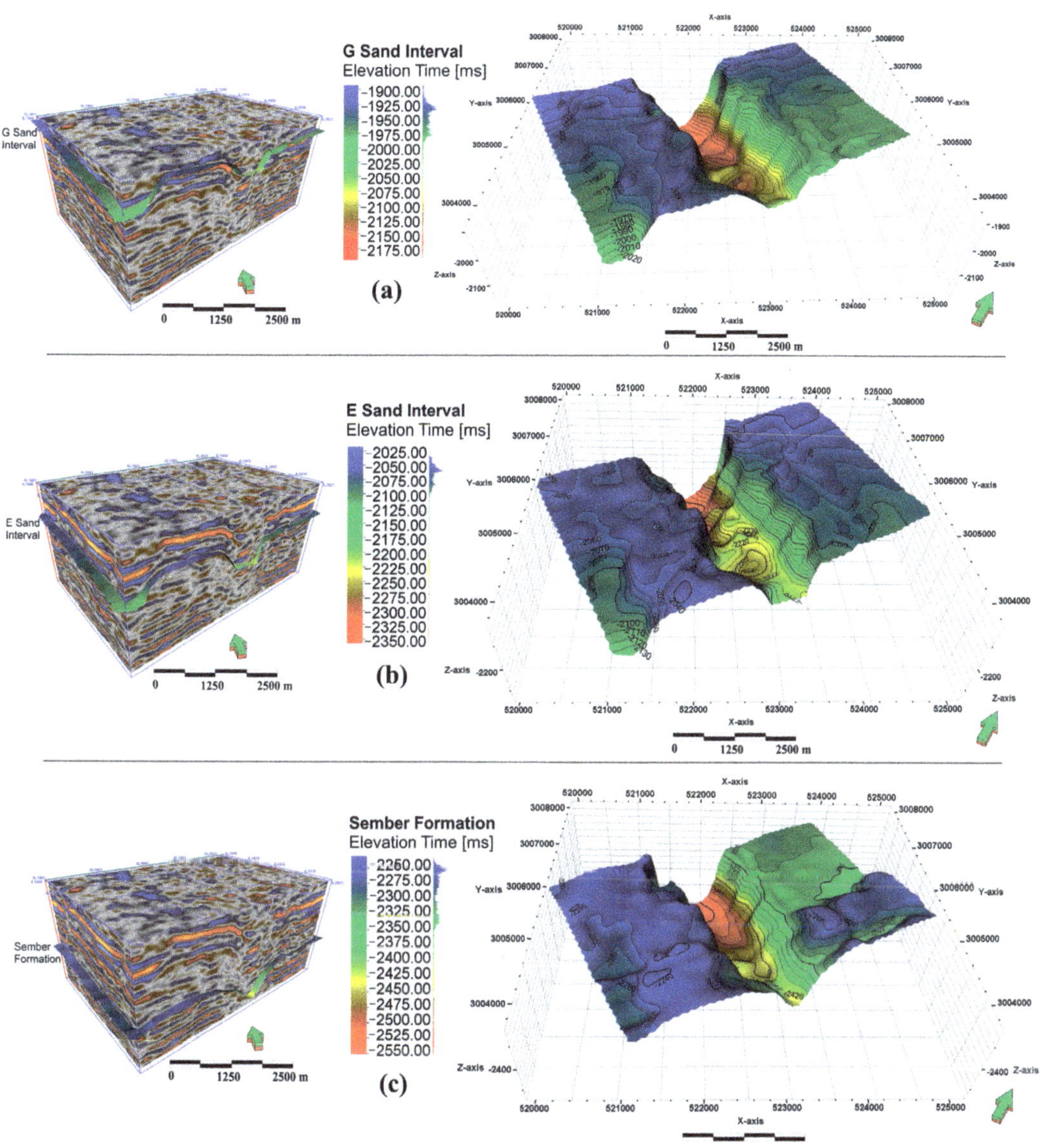

Figure 7. Three-dimensional TWT contour surfaces of the stratigraphic interfaces show different structural distributions: (**a**) TWT top surface of the G sand interval; (**b**) TWT top surface of the E sand interval; (**c**) TWT top surface of the Sembar formation in 3D space.

4.2. Three-Dimensional Fault System Models (3D FSMs)

The study area's complex and composite subsurface morphology resulted from multiple episodes of tectonic activity. Multi-stage tectonic movements contribute to the intersection of faults formed simultaneously or at different stages of tectonic movement (Figure 8a). These faults occurred during various phases of adjustment and deformation, along with the structure inversion and loss of reservoir quality in the Kadanwari field, MIB. In order to effectively represent the spatial distribution of the fault system and its influence on the fragmentation of the stratigraphic interfaces, individual 3D fault surfaces have been illustrated, along with the 3D seismic volume (Figure 8a). The orientation and spatial distribution of the fault surfaces reveal that the intact depositional environment of the Kadanwari field was influenced by the tectonic regime of the surrounding plate boundary, which has continuously influenced the stratigraphic structure (Figure 8a). The 3D seismic volume interpretation at the field scale shows that the middle part of the field has undergone greater deformation than both sides. This is why the number of faults and their complexity decrease from the center to either field side. These faults have been interpreted as normal faults, generally showing NW to SE directions, controlling the distribution of depositional facies, reservoir compartmentalization, and hydrocarbon up-dip migration in the study area. These NW–SE normal faults have governed the complex structural configuration of the stratigraphic sequence (e.g., G sand interval, E sand interval, and the Sembar Formation). The overall pattern of these faults can be regarded as a negative flower structure, which significantly increases the likelihood of the successful positioning of hydrocarbon traps. The degree of completion of these negative flower structures is positively correlated with improved hydrocarbon migration and abundance in the reservoir window (e.g., G and E sand intervals) (Figure 8b). In addition, the presence of a negative flower structure indicates the combined effects of extensional and strike-slip motions. The fault orientation results derived from the 3D dip angle models, the stereonet, the rose diagram, and the histogram show that most of the faults are oriented S30°–45° E and N25°–35° W, with an azimuth of 148°–170° and 318°–345°, and exhibiting minimum, mean and maximum dip angles of 28°, 62°, and 90° respectively (Figure 9a). The histogram in Figure 9b displays the relationship between fault frequency in percentage and dip angle in degrees, which shows that most of the faults have dips ranging between 35° and 75°. In comparison, 20% of the total fault planes have dips in the range 80°–90°, and the lowest fault dip observed is approximately 28° (Figures 8a and 9a).

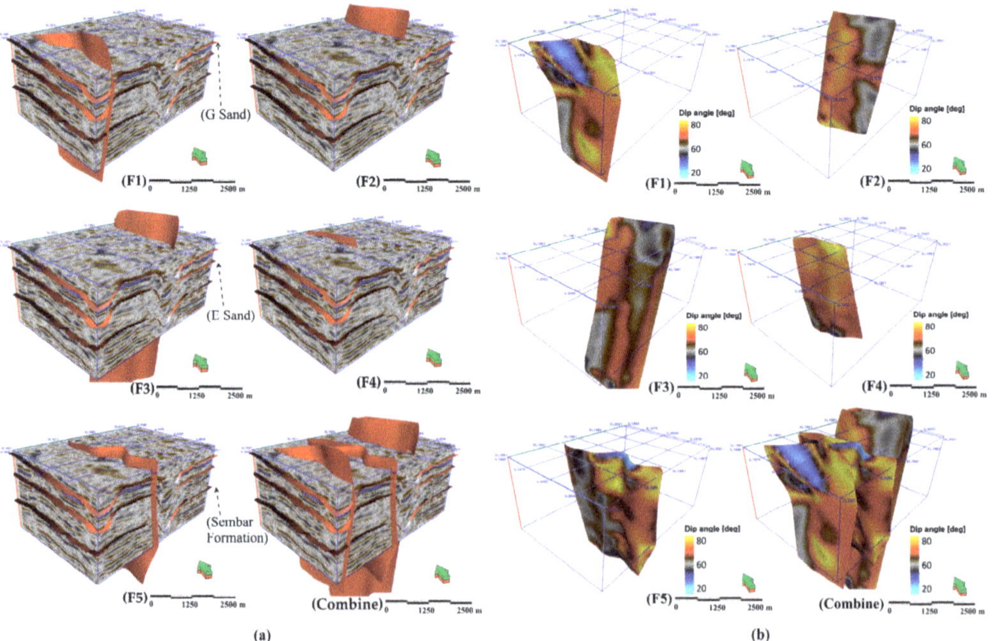

Figure 8. (a) The individual and combined distribution of tectonic extensional fault surfaces along the early Cretaceous stratigraphic sequence in the 3D seismic volume (SW-SE) and (b) 3D dip angle models (SW-SE) of the individual and combined tectonic extensional fault surfaces.

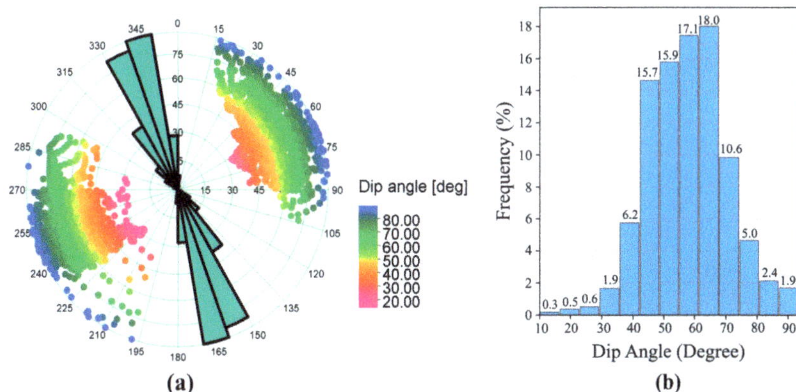

(a)   (b)

Figure 9. (**a**) Steroenet showing the distribution of faults' dips and orientations identified, along with the entire seismic survey in the Kadanwari field, MIB, Pakistan; (**b**) a histogram showing the dip angles of the interpreted faults along with their frequency.

4.3. Three-Dimensional Structural Models (3D SMs)

The 3D SMs and several 2D time-domain structural cross-sections have been derived to illustrate the detailed structural and stratigraphic setting of the sequence in the study area. Figure 10a displays the southwestern and southeastern transects of the 3D TWT model, while Figure 10b represents the 2D TWT cross-sections derived from the 3D TWT model result. Figure 11a displays the southwestern and southeastern transect of the 3D SMs derived from VBM. Moreover, the fault system is incorporated into the 2D cross-sections during model computation to constrain the horizons at each fault offset (Figure 11b).

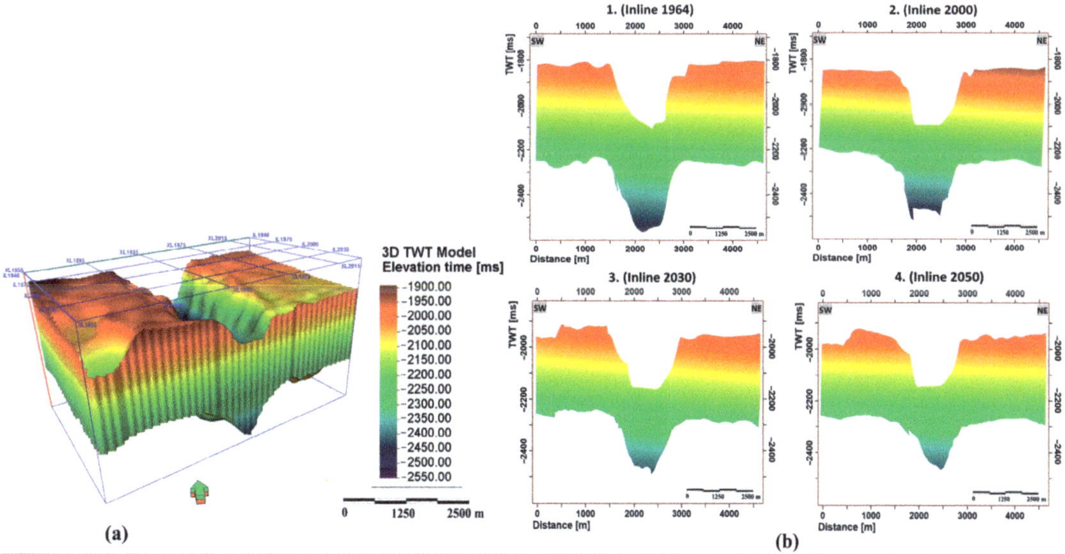

(a)   (b)

Figure 10. TWT domain structural model of the early Cretaceous stratigraphic sequence (e.g., G sand interval, E sand interval, and the Sembar Formation): (**a**) SW–SE view; (**b**) extracted inline 2D TWT cross-sections.

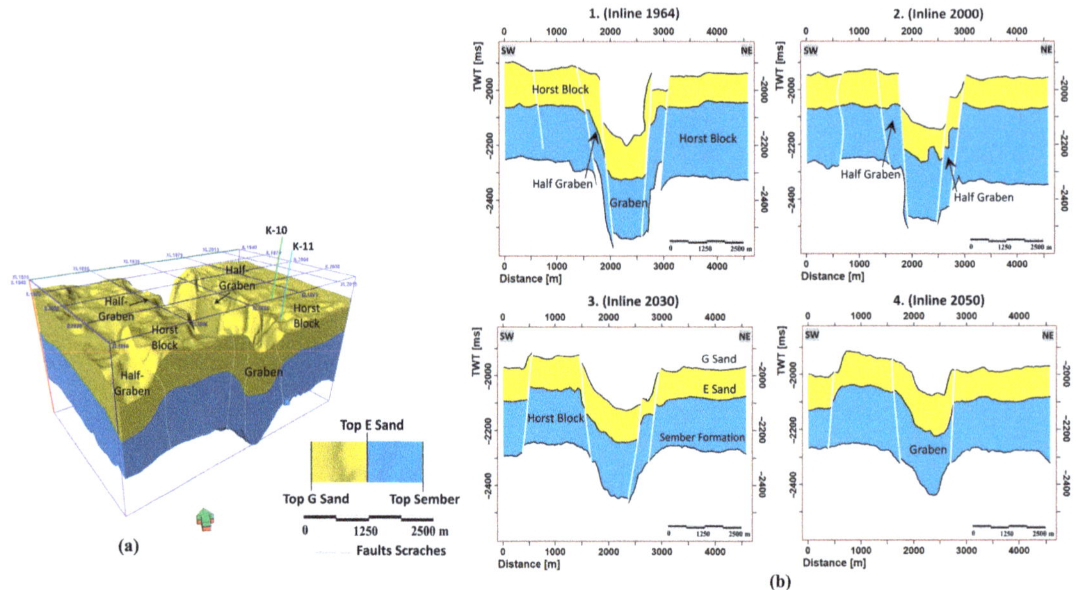

Figure 11. Final result of the 3D SM: (**a**) SW–SE view and (**b**) SW–NE oriented 2D cross-sections drawn from the 3D SM results.

Detailed structural analysis reveals that the geological structural complexity is a consequence of different tectonic phases of deformation (e.g., extensional and strike-slip deformation from the early Cretaceous to Quaternary). The complex structural mechanics of both extensional and strike-slip movements have been observed in the 3D SMs. These complex structural mechanics were controlled by the normal fault system's NW–SE dipping. The above-explained normal fault patterns (e.g., negative flower structure) show brittle deformation features, where the interpreted sequences have been displaced relative to each other (amount of displacement) (Figure 11b). These significantly control the structural domains, consisting mainly of half-graben, horst, half-graben, graben, half-graben, and horst, from SW to NE. The thickness of the sequence increases towards the NE. The thickness decreases in the central portion of the region from the top E sand interval to the top Sembar Formation. The hydrocarbon migration direction can be determined from the spatial distribution and composition of the fault system within the early Cretaceous sequences. The geometrical trend of these fault systems creates a pathway that is crucial for hydrocarbon migration in the vertical direction. This up-dip migration of the hydrocarbons from the Sembar formation towards the G and E sand intervals has resulted in hydrocarbon accumulation, validated through seismic attribute analysis and petrophysical modeling.

4.4. Seismic Attributes Interpretation
4.4.1. Variance Edge Attribute

The variance edge attribute indicates discontinuities related to faulting or stratigraphy in the vertical seismic cross-sections and is proved to help depict significant fault zones and fractures, thereby validating fault interpretation. Figure 12 shows the results of the variance edge attribute calculated from the 3D seismic volume, appropriate cross-sections (e.g., A–A′, B–B′, C–C′, D–D′, E–E′ and F–F′), and horizon slices. The horizon slices include the G sand interval, the E sand interval, and the Sembar Formation. The variance edge attribute values in the 2D variance cross-sections and horizon slices range from 0.0 to 1.0. A variance value equal to 1 indicates discontinuity (fault), while a value of 0 variances represents continuous seismic events. The darkest regions (e.g., values ranging from 0.8 to 1) in vertical strips can be interpreted as faults or fractures (Figure 12c). These faults and fractures create an essential pathway for vertical hydrocarbon migration.

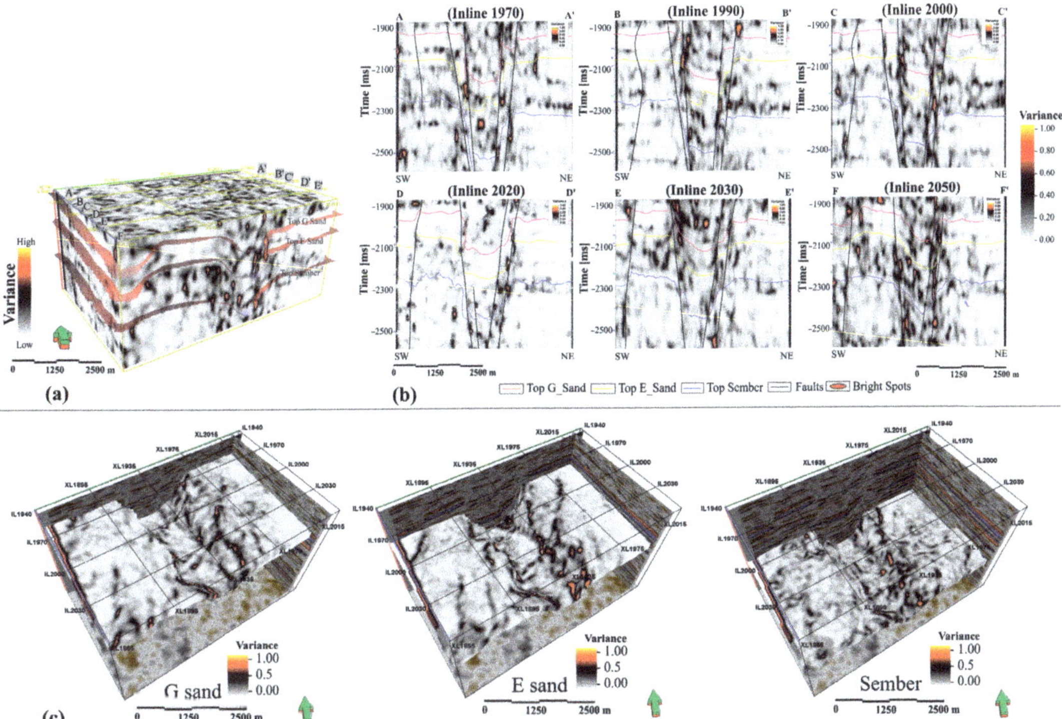

Figure 12. (**a**) 3D variance edge attribute volume, (**b**) 2D extracted variance cross-sections, and (**c**) 3D variance horizon slices of the G sand interval, the E sand interval, and the Sembar Formation showing discontinuities.

4.4.2. Sweetness Attribute

Figure 13 shows the sweetness attribute computed from the 3D seismic volume, corresponding cross-sections (e.g., A–A′, B–B′, C–C′, D–D′, E–E′ and F–F′), and sweetness horizon slices. The high sweetness anomalies (at −1900 to −2300 ms) on the 2D sweetness cross-sections (Figure 13b) and the 3D horizon slices (e.g., G sand and E sand intervals) (Figure 13c) contributed to the high amplitude and low frequency. In contrast, the low sweetness anomalies (at −2300 to −2500 ms) on the 2D sweetness cross-sections and the SW—NE parts of the 3D horizon slices (e.g., the Sembar Formation) due to the seismic reflection low amplitude and high frequency. The high amplitude (high acoustic impedance as opposed to shale) and low-frequency anomalies on the 2D sweetness cross-sections and the 3D sweetness horizon slices represent cleaner and more payable sand zones. These sweet spots suggest the presence of a high proportion of porous sand and seem to hold potential for producing gas in the SW and NE parts of the G and E sand reservoir intervals (Figure 13c). In contrast, areas with low amplitude and high-frequency anomalies within the −2100 to −2300 (ms) in the 2D sweetness cross-sections non-reservoir window (e.g., the Sembar Formation) are shale-prone; the sands here may be interbedded with shale. Although the sweetness attribute effectively distinguishes sand bodies from shale, it does so via the high acoustic impedance contrast between sand and shale.

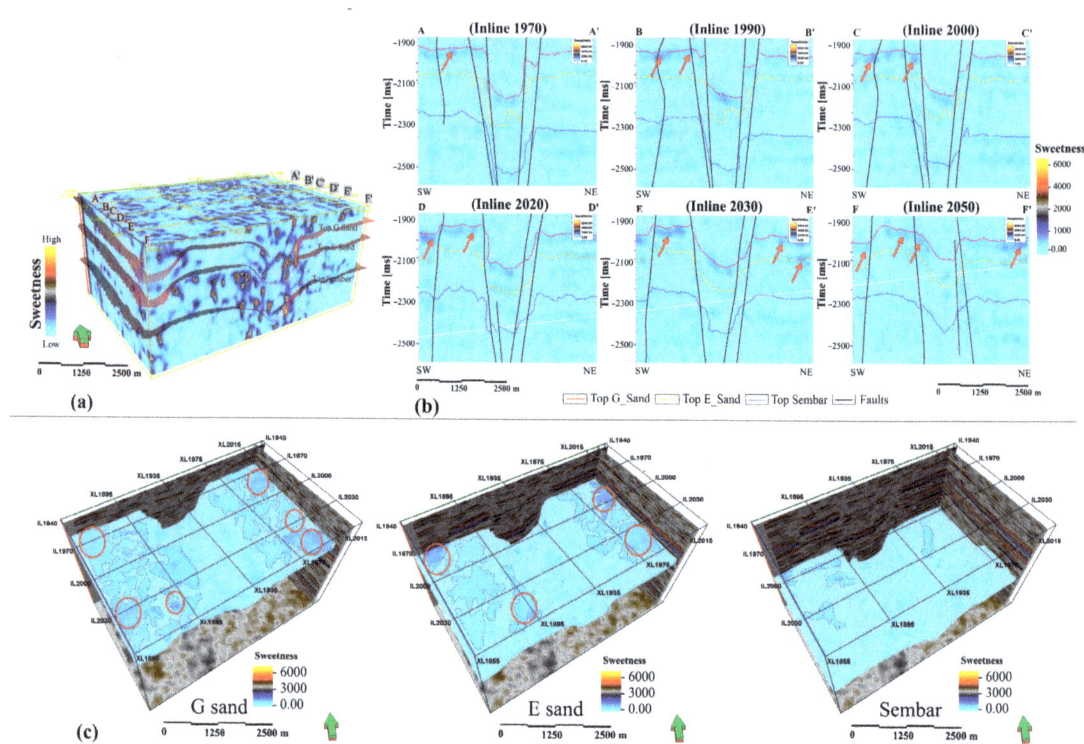

Figure 13. (**a**) 3D sweetness attribute volume, (**b**) 2D extracted sweetness cross-sections, and (**c**) 3D sweetness horizon slices of the G sand interval, the E sand interval, and the Sembar Formation.

4.4.3. RMS Amplitude Attribute

Figure 14 shows the results of RMS amplitude inferred from the 3D seismic volume, appropriate 2D cross-sections (e.g., A–A′, B–B′, C–C′, D–D′, E–E′ and F–F′), and 3D horizon slices (e.g., G sand interval, E sand interval, and the Sembar Formation). The extracted 2D RMS amplitude cross-section anomalies range from 0 to 4000 ms. These amplitude variations evaluated the structure's influence on depositional facies. The moderate to high-amplitude anomalies, e.g., between 2500 and 4000 (ms), are often associated with channel sand bodies, high porosity (porous sands), and sand-rich sand shoreward facies, especially gas saturated sand zones. The lateral and horizontal spatial distributions of facies show that faults and fractures significantly reduce the reservoir quality (G and E sand intervals). The gas-saturated sand in the SW and NE parts have high reflectivity, indicating high porosity within the G and E sand intervals (Figure 14c). These sand-rich gas-saturated zones can be considered for future gas exploration in the study area. In comparison, low amplitude anomalies, e.g., between 0 and 2000 (ms), may indicate that these zones contain sandy shale, making them unfavorable for gas production. Such unfavorable zones mainly include the Cretaceous organic-rich shales of the Sembar Formation (Figure 14c).

4.5. Petrophysical Modeling

Petrophysical modeling unveils the reservoir traits and offers suggestions of hydrocarbon-bearing zones. The average petrophysical properties such as V_{shale}, $\varnothing_{avg}$, $\varnothing_{eff}$, and S_W of the G sand interval in both wells are 36.11%, 12.5%, 7.5%, and 45%, respectively (Table 3). Similarly, the derived average petrophysical properties for the E sand interval in both wells are 30.5%, 17.4%, 12.2%, and 33.2%, respectively (Table 4). A graphical representation of these properties is presented in Figure 15. The overall description of the G and E sand reservoir intervals depends on these petrophysical properties. This may significantly influence the decision-making process in all phases of the planning and executing hydrocarbon activities in the Kadanwari field, MIIB, Pakistan.

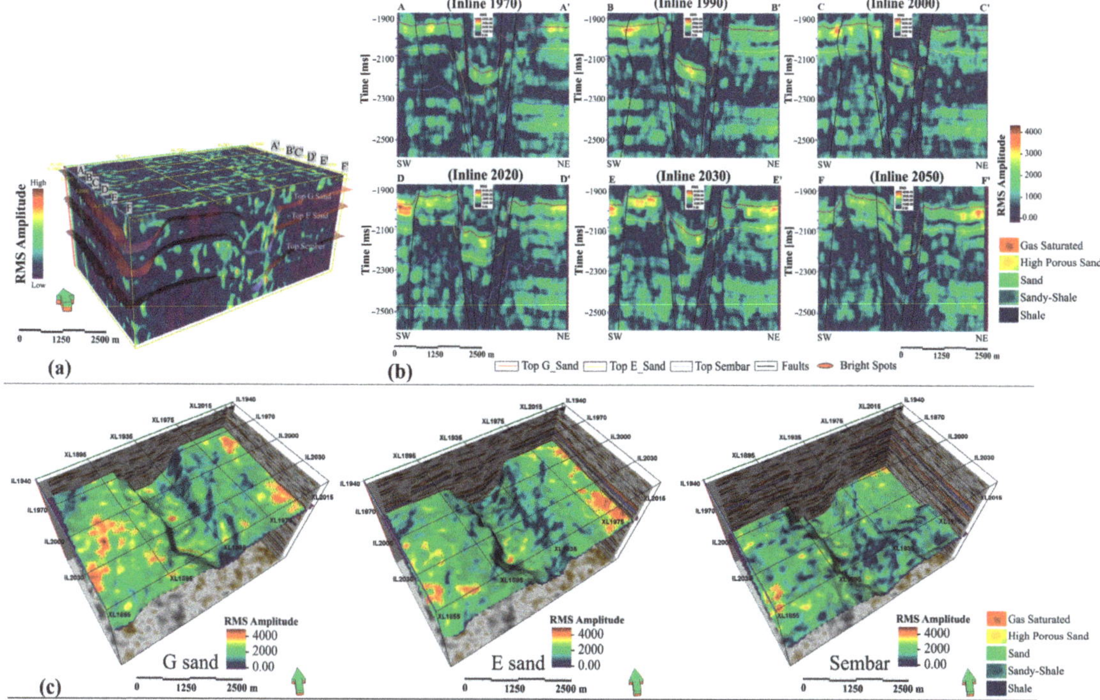

Figure 14. (**a**) 3D RMS amplitude volume, (**b**) corresponding 2D extracted cross-sections, and (**c**) 3D RMS amplitude horizon slices of G sand, E sand, and the Sembar Formation.

Table 3. Petrophysical properties of the Cretaceous G sand interval in the Kadanwari-10 and Kadanwari-11 wells.

Well No.	Intervals	Volume of Shale (V_{shale}) %	Effective Porosity ($\varnothing_{eff}$) %	Average Porosity ($\varnothing_{avg}$) %	Water Saturation (S_w) %
Kadanwari-10	G sand interval	36.11	7.8	12.2	45.4
Kadanwari-11	G sand interval	36.11	8	12.56	45.4

Table 4. Petrophysical properties of the Cretaceous E sand interval in the Kadanwari-10 and Kadanwari-11 wells.

Well No.	Intervals	Volume of Shale (V_{shale}) %	Effective Porosity ($\varnothing_{eff}$) %	Average Porosity ($\varnothing_{avg}$) %	Water Saturation (S_w) %
Kadanwari-10	E sand interval	27.02	13.2	18.1	30.09
Kadanwari-11	E sand interval	34.05	11.3	16.7	36.42

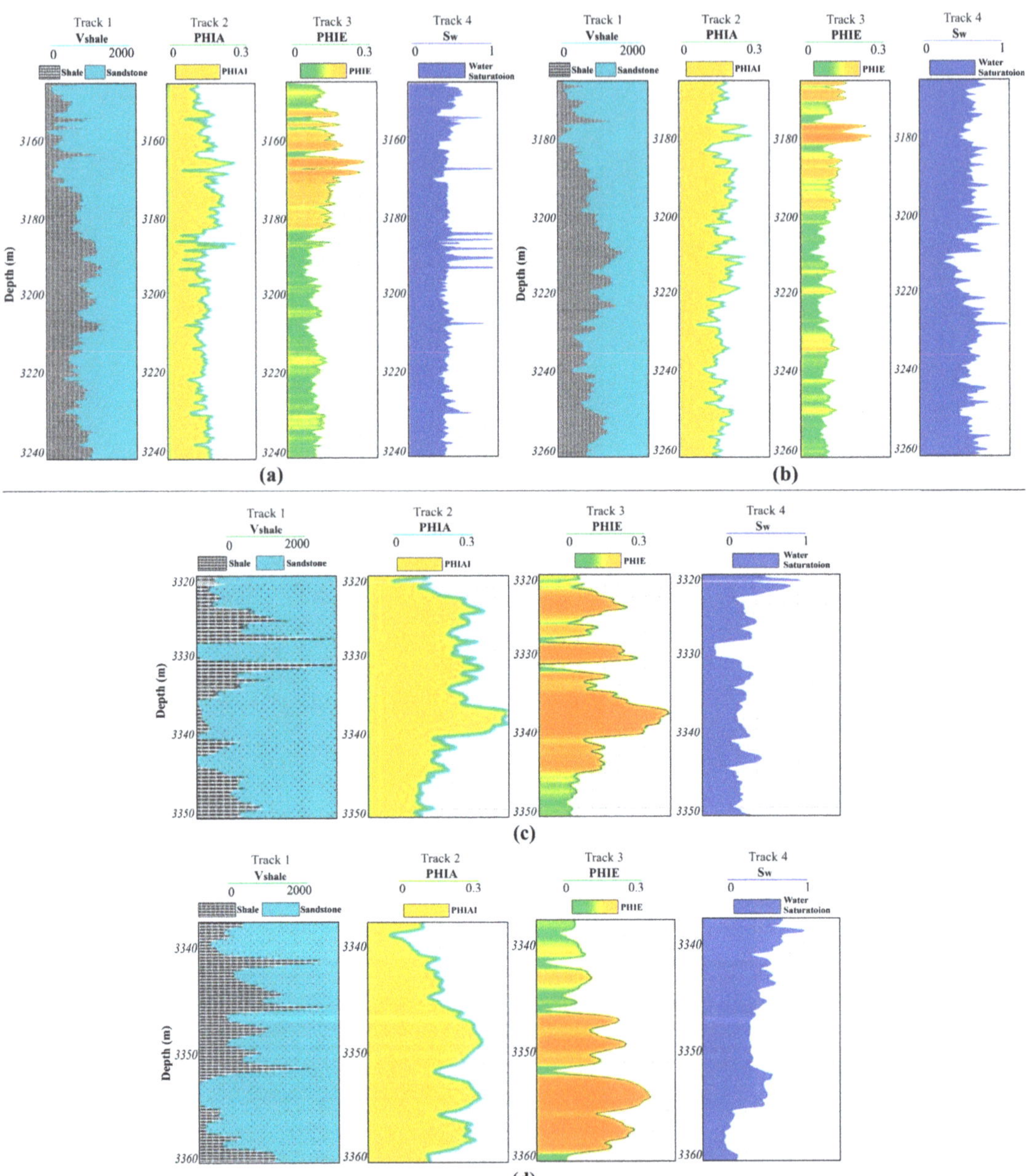

Figure 15. Relationship between the V_{shale} and volume of sand (track 1), distribution of $\varnothing_{avg}$ (track 2), distribution of $\varnothing_{eff}$ (track 3), and the relationship between S_W and S_{hc} with respect to the depth at the G sand interval in (**a**) Kadanwari 10 and (**b**) Kadanwari-11 wells, and at the E sand interval in (**c**) Kadanwari 10 and (**d**) Kadanwari-11 wells.

The GR log response is sensitive to the radioactive emissions predominantly concentrated in the clay minerals of shale and clean sand (feldspar-rich) [43]. The GR response in track 1 of each understudy well confirms the reservoir lithology to be sandstone (Figure 15a). Figure 15a represents the relationship between V_{shale} and sand content (track 1), $\varnothing_{avg}$ distribution (track 2), $\varnothing_{eff}$ distribution (track 3), and the relation between S_W and S_{hc} (track 4) at the G sand interval in the Kadanwari-10 and 11 wells. Similarly, Figure 15b represents the relationship between V_{shale} and sand content (track 1), $\varnothing_{avg}$ distribution (track 2), $\varnothing_{eff}$ distribution (track 3), and the relation between S_W and S_{hc} (track 4) at the E sand interval in the Kadanwari-10 and 11 wells.

5. Discussion

5.1. Integration of 3D SM and JGC for Hydrocarbon Evaluation

Knowledge of structural, lithological, and petrophysical characteristics is essential to reducing the uncertainties associated with reservoir description. Herein, 3D seismic data and well logs data have been critically analyzed to evaluate the complex and heterogeneous depositional environment of the early Cretaceous stratigraphic sequence, along with structural and stratigraphic characteristics, distribution of associated petrophysical properties, and spatial facies for reliable reservoir characterization. The result indicates that the early Cretaceous stratigraphic sequence appears irregularly, with faulted structural highs bounded by extension-related normal faults, heterogeneous nature of reservoir intervals, and potential gas-saturated zones.

The structural interpretation of the early Cretaceous stratigraphic sequence indicates that the interpreted horizons (e.g., G sand, E sand, and the Sembar Formation) appear normal faulted extension-related horsts, half-graben, and graben structures (Figure 11). The normal faulted inversion-related horsts, half-graben, and graben structures show how the sedimentary layer's structural features have contributed to the formation of traps and conduit mechanisms. In general, due to the exact kinematic mechanisms involved in the sequential deformation of the Kadanwari field, the structural deformation of each horizon is equal in orientation and extent (Figure 11b). However, the E sand interval seems to be more deformed than the G sand interval and the Sembar Formation. The geometric tendency of the extension-related normal fault in the early Cretaceous stratigraphic sequence is essential because it has created channels for hydrocarbon migration in the vertical direction. Hydrocarbon enrichment in the G and E sand reservoir intervals was positively related to the complexity of the internal structure of normal faults. The hydrocarbon concentration and dominance of these normal faults are more intense towards the central part of the seismic volume, where the faults are closely spaced. The normal fault plane profiles show the hanging wall and footwall cutoffs (Figure 11b). These profiles are useful for understanding the seal behavior for hydrocarbon potential and prospect evaluation.

Using 3D SM, we determined that the extension–related normal faults' geometrical trend created an essential pathway for the up-dip migration of hydrocarbons from the source rock (the Sembar Formation) towards the G and E sand reservoir intervals, resulting in hydrocarbon accumulation. These statements are validated here, as the moderate to high RMS amplitude anomalies are often associated with gas–saturated sand zones (Figure 14c). The hydrocarbon prospects have high reflectivity, indicating high porosity, which can be seen on both the sweetness (Figure 13b) and RMS amplitude (Figure 14b) attributes within the G and E sand intervals. In addition, the RMS amplitude attribute is advantageous compared with the sweetness attribute because the RMS amplitude attribute has a higher resolution when depicting the porous zones and DHIs. These results can significantly reduce the risk associated with hydrocarbon exploration and development in the Kadanwari field, MIB.

The calculated values of the S_{hc} form the basis of future production forecasts and the determination of the economic viability of a discovered reservoir. Therefore, high accuracy is needed when determining S_W, as it is used to calculate the estimated S_{hc} reserves. The low values found for V_{shale} content in the Kadanwari 10 and 11 wells indicate the cleanliness of the sandstone. Accordingly, the shale and sandstone facies were separated by a 40% cutoff value in the targeted reservoir intervals, i.e., G and E sands. The V_{shale} contents in the G and E sand intervals are influenced by clay minerals, which reduce the $\varnothing_{avg}$ and $\varnothing_{eff}$.

The high values of $\varnothing_{eff}$ refer to better volume estimation, and thus, a theoretically good reservoir, and vice versa. These high values of $\varnothing_{eff}$ indicate the amounts of connected pore spaces in the reservoir intervals [43]. The G and E sand reservoir interval properties, as derived from our petrophysical modeling, are in good agreement with the results of the regional study conducted by [40,43].

5.2. Analysis of Gas Reserve of the Kadanwari Field

Graphical representations of the S_{hc} at the G and E sand reservoir intervals are presented in track 4 in Figure 15. The petrophysical modeling results show that these intervals have good S_{hc}, as the S_W is below 60% (Table 5). A comparison between the G and E sand interval petrophysical properties shows that the E sand interval has good $\varnothing_{eff}$ and S_{hc}, with an apparent gas effect revealed by the cross-overs of the DT and NPHI logs curves (Figure 6). Therefore, it can be considered as an economically viable reservoir interval. The S_{hc} percentage of the E sand interval is satisfactory for exploration purposes.

Table 5. Depth range, thickness, and S_{hc} (%) of the Cretaceous G and E sand intervals in Kadanwari-10 and 11 wells.

Well No.	Intervals	Depth Range (m)	Thickness (m)	Hydrocarbon Saturation (S_{hc}) %
Kadanwari-10	G sand interval	3145–3240	95	54.6
	E sand interval	3320–3350	30	70.01
Kadanwari-11	G sand interval	3167–3260	93	55.02
	E sand interval	3337–3360	23	63.58

The annual report (2010–2011) of petroleum exploration and production activities in Pakistan analyzed the reserves of the Kadanwari field (Table 6) [7]. The Kadanwari field exhibits a significant amount of original recoverable gas reserves, i.e., 1110 billion cubic feet (Bcf) equivalent to 190 million barrels of oil equivalent (Mboe), respectively. From these original recoverable gas reserves, 420 (Bcf) has been extracted from the Kadanwari field. As of 30 June 2011, the balance recoverable gas reserves were 690 Bcf, equivalent to 110 Mboe. The collective original recoverable reserves of the Kadanwari field—280 Mboe—are now limited to 129 Mboe [7]. Thus, many hydrocarbon reserves are present in the Kadanwari field. The high level of gas reserves in the Kadanwari field compared to other fields (such as the Miano field) may be attributed to its complex structural configuration (e.g., negative flower structure of fault system), reservoir compartmentalization, and the up-dip migration of the hydrocarbons into the reservoir intervals (e.g., G and E sands).

Table 6. The gas reserves of the Kadanwari and Miano fields as of 30 June 2011.

Fields	Original Recoverable	Cumulative Production	Balance Recoverable
Kadanwari	1110 BCF	420 BCF	690 BCF
Miano	552 BCF	438 BCF	114 BCF

5.3. Comparative Analysis with Other Reservoirs

The detailed structural interpretation of the 3D SMs and appropriate 2D cross-sections has revealed that the geological structural complexity is a consequence of different tectonic phases of deformation (compressional regimes of the surrounding plate boundary). The structural and stratigraphic characteristics results derived from the 3D FSMs, the 3D TWT model, the 3D SMs, the dip angle models, the rose diagram, and the histogram agree with the regional studies conducted by other researchers [7,45,49]. These studies show similar structural and stratigraphic characteristics and patterns. According to [7], most faults dip towards the southwest, with an average throw in the order of about 50 m and a maximum throw of 113 m in the Kadanwari area. Wrench or strike-slip faults are absent, except for one potential one wrench fault (F3) to the south in the Kadanwari field (Figure 9a). Based on this, the possibility of strike-slip deformation may also be inferred. References [7,45,49] also stressed that the Kadanwari field is structurally essential due to the presence of fault-bounded structures, which may be considered as potential prospects. The structural characteristics of the nearby fields within the MIB and LIB, such as the Miano, Sawan, and Zamzama fields, can also be correlated with the conducted study of the Kadanwari field.

The adopted methods of 3D SM and JGC can also be utilized in the international basins having the same geology of extensional regimes featuring the horst and graben structures as a petroleum play. These include the Bach Ho oilfield in Cuu Long Basin of Vietnam [61] and the Mannar frontier sedimentary basin of Sri Lanka [62]. In addition, this study is helpful to characterize and evaluate reservoir geometrical characteristics, facies, and properties to reduce uncertainties and improve the success rate of future exploration and development plans pertaining to hydrocarbons in the regions having the same geology globally. In general, conventional methods remain inadequate in individual or integrated form. Therefore, machine learning tools may provide guidance for detailed structural and petrophysical evaluation. In future work, the manual structural interpretation, seismic attributes maps, and petrophysical properties can be used as input databases of machine learning, especially deep learning, models for subsequent automated 3D structural, facies, and petrophysical modeling.

6. Conclusions

We introduced a novel methodology of 3D structural modeling (3D SM) and joint geophysical characterization (JGC), which comprises seismic interpretation-aided 3D structural modeling, seismic attributes, and petrophysical modeling for reservoir characterization in the Kadanwari field, Middle Indus Basin (MIB), Pakistan. Our main findings are as follow:

(1) The 3D structural interpretation illustrates the complex structural mechanics, controlled by the NW–SE dipping normal faults system, operating in the early Cretaceous stratigraphic sequence. The identified features include horsts, half-graben, and graben structures. The spatial distribution of the fault system shows that the overall pattern of the interpreted fault system can be regarded as a negative flower structure. The negative flower structure incorporates the combined effects of extensional and strike-slip motion in the study area. In general, the horsts, half-graben, and graben, along with the faults, have geometrically determined reservoir (G and E sand intervals) geomorphology, up-dip hydrocarbon migration, the development of the local strata, the distribution of facies and properties, and internal structural deformation;

(2) The variance edge attribute enhanced the geometric distribution of the faults within the seismic data. The sweetness attribute distinguished the sand facies from shale, as the increased amplitude and lower frequency content represent cleaner and more payable sand zones. In contrast, areas with low amplitude and high-frequency anomalies are susceptible to shale. The RMS amplitude and sweetness attribute results indicate the hydrocarbon zones. Relatively high RMS amplitude attribute values are usually connected with lithological changes, sand-rich shoreward facies, bright spots, and especially gas-saturated sand zones. In comparison, low amplitudes anomalies indicate the zones of sandy-shale, shale, and pro-delta facies;

(3) Petrophysical modeling reveals the important parameters of G and E sand reservoir intervals. The $\varnothing_{avg}$ values calculated via the Sonic–Raymer (SR) porosity model, the RHOB log, and the NPHI log show that the G and E sand reservoir intervals have good porosities. Moreover, the E sand interval has good $\varnothing_{eff}$ and S_{hc} and displays clear signs of gas effects verified by the cross-overs of density and neutron log curves. Therefore, it can be considered an economically viable reservoir interval for future hydrocarbon production.

Author Contributions: Conceptualization, U.K. and B.Z.; methodology, U.K. and J.D.; software, U.K.; validation, B.Z., J.D., and M.K.; formal analysis, Y.T. and S.A.; investigation, I.A.; data curation, S.H.; writing—original draft preparation, U.K. and Z.J.; writing—review and editing, Y.T. and U.K.; funding acquisition, B.Z. All authors have read and agreed to the published version of the manuscript.

Funding: This study was supported by grants from the National Natural Science Foundation of China (Grant No. 42072326 and 41772348) and the National Key Research and Development Program of China (Grant No. 2019YFC1805905).

Data Availability Statement: The dataset of the current study are not publicaly availible due to a data privacy agreement we signed with the Directorate General of Petroleum Concessions (DGPC), Pakistan, but are availible from the corresponding author on reasonable request.

Acknowledgments: The authors thank Landmark Resources (LMKR) and the Directorate General of Petroleum Concessions (DGPC), Pakistan, for providing the dataset. Many thanks to Umar Ashraf (Yunnan University, China) for revising the final version of this manuscript.

Conflicts of Interest: The authors declare no conflict of interest.

References

1. Qadri, S.T.; Islam, M.A.; Shalaby, M.R.; Ali, S.H. Integration of 1D and 3D modeling schemes to establish the Farewell Formation as a self-sourced reservoir in Kupe Field, Taranaki Basin, New Zealand. *Front. Earth Sci.* **2020**, *15*, 631–648. [CrossRef]
2. Thota, S.T.; Islam, M.A.; Shalaby, M.R. A 3D geological model of a structurally complex relationships of sedimentary Facies and Petrophysical Parameters for the late Miocene Mount Messenger Formation in the Kaimiro-Ngatoro field, Taranaki Basin, New Zealand. *J. Pet. Explor. Prod. Technol.* **2021**, *11*, 1–36. [CrossRef]
3. Masters, C.D.; Root, D.H.; Attanasi, E.D.; Tedeschi, M.; Singh, S.; du Plessis, M.; Wong, S.; Redford, D.; Wightman, D.; MacGillivray, J. World resources of crude oil and natural gas. *Energy Explor. Exploit.* **1991**, *9*, 354–374.
4. Mangi, H.N.; Chi, R.; DeTian, Y.; Sindhu, L.; He, D.; Ashraf, U.; Fu, H.; Zixuan, L.; Zhou, W.; Anees, A. The ungrind and grinded effects on the pore geometry and adsorption mechanism of the coal particles. *J. Nat. Gas Sci. Eng.* **2022**, *100*, 104463. [CrossRef]
5. Mangi, H.N.; Detian, Y.; Hameed, N.; Ashraf, U.; Rajper, R.H. Pore structure characteristics and fractal dimension analysis of low rank coal in the Lower Indus Basin, SE Pakistan. *J. Nat. Gas Sci. Eng.* **2020**, *77*, 103231. [CrossRef]
6. Sheikh, N.; Giao, P.H. Evaluation of shale gas potential in the lower cretaceous Sembar formation, the southern Indus basin, Pakistan. *J. Nat. Gas Sci. Eng.* **2017**, *44*, 162–176. [CrossRef]
7. Saif-Ur-Rehman, K.J.; Lin, D.; Ehsan, S.A.; Jadoon, I.A.; Idrees, M. Structural styles, hydrocarbon prospects, and potential of Miano and Kadanwari fields, Central Indus Basin, Pakistan. *Arab. J. Geosci.* **2020**, *13*, 97.
8. Ahmad, N.; Khan, S.; Al-Shuhail, A. Seismic Data Interpretation and Petrophysical Analysis of Kabirwala Area Tola (01) Well, Central Indus Basin, Pakistan. *Appl. Sci.* **2021**, *11*, 2911. [CrossRef]
9. Khan, U.; Zhang, B.; Du, J.; Jiang, Z. 3D structural modeling integrated with seismic attribute and petrophysical evaluation for hydrocarbon prospecting at the Dhulian Oilfield, Pakistan. *Front. Earth Sci.* **2021**, *15*, 649–675. [CrossRef]
10. Bodunde, S.; Enikanselu, P. Integration of 3D-seismic and petrophysical analysis with rock physics analysis in the characterization of SOKAB field, Niger delta, Nigeria. *J. Pet. Explor. Prod. Technol.* **2019**, *9*, 899–909. [CrossRef]
11. Hossain, M.I.S.; Woobaidullah, A.; Rahman, M.J. Reservoir characterization and identification of new prospect in Srikail gas field using wireline and seismic data. *J. Pet. Explor. Prod. Technol.* **2021**, *11*, 2481–2495. [CrossRef]
12. Kargarpour, M.A. Carbonate reservoir characterization: An integrated approach. *J. Pet. Explor. Prod. Technol.* **2020**, *10*, 2655–2667. [CrossRef]
13. Islam, M.A.; Yunsi, M.; Qadri, S.T.; Shalaby, M.R.; Haque, A.E. Three-dimensional structural and petrophysical modeling for reservoir characterization of the Mangahewa formation, Pohokura Gas-Condensate Field, Taranaki Basin, New Zealand. *Nat. Resour. Res.* **2021**, *30*, 371–394. [CrossRef]
14. Li, X.; Zhou, N.; Xie, X. Reservoir characteristics and three-dimensional architectural structure of a complex fault-block reservoir, beach area, China. *J. Pet. Explor. Prod. Technol.* **2018**, *8*, 1535–1545. [CrossRef]
15. Osinowo, O.O.; Ayorinde, J.O.; Nwankwo, C.P.; Ekeng, O.; Taiwo, O. Reservoir description and characterization of Eni field Offshore Niger Delta, southern Nigeria. *J. Pet. Explor. Prod. Technol.* **2018**, *8*, 381–397. [CrossRef]
16. Haque, A.E.; Islam, M.A.; Shalaby, M.R. Structural modeling of the Maui gas field, Taranaki basin, New Zealand. *Pet. Explor. Dev.* **2016**, *43*, 965–975. [CrossRef]
17. Mutebi, S.; Sen, S.; Sserubiri, T.; Rudra, A.; Ganguli, S.S.; Radwan, A.E. Geological characterization of the Miocene–Pliocene succession in the Semliki Basin, Uganda: Implications for hydrocarbon exploration and drilling in the East African Rift System. *Nat. Resour. Res.* **2021**, *30*, 4329–4354. [CrossRef]
18. Abdeen, M.M.; Ramadan, F.S.; Nabawy, B.S.; El Saadawy, O. Subsurface Structural Setting and Hydrocarbon Potentiality of the Komombo and Nuqra Basins, South Egypt: A Seismic and Petrophysical Integrated Study. *Nat. Resour. Res.* **2021**, *30*, 3575–3603. [CrossRef]

19. Ashraf, U.; Zhu, P.; Yasin, Q.; Anees, A.; Imraz, M.; Mangi, H.N.; Shakeel, S. Classification of reservoir facies using well log and 3D seismic attributes for prospect evaluation and field development: A case study of Sawan gas field, Pakistan. *J. Pet. Sci. Eng.* **2019**, *175*, 338–351. [CrossRef]
20. Hussain, M.; Ahmed, N.; Chun, W.Y.; Khalid, P.; Mahmood, A.; Ahmad, S.R.; Rasool, U. Reservoir characterization of basal sand zone of lower Goru Formation by petrophysical studies of geophysical logs. *J. Geol. Soc. India* **2017**, *89*, 331–338. [CrossRef]
21. Moore, G.F. 3-D seismic interpretation. *Am. Geophys. Union* **2009**, *90*, 161. [CrossRef]
22. Vo Thanh, H.; Sugai, Y.; Sasaki, K. Impact of a new geological modelling method on the enhancement of the CO2 storage assessment of E sequence of Nam Vang field, offshore Vietnam. *Energy Sources Part A Recovery Util. Environ. Eff.* **2020**, *42*, 1499–1512. [CrossRef]
23. Houlding, S. *3D Geoscience Modeling: Computer Techniques for Geological Characterization*; Springer Science & Business Media: Berlin/Heidelberg, Germany, 2012.
24. Lajaunie, C.; Courrioux, G.; Manuel, L. Foliation fields and 3D cartography in geology: Principles of a method based on potential interpolation. *Math. Geol.* **1997**, *29*, 571–584. [CrossRef]
25. Mallet, J.-L. *Geomodeling*; Oxford University Press: Oxford, UK, 2002.
26. Wu, Q.; Xu, H.; Zou, X. An effective method for 3D geological modeling with multi-source data integration. *Comput. Geosci.* **2005**, *31*, 35–43. [CrossRef]
27. Zhang, B.; Chen, Y.; Huang, A.; Lu, H.; Cheng, Q. Geochemical field and its roles on the 3D prediction fo concealed ore-bodies. *Acta Petrol. Sin.* **2018**, *34*, 352–362.
28. Wang, L.; Wu, X.; Zhang, B.; Li, X.; Huang, A.; Meng, F.; Dai, P. Recognition of significant surface soil geochemical anomalies via weighted 3D shortest-distance field of subsurface orebodies: A case study in the Hongtoushan copper mine, NE China. *Nat. Resour. Res.* **2019**, *28*, 587–607. [CrossRef]
29. Thanh, H.V.; Sugai, Y.; Nguele, R.; Sasaki, K. Integrated workflow in 3D geological model construction for evaluation of CO2 storage capacity of a fractured basement reservoir in Cuu Long Basin, Vietnam. *Int. J. Greenh. Gas Control* **2019**, *90*, 102826. [CrossRef]
30. Adelu, A.O.; Aderemi, A.; Akanji, A.O.; Sanuade, O.A.; Kaka, S.I.; Afolabi, O.; Olugbemiga, S.; Oke, R. Application of 3D static modeling for optimal reservoir characterization. *J. Afr. Earth Sci.* **2019**, *152*, 184–196. [CrossRef]
31. Ayodele, O.L.; Chatterjee, T.; Opuwari, M. Static reservoir modeling using stochastic method: A case study of the cretaceous sequence of Gamtoos Basin, Offshore, South Africa. *J. Pet. Explor. Prod. Technol.* **2021**, *11*, 4185–4200. [CrossRef]
32. Jung, A.; Aigner, T.; Palermo, D.; Nardon, S.; Pontiggia, M. A Hierarchical Database on Carbonate Geobodies and Its Application to Reservoir Modelling Using Multi-Point Statistics. In Proceedings of the 72nd EAGE Conference and Exhibition Incorporating SPE EUROPEC 2010, Barcelona, Spain, 14–17 June 2010; p. cp-161-00314.
33. Li, J.; Zhang, X.; Lu, B.; Ahmed, R.; Zhang, Q. Static Geological Modelling with Knowledge Driven Methodology. *Energies* **2019**, *12*, 3802. [CrossRef]
34. Okoli, A.E.; Agbasi, O.E.; Lashin, A.A.; Sen, S. Static Reservoir Modeling of the Eocene Clastic Reservoirs in the Q-Field, Niger Delta, Nigeria. *Nat. Resour. Res.* **2021**, *30*, 1411–1425. [CrossRef]
35. Thanh, H.V.; Sugai, Y. Integrated modelling framework for enhancement history matching in fluvial channel sandstone reservoirs. *Upstream Oil Gas Technol.* **2021**, *6*, 100027. [CrossRef]
36. Vo Thanh, H.; Lee, K.-K. 3D geo-cellular modeling for Oligocene reservoirs: A marginal field in offshore Vietnam. *J. Pet. Explor. Prod. Technol.* **2022**, *12*, 1–19. [CrossRef]
37. Ali, M.; Abdelhady, A.; Abdelmaksoud, A.; Darwish, M.; Essa, M.A. 3D static modeling and petrographic aspects of the Albian/Cenomanian Reservoir, Komombo Basin, Upper Egypt. *Nat. Resour. Res.* **2020**, *29*, 1259–1281. [CrossRef]
38. Azeem, T.; Yanchun, W.; Khalid, P.; Xueqing, L.; Yuan, F.; Lifang, C. An application of seismic attributes analysis for map field, Pakistan. *Acta Geod. Geophys.* **2016**, *51*, 723–744. [CrossRef]
39. Naseer, M.T. Seismic attributes and quantitative stratigraphic simulation'application for imaging the thin-bedded incised valley stratigraphic traps of Cretaceous sedimentary fairway, Pakistan. *Mar. Pet. Geol.* **2021**, *134*, 105336. [CrossRef]
40. Ali, M.; Khan, M.J.; Ali, M.; Iftikhar, S. Petrophysical analysis of well logs for reservoir evaluation: A case study of "Kadanwari" gas field, middle Indus basin, Pakistan. *Arab. J. Geosci.* **2019**, *12*, 1–12. [CrossRef]
41. Raza, M.; Khan, F.; Khan, M.; Riaz, M.; Khan, U. Reservoir characterization of the B-interval of lower goru formation, miano 9 and 10, miano area, Lower Indus Basin, Pakistan. *Env. Earth Sci. Res. J.* **2020**, *7*, 18–32. [CrossRef]
42. Ahmad, N.; Spadini, G.; Palekar, A.; Subhani, M.A. Porosity prediction using 3D seismic inversion Kadanwari gas field, Pakistan. *Pak. J. Hydrocarb. Res.* **2007**, *17*, 95–102.
43. Khan, M.J.; Khan, H.A. Petrophysical logs contribute in appraising productive sands of Lower Goru Formation, Kadanwari concession, Pakistan. *J. Pet. Explor. Prod. Technol.* **2018**, *8*, 1089–1098. [CrossRef]
44. Dar, Q.U.; Pu, R.; Baiyegunhi, C.; Shabeer, G.; Ali, R.I.; Ashraf, U.; Sajid, Z.; Mehmood, M. The impact of diagenesis on the reservoir quality of the early Cretaceous Lower Goru sandstones in the Lower Indus Basin, Pakistan. *J. Pet. Explor. Prod. Technol.* **2021**, *11*, 1–16. [CrossRef]
45. Ahmad, N.; Chaudhry, S. Kadanwari Gas Field, Pakistan: A disappointment turns into an attractive development opportunity. *Pet. Geosci.* **2002**, *8*, 307–316. [CrossRef]
46. Kazmi, A.; Jan, M. *Geology and Tectonics of Pakistan*; Graphic Publishers: Santa Ana, CA, USA, 1997; p. 554. ISBN 9698375007.

47. Saif-Ur-Rehman, K.J.; Mehmood, M.F.; Shafiq, Z.; Jadoon, I.A. Structural styles and petroleum potential of Miano block, central Indus Basin, Pakistan. *Int. J. Geosci.* **2016**, *7*, 1145.

48. Ahmad, N.; Fink, P.; Sturrock, S.; Mahmood, T.; Ibrahim, M. Sequence stratigraphy as predictive tool in lower goru fairway, lower and middle Indus platform, Pakistan. *PAPG ATC* **2004**, *1*, 85–104.

49. Ahmed, W.; Azeem, A.; Abid, M.F.; Rasheed, A.; Aziz, K. Mesozoic Structural Architecture of the Middle Indus Basin, Pakistan—Controls and Implications. In Proceedings of the PAPG/SPE Annual Technical Conference, Islamabad, Pakistan, 26 November 2013; pp. 1–13.

50. Shakir, U.; Ali, A.; Hussain, M.; Azeem, T.; Bashir, L. Selection of Sensitive Post-Stack and Pre-Stack Seismic Inversion Attributes for Improved Characterization of Thin Gas-Bearing Sands. *Pure Appl. Geophys.* **2021**, *179*, 169–196. [CrossRef]

51. Faleide, T.S.; Braathen, A.; Lecomte, I.; Mulrooney, M.J.; Midtkandal, I.; Bugge, A.J.; Planke, S. Impacts of seismic resolution on fault interpretation: Insights from seismic modelling. *Tectonophysics* **2021**, *816*, 229008. [CrossRef]

52. Kim, M.; Yu, J.; Kang, N.-K.; Kim, B.-Y. Improved Workflow for Fault Detection and Extraction Using Seismic Attributes and Orientation Clustering. *Appl. Sci.* **2021**, *11*, 8734. [CrossRef]

53. Khan, M.; Abdelmaksoud, A. Unfolding impacts of freaky tectonics on sedimentary sequences along passive margins: Pioneer findings from western Indian continental margin (Offshore Indus Basin). *Mar. Pet. Geol.* **2020**, *119*, 104499. [CrossRef]

54. Qadri, S.T.; Islam, M.A.; Shalaby, M.R. Three-dimensional petrophysical modelling and volumetric analysis to model the reservoir potential of the Kupe Field, Taranaki Basin, New Zealand. *Nat. Resour. Res.* **2019**, *28*, 369–392. [CrossRef]

55. Taner, M.T.; Koehler, F.; Sheriff, R. Complex seismic trace analysis. *Geophysics* **1979**, *44*, 1041–1063. [CrossRef]

56. Ahmad, M.N.; Rowell, P. Application of spectral decomposition and seismic attributes to understand the structure and distribution of sand reservoirs within Tertiary rift basins of the Gulf of Thailand. *Lead. Edge* **2012**, *31*, 630–634. [CrossRef]

57. Abuamarah, B.A.; Nabawy, B.S.; Shehata, A.M.; Kassem, O.M.; Ghrefat, H. Integrated geological and petrophysical characterization of oligocene deep marine unconventional poor to tight sandstone gas reservoir. *Mar. Pet. Geol.* **2019**, *109*, 868–885. [CrossRef]

58. Iqbal, M.A.; Rezaee, R. Porosity and Water Saturation Estimation for Shale Reservoirs: An Example from Goldwyer Formation Shale, Canning Basin, Western Australia. *Energies* **2020**, *13*, 6294. [CrossRef]

59. Dolan, P. Pakistan: A history of petroleum exploration and future potential. *Geol. Soc. Lond. Spec. Publ.* **1990**, *50*, 503–524. [CrossRef]

60. Sam-Marcus, J.; Enaworu, E.; Rotimi, O.J.; Seteyeobot, I. A proposed solution to the determination of water saturation: Using a modelled equation. *J. Pet. Explor. Prod. Technol.* **2018**, *8*, 1009–1015. [CrossRef]

61. Cuong, T.X.; Warren, J. Bach ho field, a fractured granitic basement reservoir, Cuu Long Basin, offshore SE Vietnam: A "buried-hill" play. *J. Pet. Geol.* **2009**, *32*, 129–156. [CrossRef]

62. Kularathna, E.; Pitawala, H.; Senaratne, A.; Ratnayake, A. Play distribution and the hydrocarbon potential of the Mannar Basin, Sri Lanka. *J. Pet. Explor. Prod. Technol.* **2020**, *10*, 2225–2243. [CrossRef]

 *minerals*

Article

3D Multi-Parameter Geological Modeling and Knowledge Findings for Mo Oxide Orebodies in the Shangfanggou Porphyry–Skarn Mo (–Fe) Deposit, Henan Province, China

Zhifei Liu [1], Ling Zuo [1], Senmin Xu [2], Yaqing He [3], Chunyi Wang [3], Luofeng Wang [2], Tao Yang [2], Gongwen Wang [1,4,5,*], Linggao Zeng [6], Nini Mou [1] and Wangdong Yang [1]

[1] School of Earth Sciences and Resources, China University of Geosciences (Beijing), Beijing 100083, China; zfliu@cugb.edu.cn (Z.L.); 3001200120@cugb.edu.cn (L.Z.); nini@cugb.edu.cn (N.M.); yangwdcugb@163.com (W.Y.)
[2] Luoyang Fuchuan Mining Co., Ltd., Luoyang 471500, China; xusm@cn.cmoc.com (S.X.); wangluofeng1@126.com (L.W.); kinkye@163.com (T.Y.)
[3] China Molybdenum Co., Ltd., Luoyang 471500, China; heyq@cn.cmoc.com (Y.H.); wangcy@cn.cmoc.com (C.W.)
[4] MNR Key Laboratory for Exploration Theory & Technology of Critical Mineral Resources, China University of Geosciences, Beijing 100083, China
[5] Beijing Key Laboratory of Land and Resources Information Research and Development, Beijing 100083, China
[6] No. 9 Geological Party, Bureau of Geo-Exploration and Mineral Development of Xinjiang Province, Urumqi 830000, China; lgzeng100@163.com
* Correspondence: gwwang@cugb.edu.cn

Citation: Liu, Z.; Zuo, L.; Xu, S.; He, Y.; Wang, C.; Wang, L.; Yang, T.; Wang, G.; Zeng, L.; Mou, N.; et al. 3D Multi-Parameter Geological Modeling and Knowledge Findings for Mo Oxide Orebodies in the Shangfanggou Porphyry–Skarn Mo (–Fe) Deposit, Henan Province, China. *Minerals* **2022**, *12*, 769. https://doi.org/10.3390/min12060769

Academic Editor: Yosoon Choi

Received: 13 May 2022
Accepted: 12 June 2022
Published: 17 June 2022

Publisher's Note: MDPI stays neutral with regard to jurisdictional claims in published maps and institutional affiliations.

Abstract: The Shangfanggou Mo–Fe deposit is a typical and giant porphyry–skarn deposit located in the East Qinling–Dabie molybdenum (Mo) polymetallic metallogenic belt in the southern margin of the North China Block. In this paper, three-dimensional (3D) multi-parameter geological modeling and microanalysis are used to discuss the mineralization and oxidation transformation process of molybdenite during the supergene stage. Meanwhile, from macro to micro, the temporal–spatial–genetic correlation and exploration constraints are also established by 3D geological modeling of industrial Mo orebodies and Mo oxide orebodies. SEM-EDS and EPMA-aided analyses indicate the oxidation products of molybdenite are dominated by tungsten–powellite at the supergene stage. Thus, a series of oxidation processes from molybdenite to tungsten–powellite are obtained after the precipitation of molybdenite; eventually, a special genetic model of the Shangfanggou high oxidation rate Mo deposit is formed. Oxygen fugacity reduction and an acid environment play an important part in the precipitation of molybdenite: (1) During the oxidation process, molybdenite is first oxidized to a $MoO_2 \cdot SO_4$ complex ion and then reacts with a carbonate solution to precipitate powethite, in which W and Mo elements can be substituted by complete isomorphism, forming a unique secondary oxide orebody dominated by tungsten–powellite. (2) Under hydrothermal action, Mo^{4+} can be oxidized to jordisite in the strong acid reduction environment at low temperature and room temperature during the hydrothermal mineralization stage. Ilsemannite is the oxidation product, which can be further oxidized to molybdite.

Keywords: oxidation of molybdenite; ore-forming process; 3D multi-parameter geological modeling; microanalysis; Shangfanggou

1. Introduction

Three-dimensional (3D) geological modeling is a prominent technology that can be used to calculate or extract key parameters from 3D information on ore deposits [1]. Applying the theory of a metallogenic system, combined with multi-parameter or multi-source (geological, geophysical, geochemical, and hyperspectral) and multi-method modeling and analysis of a 3D ore-forming geologic body, the capabilities of quantitative and big data collation, mineral exploration, and mining are greatly improved [2,3]. It has been proved that 3D GIS or modeling packages (e.g., Micromine, GOCAD, and Surpac) are excellent means of data representation and interpretation [4].

3D geological modeling provides distinct advantages in assisting geologists in determining the geological background, mineralization, and mineral exploration in a comprehensive and effective manner [5–18].

Porphyry–skarn-type molybdenum (Mo) deposits are one of the strategic and economic resources in China. The East Qinling Mo belt (EQMB; Figure 1B) is referred to as one of the most significant Mo provinces in the world, containing reserves of about 6 Mt Mo [19,20]. The Shangfanggou Mo–Fe deposit is an important deposit in the East Qinling Mo polymetallic metallogenic belt. It is adjacent to the Nannihu–Sandaozhuang Mo–W polymetallic deposit in the northeast (Figure 1D), which together constitute the main body of the Nannihu Mo ore field in Luanchuan district, China. In 2000, the estimated Mo reserves of the three major deposits (Nannihu, Sandaozhuang, and Shangfanggou) were about 2.4×10^6 tons (with an average Mo grade of 0.109%) [21], while the Shangfanggou deposit contained 0.72 million tons of Mo and 59.91 million tons of Fe metal, with average grades of 0.135% Mo [22,23] and 30.14% Fe [24].

Although there have been many studies on 3D geological modeling of the Luanchuan ore district [25,26], they have been insufficient to construct 3D multi-parameter geological modeling of the study area. Moreover, the formation of the Shangfanggou porphyry–skarn deposit is mainly controlled by strong tectonic magmatic movement in the Mesozoic. The complex geological background, multiple geological factors, and multi-scale and multi-format data sources pose challenges to 3D geological modeling and deep prediction.

Currently, the industrial Mo ores used by the mine are mainly the primary molybdenite; the onefold Mo oxide ores nearly cannot be recycled and utilized because they are difficult to beneficiate, and the ore-forming process (supergene oxidation process) of this kind of ore type also lacks research. In recent years, global Mo ore output has tended to decline, and the situation of rapid economic and social development has led to a demand for Mo. Therefore, it has become an important and urgent topic to study the genesis and prospecting of Mo ores in the oxidation zone and to utilize Mo oxide ores comprehensively and efficiently. On the basis of previous studies, this paper systematically discusses the mineralization and oxidation process of Mo orebodies in the Shangfanggou porphyry–skarn Mo–Fe deposit. The temporal–spatial–genetic correlation and exploration constraints of magnetite and gangue minerals to Mo and Mo oxide orebodies are discovered, and the distribution characteristics of the oxidation zone are further investigated (Figure 2). These findings provide a reference for the genesis and exploration, as well as the recovery and utilization, of oxidized ores for mining and beneficiation of the mine.

2. Geological Setting

2.1. Regional Geology

The EQMB is part of the Central China orogenic belt which is bounded by the San-Bao Fault to the north and the Shang-Dan Fault to the south [27,28] (Figure 1B). As the "Mo capital of China", more than 30 polymetallic deposits and ore spots are densely distributed in the Luanchuan ore district. This is consistent with the horizontal zoning of geochemical anomaly elements in the district. Namely, they are regularly distributed from inside to outside around the porphyry pluton: the porphyry–skarn Mo–W deposits and skarn sulfur polymetallic deposits mainly developed in the central zone of geochemical anomalies, while the hydrothermal vein Pb–Zn–Ag deposits are mainly distributed in the middle zone of geochemical anomalies [29]. Of course, both the ore district scale and the deposit scale have favorable prospecting potential at depth.

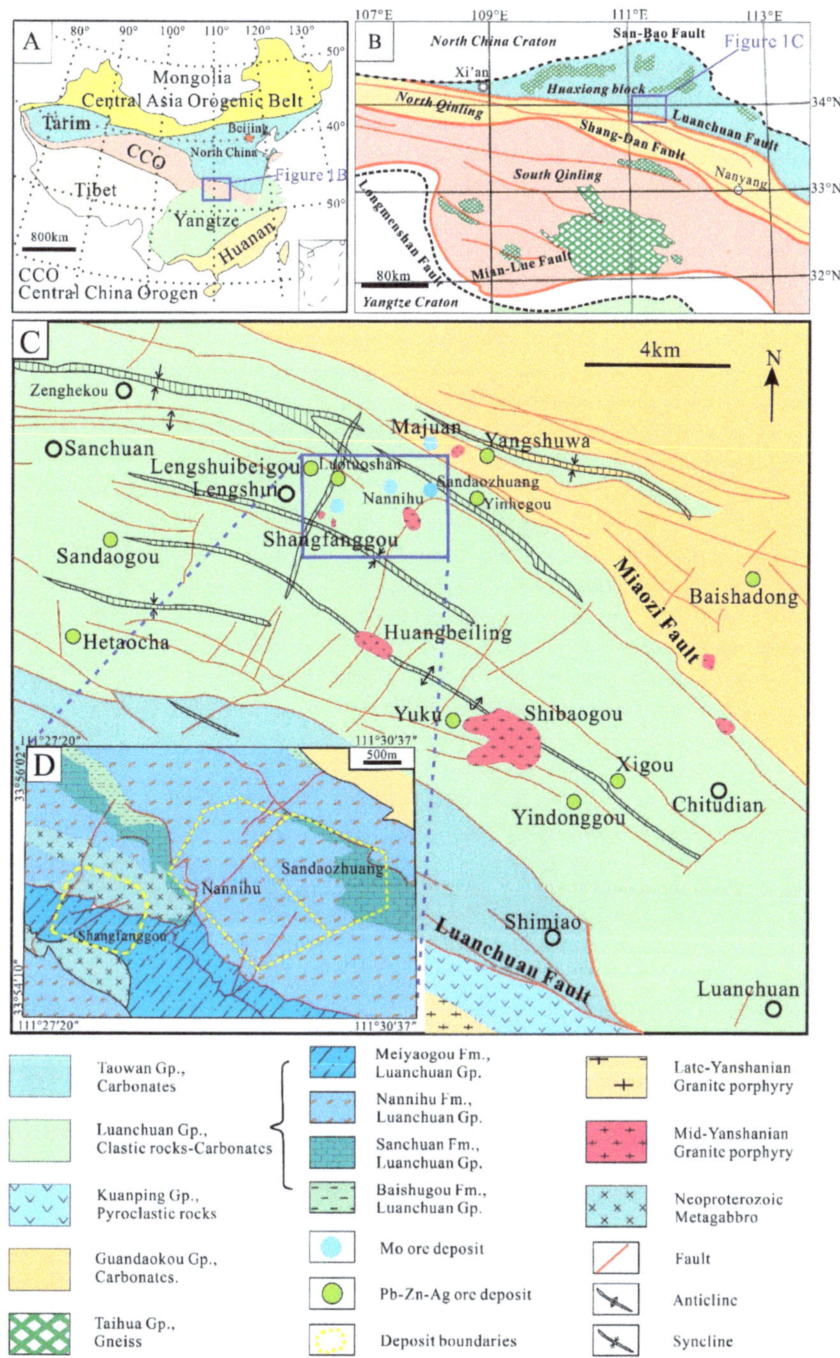

Figure 1. (**A**) Tectonic map of China, showing the location of the Qinling Orogen Belt; (**B**) tectonic subdivision of the Qinling Orogen Belt, showing the location of the Luanchuan ore district; (**C**) geological map of the Luanchuan ore district, showing the Shangfanggou Mo–Fe deposit (modified after [39]); (**D**) sketch of the Shangfanggou, Nannihu, and Sandaozhuang deposits (modified after [22,23]).

The Luanchuan ore district has a typical double-crust structure composed of basement and sedimentary cover. The basement is the Neoarchean Taihua Group deep metamorphic rock series, and the caprock series is mainly the "trident-rift" volcanic formation of the Middle Proterozoic Xiong'er Group [30], the Guandaokou group of the Middle Proterozoic, the passive continental clastic rock formation of the Luanchuan Group of the New Proterozoic, and the clastic rock–carbonates sedimentary formation of the slope facies of the Taowan Group, Lower Paleozoic [31]. The Nannihu Mo ore field, discovered in the 1960s and 1970s, is located in the north of the ore district, which is one of Mo's major places of production in China and even the world. The Nannihu Mo ore field, the Shangfanggou Mo–Fe deposit included, is mainly hosted in the Luanchuan Group.

Magmatism mainly concentrated in the Neoarchean, early Mesoproterozoic, and Mesozoic. Thereinto, the Mesozoic (Yanshanian) mineralization related to small porphyry was intense [32]. The main Mo mineralization is involved with the late Mesozoic granitic magmatism in the EQMB [33–36], forming giant porphyry–skarn Mo deposits, including the Nannihu, Shangfanggou mine, etc.

The main regional faults are the Luanchuan and Machaoying faults, which are distributed in a NWW-trending, and there are NE-trending secondary faults. These faults control the distribution of magmatic rocks and ore deposits, and giant deposits are often formed at the intersection of them [37,38]. That is, a series of NW-trending thrusts at a regional scale controlled the formation of major Mo deposits, which were formed by the regional Indosinian Qinling orogenic events. The secondary NW-trending folds and NE-trending faults and intrusive stock structures were formed by the thrusts during the Caledonian–Indosinian orogenic event. They are ore-bearing belts and ore-forming structures [2].

2.2. Deposit Geology

The Shangfanggou porphyry–skarn Mo–Fe deposit, located in the northwest of Luanchuan County, is world-renowned for its large scale, high grade, and suitability for open-pit mining. It is adjacent to the Sandaozhuang–Nannihu Mo–W polymetallic deposit in the northeast, constituting the main part of the Nannihu ore field in Luanchuan, a part of the world-famous district (Figure 1D). The deposit is located in the Sanchuan–Luanchuan fold belt in the southern margin of the North China Block, controlled by the Zenghekou–Shibaogou syncline. The genesis of the deposit is closely related to structure, granite porphyry pluton, wall rocks, strata (dolomite marble, skarn), and alteration, which are mainly distributed in NW- and NWW-trending (Figure 1C,D and Figure 2).

Metasediment of the upper Nannihu and the Meiyaogou Formation of the Luanchuan Group mainly crop out in the area, which belongs to the sedimentary environment of a shallow sea shelf-confined platform supratidal–subtidal zone. Under the influence of contact thermometamorphism and contact metasomatism caused by the intrusion of the Shangfanggou pluton, various types of hornfels and skarnization developed, respectively (for details of the lithology, see Section 3.1). Moreover, the alteration of wall rocks is composed of potassic alteration, silicification, and phlogopitization (Figures 2 and 3). With the intrusion of the pluton, the magnesian skarn formed from the dolomite marble of the middle Meiyaogou Formation (divided into three layers), which is the predominant ore-hosted position.

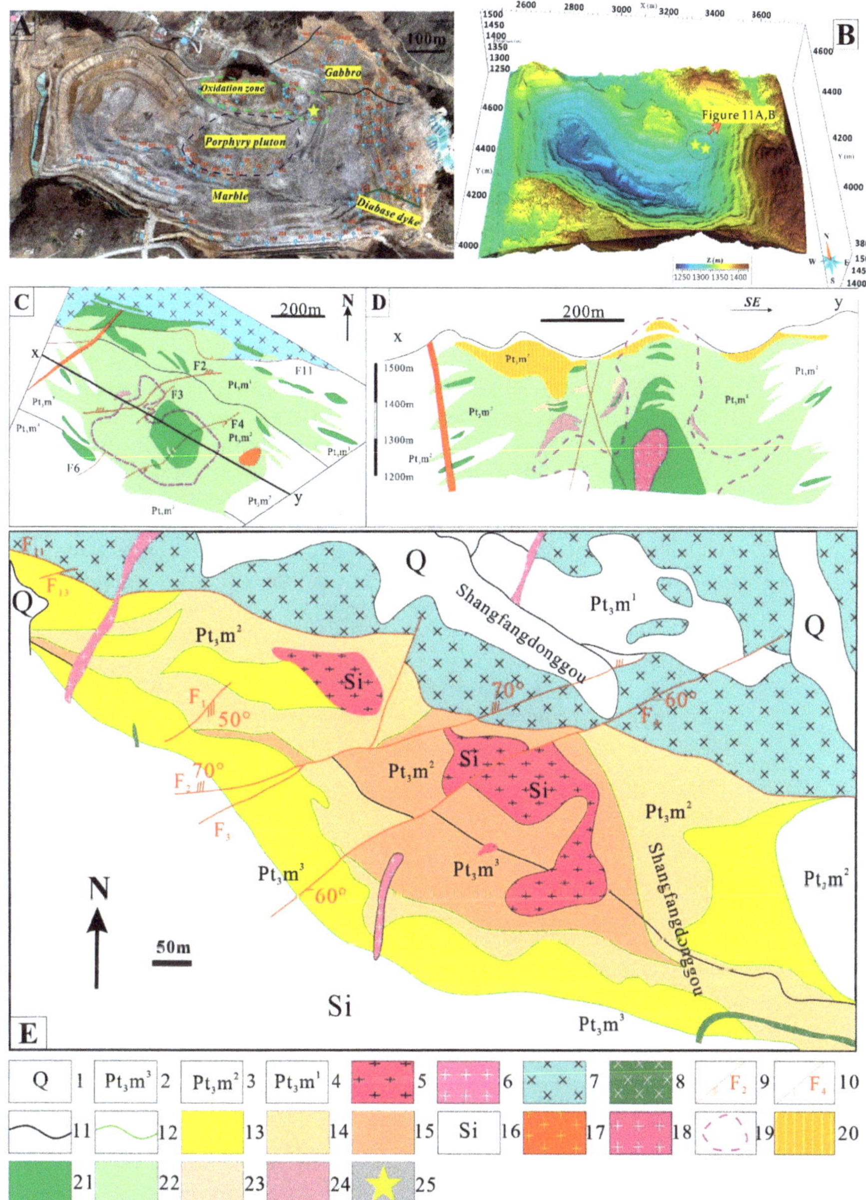

1. Quaternary; 2. Upper Meiyaogou Fm.; 3. Middle Meiyaogou Fm; 4. Lower Meiyaogou Fm.; 5. Granite p-
-orphyry; 6. Granite porphyry dike; 7. Gabbro; 8. Diabase; 9. Compression fault; 10. Transtension fault; 11.
Geological boundary; 12. Alteration zone boundary; 13. Magnetite-tremolitization-diopsidization zone; 14.
Serpentinization zone; 15. Phlogopitization-actinolitization zone; 16. Strong silicification zone; 17. Post--
-ore porphyry; 18. Mineralized syenogranite porphyry; 19. Border of porphyritic biotite granite; 20. Oxidi-
-zed Mo orebody; 21. Low grade Mo orebody; 22. High grade Mo orebody; 23. Low grade Fe orebody; 24.
High grade Fe orebody; 25. Sample location.

Figure 2. Geological sketch and location of Shangfanggou Mo–Fe deposit: (**A**) the open pit (red points, sample spots collected and analyzed in the fieldwork); (**B**) DEM of the mine; (**C,D**) geologic profile of the deposit, showing the Mo oxide orebodies and geology at 1132 m level and prospecting line NO. 5 (modified after [22,23,40–42]); (**E**) geological map of the deposit (after [43]).

Magmatism in the mining area is controlled by the fault structure, which mainly includes Caledonian metagabbro (in the south and north), syenite porphyry, and Yanshanian intermediate–acid granite porphyry (central part of the mining area). The metagabbro dikes yield K–Ar ages of ca. 743 Ma [44] and zircon SHRIMP U-Pb ages of ca. 830 Ma [45], which are associated with the rifting background at the southern margin of the North China Continent [46]. Therefore, the intrusive, alkali–feldspar granite porphyry in the middle Yanshanian, a time when the predominate mineralization was undergoing (Shangfanggou pluton, distributed in the center of the study area), is related to metallogenesis.

The deposit is controlled by the Shangfanggou syncline, which is located in the northern wing of the Shangfanggou syncline in the Sanchuan–Luanchuan County depression fold fault belt. NW–NWW- and NE–NNE-trending faults are developed, and the former controls the distribution characteristics of the plutons in the area, while the latter provides the conditions for migration and accumulation of Yanshanian ore-bearing hydrothermal fluid as a migration pathway. The NW–NWW trending dominates in magmatic rocks and various structures in the area, superimposing the later NNE–NE-trending faults. The NWW–NW-trending structure is mainly composed of the Zenghekou–Shibaogou syncline and a series of thrust nappe faults with NWW trending, in which the thrusts with southward thrusting and NE dip are developed. NNE–NE faults are intruded by the granitic porphyry dikes in both directions that control the formation of the Mo–Fe deposit and the intersection area of the two different directions is the occurrence site of the Shangfanggou orebody.

The Shangfanggou Mo–Fe polymetallic deposit occurs in the inner and outer contact zone of the alkali–feldspar granite porphyry. That is, the Mo ore bodies are hosted not only in the magnesium skarn but in the porphyry pluton and adjacent metagabbro and hornfels, which shows the controlling effect of the pluton on the deposit, while the Fe ore bodies occur in the former more often.

The ores can be divided into skarn, granite porphyry, hornfels and metagabbro types, according to the different host rocks (Figure 3A–G). The ore minerals consist of molybdenite, magnetite, and pyrite, followed by pyrrhotite, chalcopyrite, sphalerite, galena, etc. The gangue minerals are mainly quartz, diopside, garnet, tremolite, talc, serpentine, chlorite, feldspar, followed by sericite, carbonates, fluorite, etc. (Table 1; after [47]).

Table 1. Stages of mineralization and paragenesis for the Shangfanggou Mo-Fe deposit.

Stages	Minerals
Protolith	Dolomite, K-feldspar, plagioclase, quartz, biotite, ilmenite, wollastonite
Late magmatism (silicification and potassic alteration) stage (1)	K-feldspar, quartz, biotite …
Early skarn stage (2)	Garnet, diopside, forsterite …
Late skarn stage (3)	Tremolite, magnetite, phlogopite, serpentine, calcite, actinolite, chlorite, talc, fluorite, serpentine …
Hydrothermal stage (4)	Quartz, pyrite, molybdenite, K-feldspar, chalcopyrite, galena, sphalerite, epidote, scheelite, fluorite, pyrrhotite …
Supergene stage (5)	Molybdite, ilsemannite, scheelite, limonite, tungsten–powellite …

The gabbro in the northern part of the mine serves as part of the main orebody that is affected by mineralization. The Mo orebody occurs in (meta-)gabbro as well, accompanied by magnetite. In this kind of ore, pyroxene alteration is developed and widespread, which can be altered into amphibole and chlorite, giving rise to the partially argillated surface, and pyrite is mostly weathered into limonite (Figure 3G,H).

Figure 3. Photographs of alteration and mineralization characteristics in different ore types: (**A**) porphyry Mo ore and potassic alteration; (**B**) molybdenite with leaf-like texture unevenly embedded in gangue minerals such as quartz and K-feldspar or in veinlets in granite porphyry; (**C**) quartz veins (silicification) and potassic alteration in the granitic porphyry; (**D**) marble-type Mo ore; (**E**) skarn-type Mo ore with disseminated structure; (**F**) garnet diopside skarn-type Mo ore with epidotization; (**G**) metagabbro-type Mo ore with ferritization; (**H**) the alteration of pyroxene to amphibole in metagabbro and contiguous magnetite developed; (**I**) phlogopitization (colorful interference color of phlogopite).

The skarn-type Mo ores are distributed around the granite porphyry in the hanging wall (south) and on both sides of the east and west of the granite porphyry pluton, with a small amount of rock footwall, which is the dominant ore type of the deposit. Stellate and disseminated Mo mineralization can be observed, showing obvious metasomatism among this kind of ore. Molybdenite often occurs in the intergranular or fissure of skarn minerals such as diopside and garnet (Figure 3E,F).

The ore is dominated by a veinlet stockwork structure, followed by a disseminated structure. The veinlet structure is the most widespread and predominant structure in the area; the veinlet is not straight and has poor continuity. The ore texture is dominated by a granular texture, followed by a metasomatic relict texture (Figure 4).

Figure 4. Photographs of ore petrography of the Shangfanggou Mo–Fe deposit: (**A**) quartz–molybdenite veinlet and pyrite veinlets crosscut by the late quartz veinlet; (**B**) banded structure; (**C**) membranous structure of the Mo oxide ore; (**D**) stockwork structure showing multistage mineralization; (**E**) disseminated Mo mineralization in the garnet-skarns; (**F**) the curved lepidosome molybdenite and paragenetic pyrite, magnetite, and chalcopyrite unevenly distributed between gangue minerals. Abbreviations: Q, quartz; Mo, molybdenite; Py, pyrite; Grt, garnet; Mt, magnetite; Ccp, chalcopyrite.

3. 3D Geological Modeling

Construct 3D geological model at the deposit scale using a variety of datasets such as strata, lithology, regional and deposit structures, geological section, boreholes, ore grade, and so on. 3D geological modeling usually consists of the following three steps: (i) geoscience data extraction and processing, (ii) 3D geological modeling, and (iii) interpretation and verification [1].

3.1. 3D Lithology Modeling

In GOCAD software, a 3D model of the lithology is established based on the datasets of geological section and stratum histogram. In Figure 5, the lithology of the three strata of the Luanchuan Group's Meiyaogou Formation can be visually recognized: the lower member is quartzite intercalated with mica schist and marble, etc., the middle part is thick dolomite marble, and the upper member is mainly dolomite marble. Granite porphyry pluton is distributed in the central part of the mining area, which is irregularly divided by faults and wall rocks on the surface and has a distinct contact boundary with the wall rocks.

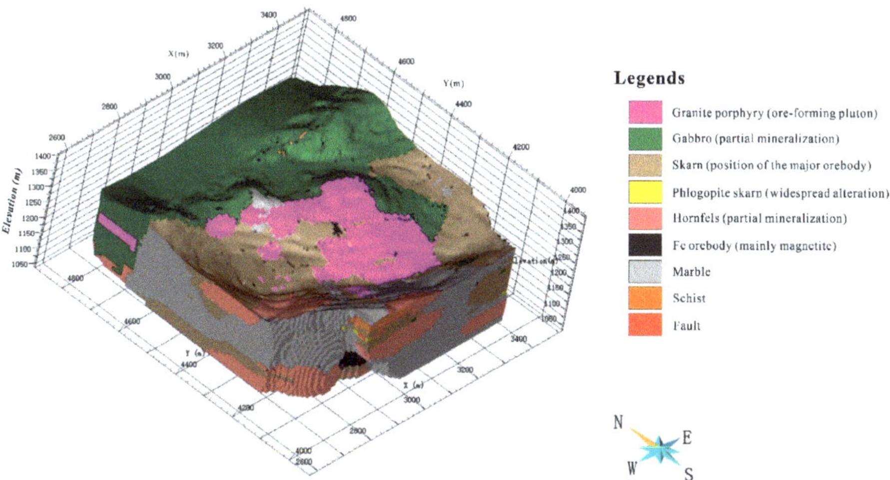

Figure 5. 3D lithology model of the Shangfanggou deposit.

3.2. 3D Structural Model

A structural model is established mainly based on the profile of prospecting lines and adjusted and verified by the horizontal geologic map. The software is mainly divided into the following four parts: fault, outline, contact, and surface. It is important to note that the definition of fault properties is critical.

The NWW–NW-trending faults dominating the study area and a series of NW-trending thrust structures control the formation of the main Mo orebodies (Figure 6). They are ore-bearing belts and ore-forming structures [2]. Moreover, NNE–NE trending also controls the formation of the deposit.

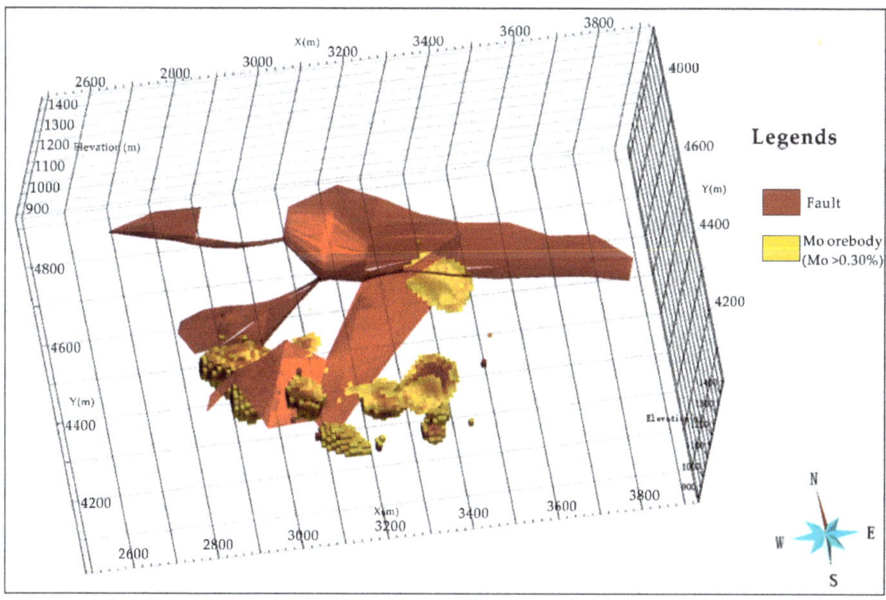

Figure 6. Spatial relationship between rich orebody and main faults (superposition of main faults with ore shoot).

3.3. 3D Orebodies and Grade Modeling

3D orebody geological modeling involves the delineation of the alteration zone, the identification of centers and boundaries of the orebody, and the mapping of fractures and faults. 3D orebody grade modeling is based on the borehole dataset through the application of indicator kriging (IK), a geostatistical interpolation tool from the GOCAD software [48]. In the case of the datasets with a favorable structure and a large amount of data, kriging interpolation with variogram function is more advantageous. The variation function has a good effect on analyzing the continuity of spatial data. Indeed, Mo orebodies are controlled by the diverse structures in the deposit, and the 3D orebody grade model is beneficial for inferring the geological events associated with mineralization. In addition, 3D orebody grade modeling by IK is helpful in identifying and extracting the different orebodies in one deposit [49,50].

The primary orebody is distributed in the form of an 'inverted cup,' with Mo oxide orebodies on the shallow surface. The orebody's border is uneven, and there are several branchings on the east and west sides. The thickness of the orebody varies on both sides, with the southwest being thinner and the northeast being thicker. The Mo orebodies are basically distributed above 600 m elevation but are mainly concentrated above 900 m, and the Mo oxide orebodies are typically found over 1300 m.

The skarn-type Fe orebodies are distributed in lenticular, irregular cystic, and saddle forms, with the shape determined by the occurrence of the pluton's contact zone. Analyzing both Figure 5 and Figure 8, the conclusion that can be drawn is that the depression and the roof of the pluton are favorable places for the enrichment of Fe mineralization. In the mine, the Fe orebodies are mostly found above 1000 m, with a downward trend, which is similar to the distribution characteristics of the Mo orebodies (Figures 7 and 8). Therefore, through the superposition of Mo mineralization, Fe, as an associated element, constitutes the Mo–Fe orebodies together with the Mo orebodies, indicating the spatial correlation between the Fe orebodies and the Mo orebodies and further speculating that the two are also related in genesis. The relative grade information of the two is shown in Figures 9 and 10.

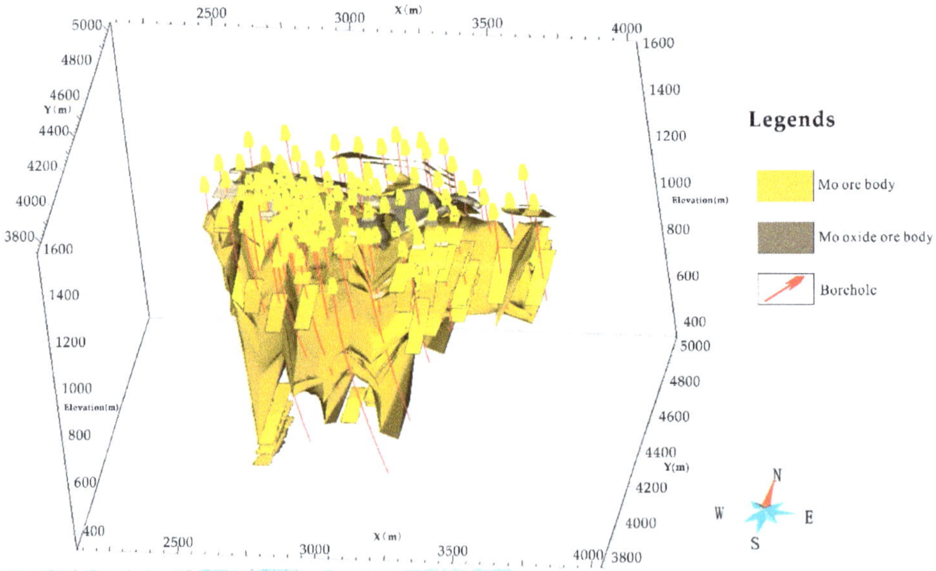

Figure 7. 3D model of Mo orebodies and Mo oxide orebodies.

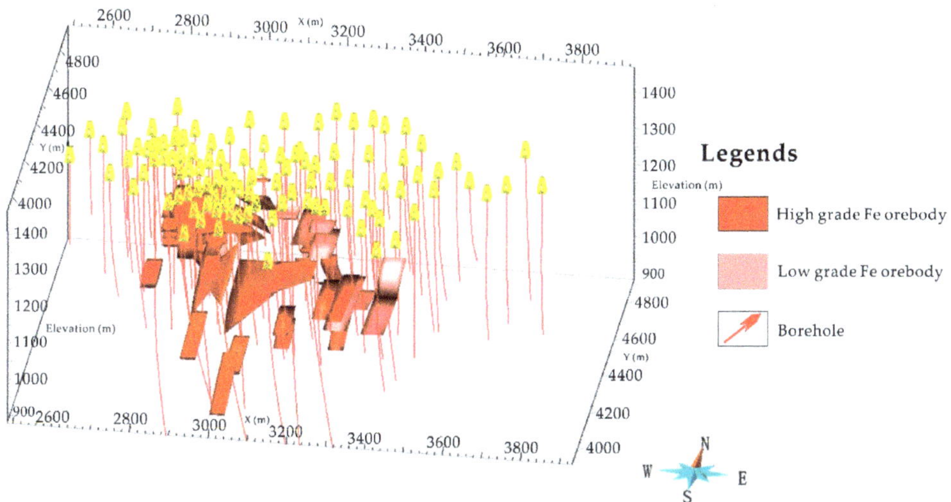

Figure 8. 3D model of Fe orebodies.

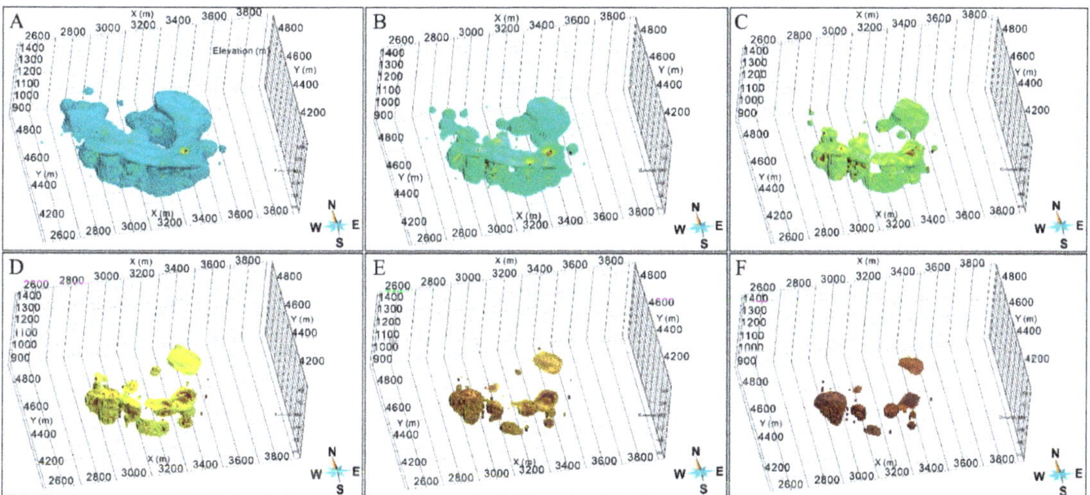

Figure 9. Distribution of Mo orebodies with different grades: (**A**) the grade > 0.10%; (**B**) the grade > 0.15%; (**C**) the grade > 0.20%; (**D**) the grade > 0.25%; (**E**) the grade > 0.30%; (**F**) the grade > 0.35%.

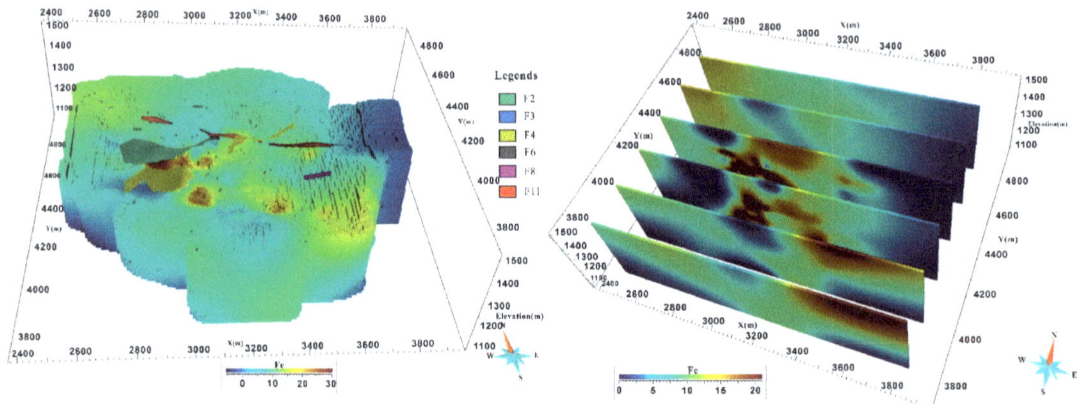

Figure 10. 3D grade model of Fe orebodies.

Figures 5, 7 and 9 illustrate that there is no ore in the core of the pluton. Although the orebodies are found on both sides, those toward the southwest are thick, concentrated, and of high grade.

The average grade of total Fe (TFe) in the Fe orebodies in the area is normally 20–30%, which is mostly concentrated in the hanging wall of the pluton or the top wall rock (magnesian skarn) within the range of 100–150 m from the intrusion contact zone of the pluton (Figure 10).

4. Sampling and Analysis Methods

4.1. Sampling

Mo oxide ores were collected from the oxide zone of the deposit for microscopic identification (Figure 2), SEM (scanning electron microscope)-EDS (energy dispersive X-ray spectroscopy), and EPMA (electron microprobe analysis) composition analysis.

The determination of major elements and high-resolution surface scanning and backscattering images were carried out at the Central Laboratory of Genetic Minerals, China University of Geosciences, Beijing, using FE-SEM Tescan. In addition, EPMA was performed at the Beijing Institute of Geology of Nuclear Industry, Beijing, China.

4.2. SEM-EDS

SEM-EDS was utilized to conduct a semi-quantitative analysis of molybdenite and its oxides, as well as the gangue minerals associated with them, in order to preliminarily explore their composition, content, and distribution (e.g., Figure 11D–I). FE-SEM Tescan: MIRA 3 XMU field emission scanning electron microscope equipped with OXFORD X-Max 20 mm^2 X-ray energy spectrometer; backscattering electron image resolution was 2.0 nm@30 KV with beam drift $\leq$ 0.2% h; and energy dispersive spectroscopy was equipped with an electric refrigeration detector and peak to back ratio was better than 20,000:1. Moreover, Mn-ka is superior to 127 eV with an energy resolution of 20,000 CPS. The data were analyzed and processed using an Inca system with Inca X-Stream and Inca Mics microanalysis processors. The conditions of point analysis were as follows: high voltage, 20 KV; emission current, 83 μA; electron beam intensity, 18.36; absorption current, 4.0 nA; spot size, 85 nm; acquisition time, 30 s; process time, 3; dead time, less than 30%; count rate, higher than 20 Kcps.

4.3. EPMA

Representative measuring points were selected for more accurate component analysis of the samples preliminarily analyzed by SEM-EDS to determine the oxidation products of molybdenite: instrument (electron microprobe analyzer) model, JXA-8100. The implementation standard is GB/T 15074-2008 "General Principles of Quantitative Analysis Methods for Electron Probe", China. In addition, the analysis conditions are as follows: an acceleration voltage of 20 kV was used for the experiment while the beam current was 1×10^{-8} A and the exit angle was 40°. Meanwhile, the spectral analysis method and ZAF correction method were adopted.

5. Discussion

5.1. Compositions

5.1.1. EPMA

The composition data are listed in Table 2. The content of WO3 ranges from 11.21 to 32.40, with an average of 22.35 (unit: wt.%, and all of the following). The concentration of MoO_3 varies from 39.80 to 60.03, with an average of 49.64; it contains 24.18–27.08 CaO, with an average of 25.70 and a relatively uniform distribution; and the SO_3 content ranges from 1.18 to 1.93, with an average of 1.42.

Table 2. Representative analysis results (wt.%) of Mo oxide ores.

Points	1	2	3	4	5	6	7	8	9	10	11	12	13	14	15
F	/	/	/	/	/	/	/	/	/	/	/	/	/	/	/
SiO_2	/	/	/	/	/	/	/	/	/	/	/	/	/	/	/
WO_3	12.59	32.40	32.24	14.11	23.93	17.32	11.21	17.67	28.22	30.08	27.38	18.55	29.50	25.17	31.76
SO_3	1.66	1.26	1.41	1.49	1.33	1.64	1.48	1.49	1.18	1.24	1.19	1.45	1.34	1.35	1.93
Al_2O_3	/	/	/	0.02	/	/	/	/	/	/	/	/	/	0.02	/
MgO	/	/	/	/	/	/	/	/	/	/	/	0.13	/	/	/
MoO_3	58.51	40.91	40.65	57.55	48.86	53.66	60.03	54.21	43.81	42.54	45.56	51.85	43.74	47.59	39.80
CaO	26.73	24.35	24.37	26.36	25.07	26.69	27.08	26.05	25.97	25.29	25.03	26.77	24.18	25.17	24.80
SeO_2	0.10	0.63	0.66	0.14	0.51	0.26	0.13	0.31	0.44	0.77	0.64	0.34	0.54	0.60	0.63
FeO	0.07	0.10	0.05	0.03	0.12	0.03	/	0.06	0.11	0.07	0.07	0.14	0.20	0.09	0.06
Cl	0.16	0.08	0.11	0.19	0.15	0.14	0.13	0.16	0.14	0.10	0.12	0.13	0.14	0.12	0.07
MnO	/	/	0.04	/	/	/	/	/	0.04	0.07	0.03	0.04	/	/	/
P_2O_5	/	/	/	/	/	/	0.06	/	/	/	/	/	/	/	/
CuO	/	/	/	/	/	/	/	/	/	/	/	/	/	/	/
PbO	/	/	/	/	/	/	/	/	/	/	/	/	/	/	/
TeO_2	/	/	/	/	/	/	/	/	/	/	/	/	/	/	/
Total	99.82	99.73	99.53	99.89	99.97	99.74	100.12	99.95	99.91	100.16	100.02	99.40	99.64	100.11	99.05

/: Not detected.

Within the detection limit of the instrument, the mineral does not contain Cu, Pb, Zn, Si, F, Te, etc. Mn, Al, etc. were found on occasion. In terms of the contents, Mo is the predominant element in the oxide mineral granule, while Mo elements are mainly present in the tungsten–powellite.

The relative contents of Mo and Ca at the seventh site are the highest, which are 60.03% and 27.08%, respectively, while the content of W is the lowest at only 11.21%. A minor quantity of P is contained, and Fe, Mn, etc. are not included.

5.1.2. SEM-EDS

The tungsten–powellite is finely granular and is irregularly dispersed in the oxide ores (Figure 12).

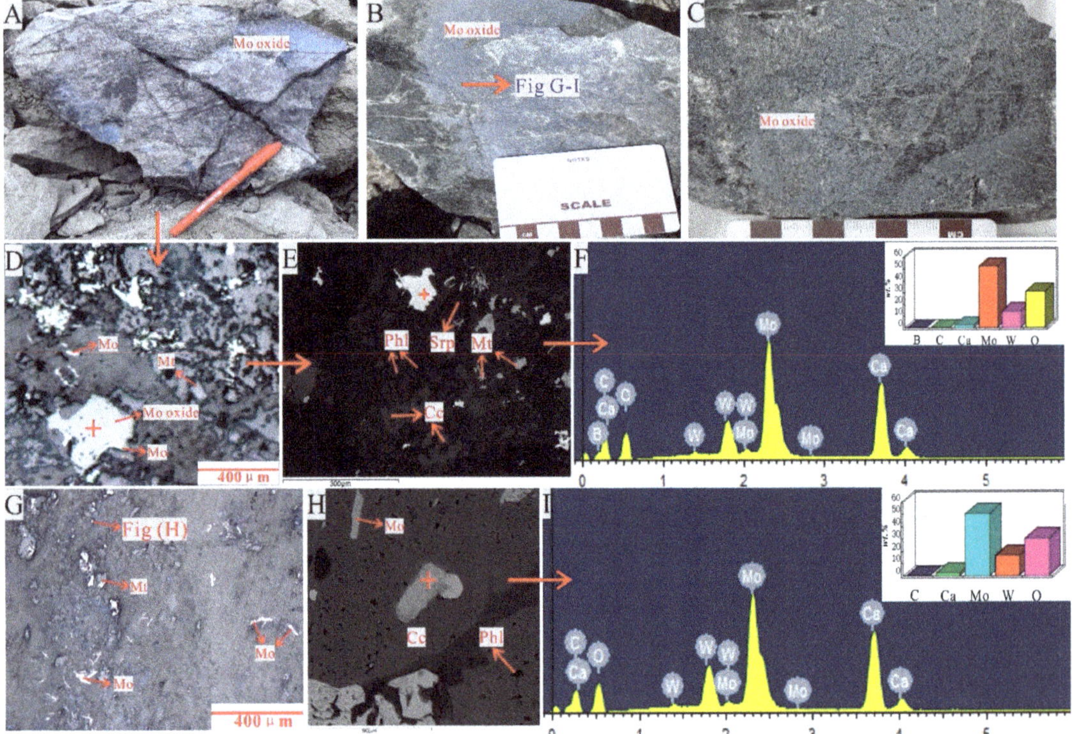

Figure 11. Photographs showing the characteristics of field and hand specimens of Mo oxide ores as well as SEM analysis results: (**A–C**) Mo oxide ores; (**D,G**) molybdenite and associated with magnetite; (**E,H**) the test points of EDS and BSE of the Mo oxide minerals and accompanying gangue minerals; (**F,I**) results of EDS. Abbreviations: Mo, molybdenite; Mt, magnetite; Cc, calcite; Phl, phlogopite; Srp, serpentine.

Mo oxide minerals are primarily found in the skarnized marble, which is usually associated with phlogopitization and serpentinization. The tungsten–powellite is found between the grains and veins of calcite or hydrothermal alteration minerals such as phlogopite and serpentine, indicating a later stage of tungsten–powellite development rather than the skarn stage or earlier.

In Figure 12, compared with Mo, the distribution of the W element is not uniform, and obvious Mo and W bands are visible, while for the element, Mo distribution is relatively uniform, and the Mo content of molybdenite in the (almost) unoxidized margin is higher than that in the middle of the mineral.

The element distribution of Ca, Fe, and O is uniform, while the content of Fe is significantly lower than that of Ca. In addition, S is mainly distributed in the margin molybdenite. All of these are consistent with the results in Table 2 and the characteristics of tungsten–powellite, indicating the incomplete oxidation or dynamic oxidation process of molybdenite. The contents of Mo, Ca, and W are higher than that of O, Mg, and Si, which further confirms the oxidation of molybdenite. The characteristics of the zonal concentration of W content also demonstrate the distribution of a W-bearing hydrothermal solution.

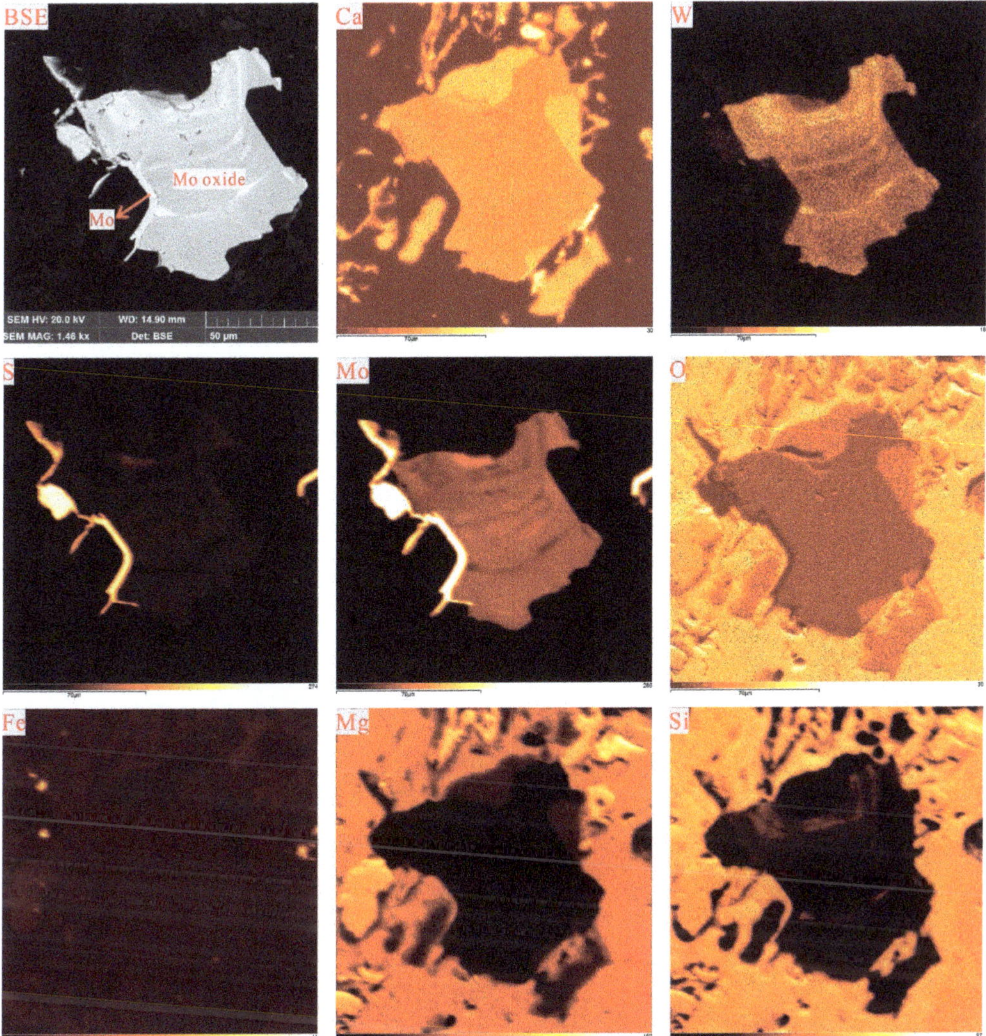

Figure 12. Mapping of the elements scanning results based on SEM. Abbreviation: Mo, molybdenite.

When molybdenite is not completely oxidized, it can be observed that the pseudomorph of molybdenite is retained in the granules and distributed along the edge of tungsten–powellite (Figures 11D and 12). Moreover, unoxidized molybdenite is sporadically present in the samples. Magnetite is disseminated and indicates a high oxygen fugacity environment.

5.2. Mineralization and Supergene Oxidation Process

Mineralization is intimately linked to temperature and oxygen fugacity ($f(O_2)$). Previous research has revealed that the ore-forming fluids of many of the East Qinling's magmatic–hydrothermal deposits are strong oxidizing fluids with high temperature, salinity, and oxygen fugacity [51–53].

The crystallization temperature and $f(O_2)$ of biotite in the Shangfanggou pluton are 750 °C–860 °C and −8.0−−6.5, respectively, indicating a relatively high temperature and f(O_2) environment. The crystallization temperature and f(O_2) of biotite related to Mo mineralization are significantly greater than those of other ore-forming plutons. It is speculated that the pluton related to Mo mineralization also originates from a high temperature and f(O_2) environment [54].

The K–Ar, Rb–Sr, and zircon U–Pb ages of the Shangfanggou porphyry pluton are all between 134 and 158 Ma [33,55–58], similar to the Re–Os isotopic dating of molybdenite (metallogenic age, 143.8~145.8 Ma) [59], suggesting the deposit is of magmatic–hydrothermal genesis in Yanshanian. The Yanshanian magmatism provides hydrothermal and metallogenic material sources for Mo polymetallic mineralization in the study area [54]. Meanwhile, the magma source is mixed crust and mantle-derived (with I-type granite characteristics, rich in Mg and poor in Fe) [60].

The ore-forming fluid of the Shangfanggou Mo–Fe deposit is mostly magmatic water, with a tiny quantity of metamorphic water in the strata, according to H–O isotope research. The S isotope composition is rather consistent and is characterized by deep source sulfur, which might come from the pluton. The Pb isotopic composition of the Nannihu ore field granite porphyry is notably different from that of the Shangfanggou pluton, which may inherit the Pb isotopic composition features of the Shangfanggou pluton [61].

In general, during the mineralization stage, the fluid system evolved from oxidation to reduction, and the composition changed from complex to simple [51]. A simplified deposit exploration model is described in Figure 13.

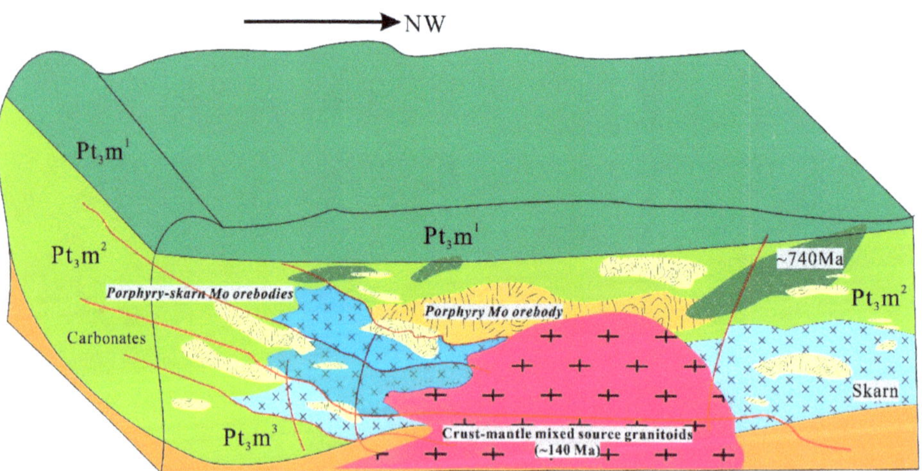

Figure 13. Sketch map of the Shangfanggou deposit exploration model.

To sum up, in such a metallogenic environment, a decrease in f(O_2) and an increase in sulfur fugacity play an important role in molybdenite precipitation. With local redox reaction or neutralization on the contact zone between the hydrothermal solution and carbonates, plenty of sulfides began to precipitate. At the high to medium temperature stage, a substantial quantity of molybdenites developed, resulting in high-grade orebodies dominated by molybdenite.

Under hydrothermal action, molybdenite precipitates under acidic conditions; that is, molybdenite is the most stable under acidic conditions. Under the condition of a supergene environment, molybdenite is oxidized, leached, and migrated with the medium in the state of a $MoO_2 \cdot SO_4$ complex ion, and then contacted with carbonate solution to precipitate powethite. Therefore, W and Mo elements can be completely replaced by complete isomorphism, forming unique secondary orebodies dominated by tungsten–powellite with a high oxidation rate in the Shangfanggou Mo–Fe deposit. The basic reaction process is as follows:

$$2MoS_2 + 9O_2 + 2H_2O = 2 \ (MoO_2 \cdot SO_4) + 2H_2SO_4; \tag{1}$$

$$MoO_2 \cdot SO_4 + Ca \ (HCO_3)_2 = CaMoO_4 + H_2SO_4 + 2CO_2 \tag{2}$$

Moreover, powethite also metasomatized tungsten–powellite. The light and dark bands alternately formed by Mo-rich or W-rich bands in the mineral grains can be observed in SEM (Figure 12); the percentage of tungsten–powellite to total Mo changes depending on the oxidation degree of molybdenite.

While under hydrothermal action, Mo^{4+} can be oxidized to jordisite in a strong acid reduction environment at low temperature and room temperature during the metallogenic epoch. Ilsemannite is the oxidation product, which can be further oxidized to molybdite.

6. Conclusions

Combined with the 3D model, (1) it can be confirmed that the surrounding rocks of the gently inclined upper contact zone and the top of the pluton are more conducive to the formation of high-grade orebodies, the form of Mo orebodies is mostly related to spatial morphology and the occurrence of small plutons, and there is no ore in the core of the pluton. (2) Molybdenite began to be generated after the precipitation of magnetite, resulting in the formation of Fe orebodies earlier than that of Mo, while the Fe orebodies normally formed Mo–Fe orebodies caused by the superimposed Mo mineralization. In general, magnetite-dominated Fe orebodies have a temporal, spatial, and genetic relevance and/or control effect on Mo and its oxide orebodies.

The Shangfanggou Mo–Fe deposit was formed in an environment of high temperature and high oxygen fugacity in Yanshanian, provided by a crust–mantle mixed source and magmatic–hydrothermal and ore-forming source. The ore-forming fluid is given priority to magmatic water, with the possible addition of a small quantity of metamorphic water from the strata. With the precipitation of molybdenite, it experiences the following two different oxidation processes with a decrease in temperature, oxygen fugacity, and acidity: (1) Supergene oxidation (Stage 5)—molybdenite-$MoO_2 \cdot SO_4$ complex ion–Mo oxide minerals aggregate dominated by powethite–tungsten–powellite. During the oxidation process, molybdenite is first oxidized to a $MoO_2 \cdot SO_4$ complex ion and then reacts with a carbonate solution to precipitate powethite, in which W and Mo elements can be substituted by complete isomorphism, forming unique secondary oxide orebodies dominated by tungsten–powellite. (2) Metallogenic epoch (Stage 4)—when the solution is strongly acidic, molybdenite is first transformed into jordisite, and the oxidation product is ilsemannite, which finally can be completely oxidized to molybdite.

Author Contributions: Conceptualization, G.W.; methodology, Z.L. and L.Z. (Ling Zuo); software, N.M. and W.Y.; validation, S.X., C.W., Y.H., T.Y. and L.W.; formal analysis, Z.L.; investigation, Z.L. and G.W.; project administration, L.Z. (Ling Zuo) and L.Z. (Linggao Zeng); data curation, Z.L. and G.W.; writing—original draft preparation, Z.L.; writing—review and editing, Z.L. and G.W. All authors have read and agreed to the published version of the manuscript.

Funding: This research was supported by Fuchuan Mining Co., Ltd., China Geological Survey (Grant No. DD20190570).

Data Availability Statement: All the data are presented in the article.

Acknowledgments: We are grateful to Yi Cao for his constructive comments and discussion. Many thanks are given to Xiaojiang Zhou from Luoyang Fuchuan Co., Ltd. for his kind assistance in the fieldwork. We also thank Leilei Huang and Yulin Chang for their help in data collection and processing, and for the laboratory assistance provided by Peipei Li of the Central Laboratory of Genetic Minerals, China University of Geosciences (Beijing) during the SEM. The valuable comments and suggestions from the four reviewers considerably improved the manuscript.

Conflicts of Interest: The authors declare no conflict of interest.

References

1. Wang, G.W.; Zhang, S.T.; Yan, C.H.; Song, Y.W.; Sun, Y.; Li, D.; Xu, F.M. Mineral potential targeting and resource assessment based on 3D geological modeling in Luanchuan region, China. *Comput. Geosci.* **2011**, *37*, 1976–1988. [CrossRef]
2. Wang, G.W.; Ma, Z.B.; Li, R.X.; Song, Y.W.; Qu, J.A.; Zhang, S.T.; Yan, C.H.; Han, J.W. Integration of multi-source and multi-scale datasets for 3D structural modeling for subsurface exploration targeting, Luanchuan Mo-polymetallic district, China. *J. Appl. Phys.* **2017**, *139*, 269–290. [CrossRef]
3. Alizadeh, A.A.; Guliyev, I.S.; Kadirov, F.A.; Eppelbaum, L.V. Geosciences in Azerbaijan. In *Economic Minerals and Applied Geophysics*, 2nd ed.; Springer: Heidelberg, Germany, 2017; Volume 2, p. 340.
4. Smirnoff, A.; Boisvert, E.; Serge, J. Support vector machine for 3D modelling from sparse geological information of various origins. *Comput. Geosci.* **2008**, *34*, 127–143. [CrossRef]
5. Arias, M.; Nunez, P.; Arias, D.; Gumiel, P.; Castanon, C.; Fuertes-Blanco, J.; Martin-Izard, A. 3D Geological Model of the Touro Cu Deposit, A World-Class Mafic-Siliciclastic VMS Deposit in the NW of the Iberian Peninsula. *Minerals* **2021**, *11*, 85. [CrossRef]
6. Olivier, K.; Thierry, M. 3D geological modelling from boreholes, cross-sections and geological maps, application over former natural gas storages in coal mines. *Comput. Geosci.* **2008**, *34*, 278–290. [CrossRef]
7. Fallara, F.; Legault, M.; Rabeau, O. 3-D Integrated Geological Modeling in the Abitibi Subprovince (Quebec, Canada): Techniques and Applications. *Explor. Min. Geol.* **2006**, *15*, 27–43. [CrossRef]
8. Calcagno, P.; Chilès, J.P.; Courrioux, G.; Guillen, A. Geological modelling from field data and geological knowledge: Part I. Modelling method coupling 3D potential-field interpolation and geological rules. *Phys. Earth Planet. Inter.* **2008**, *171*, 147–157. [CrossRef]
9. Caumon, G.; Collon-Drouaillet, P.; Le Carlier de Veslud, C.; Viseur, S.; Sausse, J. Surface-Based 3D Modeling of Geological Structures. *Math. Geosci.* **2009**, *41*, 927–945. [CrossRef]
10. Sprague, K.; De Kemp, E.; Wong, W.; McGaughey, J.; Perron, G.; Barrie, T. Spatial targeting using queries in a 3-D GIS environment with application to mineral exploration. *Comput. Geosci.* **2006**, *32*, 396–418. [CrossRef]
11. De Kemp, E.; Monecke, T.; Sheshpari, M.; Girard, E.; Lauzière, K.; Grunsky, E.; Schetselaar, E.; Goutier, J.; Perron, G.; Bellefleur, G. 3D GIS as a support for mineral discovery. *Geochem. Explor. Environ. Anal.* **2011**, *11*, 117–128. [CrossRef]
12. Yunsel, T.Y.; Ersoy, A. Geological Modeling of Gold Deposit Based on Grade Domaining Using Plurigaussian Simulation Technique. *Nat. Resour. Res.* **2011**, *20*, 231–249. [CrossRef]
13. Mammo, T. Geophysical Models for the Cu-Dominated VHMS Mineralization in Katta District, Western Ethiopia. *Nat. Resour. Res.* **2013**, *22*, 5–18. [CrossRef]
14. Wang, G.W.; Pang, Z.S.; Boisvert, J.B.; Hao, Y.L.; Cao, Y.X.; Qu, J.N. Quantitative assessment of mineral resources by combining geostatistics and fractal methods in the Tongshan porphyry Cu deposit (China). *J. Geochem. Explor.* **2013**, *134*, 85–98. [CrossRef]
15. Hamedani, M.L.; Plimer, I.R.; Xu, C. Orebody Modelling for Exploration: The Western Mineralisation, Broken Hill, NSW. *Nat. Resour. Res.* **2012**, *21*, 325–345. [CrossRef]
16. Lindsay, M.D.; Aillères, L.; Jessell, M.W.; De Kemp, E.; Betts, P.G. Locating and quantifying geological uncertainty in three-dimensional models: Analysis of the Gippsland Basin, southeastern Australia. *Tectonophysics* **2012**, *546–547*, 10–27. [CrossRef]
17. Mejía-Herrera, P.; Royer, J.J.; Caumon, G.; Cheilletz, A. Curvature Attribute from Surface-Restoration as Predictor Variable in Kupferschiefer Copper Potentials. *Nat. Resour. Res.* **2015**, *24*, 275–290. [CrossRef]
18. Zanchi, A.; Francesca, S.; Stefano, Z.; Simone, S.; Graziano, G. 3D reconstruction of complex geological bodies: Examples from the Alp. *Comput. Geosci.* **2009**, *35*, 49–69. [CrossRef]
19. Li, N.; Chen, Y.J.; Zhang, H.; Zhao, T.P.; Deng, X.H.; Wang, Y.; Ni, Z.Y. Molybdenum deposits in east Qinling. *Earth Sci. Front.* **2007**, *14*, 186–198, (In Chinese with English Abstract).
20. Chen, Y.; Zhai, M.; Jiang, S. Significant achievements and open issues in study of orogenesis and metallogenesis surrounding the North China continent. *Acta Petrol. Sin.* **2009**, *25*, 2675–2726, (In Chinese with English Abstract).
21. *Subsurface Mineral Prospecting Report in the Luanchuan Ore District*; Henan Institute of Geological Survey: Zhengzhou, China, 2015. (Unpublished materials not intended for publication). (In Chinese)
22. *Exploration Report of the Sandaozhuang Mo–W Deposit, Luanchuan County*; Geological Survey Team 1 of Henan Bureau of Geology: Luoyang, China, 1980. (Unpublished materials not intended for publication). (In Chinese)
23. *Exploration Report of the Nannihu Mo (-W) Deposit, Luanchuan County*; Geological Survey Team 1 of Henan Bureau of Geology: Luoyang, China, 1985. (Unpublished materials not intended for publication). (In Chinese)
24. Jiang, D.M.; Fu, Z.G.; Gao, S.H.; Kong, D.C. The analysis of occurrence state for the ore-forming material in Nannihu molybdenum mine field. *China Molybdenum Ind.* **2008**, *32*, 23–31, (In Chinese with English Abstract).
25. Wang, G.W.; Zhu, Y.Y.; Zhang, S.T.; Yan, C.H.; Song, Y.W.; Ma, Z.B.; Hong, D.M.; Chen, T.Z. 3D geological modeling based on gravitational and magnetic data inversion in the Luanchuan ore region, Henan Province, China. *J. Appl. Geophys.* **2012**, *80*, 1–11. [CrossRef]
26. Zhang, Z.Q.; Wang, G.W.; Ma, Z.B.; Gong, X.Y. Interactive 3D Modeling by Integration of Geoscience Datasets for Exploration Targeting in Luanchuan Mo Polymetallic District, China. *Nat. Resour. Res.* **2018**, *27*, 315–346. [CrossRef]
27. Chen, Y.J.; Li, C.; Zhang, J.; Li, Z.; Wang, H.H. Sr and O isotopic characteristics of porphyries in the Qinling molybdenum deposit belt and their implication to genetic mechanism and type. *Sci. China Earth Sci.* **2000**, *43*, 82–94. [CrossRef]

28. Chen, Y.J.; Pirajno, F.; Li, N.; Guo, D.S.; Lai, Y. Hide Researcher Isotope systematics and fluid inclusion studies of the Qiyugou breccia pipe-hosted gold deposit, Qinling Orogen, Henan province, China: Implications for ore genesis. *Ore Geol. Rev.* **2009**, *35*, 245–261. [CrossRef]

29. Ye, H.S. The Mesozoic Tectonic Evolution and Pb-Zn-Ag Metallogeny in the South Margin of North China Craton. Ph.D. Thesis, Chinese Academy of Geological Sciences, Beijing, China, 2006.

30. Yan, C.H. *Internal Structure of Pb-Zn-Ag Metallogenic System in East Qinling*, 1st ed.; Geological Publishing House: Beijing, China, 2004; pp. 1–11. (In Chinese)

31. Shi, Q.Z.; Tao, Z.Q.; Pang, J.Q.; Qu, M.X. Study on Luanchuan Group in the South Margin of North China Plate. *J. Geol. Miner. Res. North CHN* **1996**, *1*, 51–59, (In Chinese with English Abstract).

32. Yan, C.H.; Liu, G.Y.; Peng, Y.; Song, Y.W.; Wang, J.Z.; Zhao, R.J.; Zeng, X.Y.; Lv, W.D.; Yao, X.N.; Ma, H.W.; et al. *Metallogenic Regularity of Pb-Zn-Ag in Southwest Henan Province*, 1st ed.; Geological Publishing House: Beijing, China, 2009; pp. 70–72. (In Chinese)

33. Bao, Z.W.; Zeng, Q.S.; Zhao, T.P.; Yuan, Z.L. Geochemistry and petrogenesis of the ore-related Nannihu and Shangfanggou granite porphyries from east Qinling belt and their constaints on the molybdenum mineralization. *Acta Petrol. Sin.* **2009**, *25*, 2523–2536, (In Chinese with English Abstract).

34. Bao, Z.W.; Sun, W.D.; Zartman, R.E.; Yao, J.M.; Gao, X.Y. Recycling of subducted upper continental crust: Constraints on the extensive molybdenum mineralization in the Qinling-Dabie orogen. *Ore Geol. Rev.* **2017**, *81*, 451–465. [CrossRef]

35. Li, N.; Chen, Y.J.; Pirajno, F.; Gong, H.J.; Mao, S.D.; Ni, Z.Y. LA-ICP-MS zircon U-Pb dating, trace element and Hf isotope geochemistry of the Heyu granite batholith, eastern Qinling, central China: Implications for Mesozoic tectono-magmatic evolution. *Lithos* **2012**, *142–143*, 34–47. [CrossRef]

36. Li, N.; Pirajno, F. Early Mesozoic Mo mineralization in the Qinling Orogen: An overview. *Ore Geol. Rev.* **2017**, *81*, 431–450. [CrossRef]

37. Zhang, G.W.; Zhang, B.R.; Xiao, Q.H. *Qinling Orogenic Belt and Continental Dynamics*, 1st ed.; Science Press: Beijing, China, 2001; pp. 162–320.

38. Duan, S.G.; Xue, C.J.; Yan, C.H.; Liu, G.Y.; Song, Y.W.; Zhang, D.H. Comparison of Pb-Zn metallogenic characteristics between northern and southern margins of North China Landmass. *Miner. Depos.* **2008**, *3*, 383–398.

39. Qi, J.P.; Chen, Y.J.; Ni, P.; Lai, Y.; Ding, J.Y.; Song, Y.W.; Tang, G.J. Fluid inclusion constraints on the origin of the Lengshuibeigou Pb-Zn-Ag deposit, Henan province. *Acta Petrol. Sin.* **2007**, *23*, 2119–2130, (In Chinese with English Abstract).

40. Luo, M.J.; Zhang, F.M.; Dong, Q.Y.; Xu, Y.R.; Li, S.M.; Li, K.H. *Molybdenum Deposits in China*; Henan Sci. & Tech. Press: Zhengzhou, China, 1991; p. 452. (In Chinese)

41. Weng, J.C.; Gao, S.H.; Shi, C.; Wang, Y.M.; Xu, C.Z. Study on the alteration zoning law of Shangfanggou huge-large molybdenum ore deposit. *China Molybdenum Ind.* **2008**, *3*, 16–24, (In Chinese with English Abstract).

42. Mao, J.W. Mesozoic molybdenum deposits in the east Qinling–Dabie orogenic belt: Characteristics and tectonic settings. *Ore Geol. Rev.* **2011**, *43*, 264–293. [CrossRef]

43. Jian, X.L.; Zhang, S.Y.; Pang, Y.C. Henan Province. Verification Report of Mo-Fe Mineral Resource Reserve in the Shangfanggou Mine Area, Luanchuan County, 2012, (Unpublished materials not intended for publication) (In Chinese)

44. Jiang, G.Q.; Zhou, H.R. Sequence, sedimentary environment and tectonic palaeogeographic significance of Luanchuan Group in Luanchuan area, western Henan, Province. *Geoscience* **1994**, *4*, 430–440, (In Chinese with English Abstract).

45. Wang, X.L.; Jiang, S.Y.; Dai, B.Z.; Griffin, W.L.; Dai, M.N.; Yang, Y.H. Age, geochemistry and tectonic setting of the Neoproterozoic (ca 830 Ma) gabbros on the southern margin of the North China Craton. *Precambrian Res.* **2011**, *190*, 35–47. [CrossRef]

46. Chen, Y.J.; Fu, S.G. *Metallogenic Regularity of Gold Deposits in Western Henan, Province*, 1st ed.; Seismological Press: Beijing, China, 1992; pp. 1–6. (In Chinese)

47. Cao, Y.; He, Y.Q.; Zhang, H.; Wang, D.; Hou, S.Z.; Li, Y.T. Geological Characteristics, Mineralization Enrichment Regularity and Prospecting Direction of Shangfanggou Molybdenum Iron Mine in Luanchuan County. *China Molybdenum Ind.* **2021**, *45*, 16–21, (In Chinese with English Abstract).

48. Deutsch, C.V. *Geostatistical Reservoir Modeling*; Oxford University Press: New York, NY, USA, 2002; p. 336.

49. Afzal, P.; Alghalandis, Y.F.; Khakzad, A.; Moarefvand, P.; Omran, N.R. Delineation of mineralization zones in porphyry Cu deposits by fractal concentration–volume modeling. *J. Geochem. Explor.* **2011**, *108*, 220–232. [CrossRef]

50. Qu, J.N. Mineral Resources Quantitative Evaluation Based on 3D Geological Modeling: A Case Study on Nannihu Ore Field. Master's Dissertation, China University of Geosciences, Beijing, China, 2013.

51. Yang, Y.; Zhang, J.; Yang, Y.F.; Shi, Y.X. Characteristics of fluid inclusions and its geological implication of the Shangfanggou Mo deposit in Luanchuan county, Henan province. *Acta Petrol. Sin.* **2009**, *25*, 2563–2574, (In Chinese with English Abstract).

52. Yang, Y.F.; Li, N.; Yang, Y. Fluid inclusion study of the Nannihu porphyry Mo-W deposit, Luanchuan county, Henan province. *Acta Petrol. Sin.* **2009**, *25*, 2550–2562, (In Chinese with English Abstract).

53. Deng, X.H.; Yao, J.M.; Li, J.; Liu, G.F. Fluid inclusion and Re-Os isotopic constraints on the timing and origin of the Shimengou Mo deposit, Xixia County, Henan Province. *Acta Petrol. Sin.* **2011**, *27*, 1439–1452, (In Chinese with English Abstract).

54. Zhang, S.T.; Wang, G.W.; Tang, L.; Guo, B.; Han, J.W.; Li, D. *Metallogenic Background Process and Quantitative Evaluation of Large-Super Large Deposit in Luanchuan Mo-Pb-Zn Polymetallic Ore District*, 1st ed.; Geological publishing house: Beijing, China, 2021; pp. 38–55. (In Chinese)

55. Institute of Geological Sciences; Henan Bureau of Geology and Mineral Resources; Geological Survey Team 1 of Henan Bureau of Geology; Luoyang, Henan Province; Department of Earth Sciences; Nanjing University. *Geology and Mineralization of the North and South China Paleo-Plate Convergence Zone: A Case Study of the East Qinling-Tongbai Area*, 1st ed.; Nanjing University Press: Nanjing, China, 1988; pp. 442–465. (In Chinese)
56. Li, X.Z.; Yan, Z.; Lu, X.X. *Qinling-Dabie Mountain Granite*; Geological Publishing House: Beijing, China, 1993; pp. 10–27. (In Chinese)
57. Ma, Z.D. A preliminary research on the genesis and the tectonic setting of molybdenum metallogenetic zone in the East Qinling, western Henan in accordance with lead-isotopic composition characters. *Earth Sci.* **1984**, *4*, 57–64, (In Chinese with English Abstract).
58. Mao, J.W.; Ye, H.S.; Wang, R.Y.; Dai, J.Z.; Jian, W.; Xiang, J.F.; Zhou, K.; Meng, F. Mineral deposit model of Mesozoic porphyry Mo and vein-type Pb-Zn-Ag ore deposits in the eastern Qinling, Central China and its implication for prospecting. *Geol. Bull. China* **2009**, *1*, 72–79, (In Chinese with English Abstract).
59. Li, Y.F.; Mao, J.W.; Bai, F.J.; Li, J.P.; He, Z.J. Re-Os Isotopic dating of Molybdenites in the Nannihu Molybdenum (Tungsten) orefield in the Eastern Qinling and its geological significane. *Geol Rev.* **2003**, *6*, 652–660, (In Chinese with English Abstract).
60. Li, Y.F.; Mao, J.W.; Hu, H.B.; Guo, B.J.; Bai, F.J. Geology, distribution, types and tectonic settings of Mesozoic molybdenum deposits in East Qinling area. *Mineral. Depos.* **2005**, *3*, 292–304, (In Chinese with English Abstract).
61. Tang, C.H. The Comparative Study of Petrogenesis and Metallogeny for Nannihu and Shangfanggou in Luanchuan Ore-Concentration Area. Master's Thesis, China University of Geosciences, Beijing, China, 2015.

Review

Micro-Mechanisms and Implications of Continental Red Beds

Wang He [1], Zhijun Yang [1,*], Hengheng Du [1], Jintao Hu [1], Ke Zhang [1], Weisheng Hou [1] and Hongwei Li [2]

[1] School of Earth Science and Engineering, Sun Yat-sen University, Zhuhai 519080, China; hewang7@mail2.sysu.edu.cn (W.H.); duhh5@mail2.sysu.edu.cn (H.D.); hujt23@mail2.sysu.edu.cn (J.H.); eeszke@mail.sysu.edu.cn (K.Z.); houwsh@mail.sysu.edu.cn (W.H.)

[2] Guangdong Geological Survey Institute, Guangzhou 510080, China; lihongw1981@126.com

* Correspondence: yangzhj@mail.sysu.edu.cn

Abstract: Continental red beds, widely formed at various geologic timescales, are sedimentary rocks and sediments with red as the main color. Geoscientists have analyzed the geomorphology, paleomagnetism, paleoenvironments, paleontology, energy, and minerals in continental red beds. Despite the agreement that fine-grained hematite is closely related to the color of continental red beds, controversies and problems still exist regarding the micro-mechanism of their formation. As a review, this paper details the composition and color properties of pigmentation in red beds, analyzes the existence and distribution of authigenic hematite, and summarizes the iron sources and the formation of hematite. In addition, we introduce the fading phenomenon observed in continental red beds, including three types of secondary reduction zones: reduction spots, reduction strips, and reduction areas. Lastly, this paper summarizes the evolution of color in continental red beds, emphasizes the relationship between authigenic hematite and the diagenetic environment, and proposes possible research directions for future red bed-related issues.

Keywords: continental red beds; hematite; secondary reduction zone; Danxia landform

Citation: He, W.; Yang, Z.; Du, H.; Hu, J.; Zhang, K.; Hou, W.; Li, H. Micro-Mechanisms and Implications of Continental Red Beds. *Minerals* **2022**, *12*, 934. https://doi.org/10.3390/min12080934

Academic Editor: Luca Aldega

Received: 29 June 2022
Accepted: 22 July 2022
Published: 25 July 2022

Publisher's Note: MDPI stays neutral with regard to jurisdictional claims in published maps and institutional affiliations.

1. Introduction

Red beds can be classified as continental or oceanic red beds according to the depositional environment in which they formed. At present, most are continental; oceanic ones are few due to their special formation conditions [1–3]. Continental red beds are not lithologically different from ordinary clastic sedimentary rocks, but they are of interest to geologists and geomorphologists because of their bright colors and unique landform types. According to erosion characteristics especially in China, red bed landforms are divided into Danxia landforms, red bed mountains, red bed hills, and red bed plateaus or platforms [4].

Research on continental red beds can be traced back to the 19th century [5–7]. Geologists have conducted a wide range of research regarding geomorphology, paleomagnetism, paleontology, energy, and minerals [8–15]. However, debates on the formation mechanism of their color still exist. Early geologists held two views on the origin of red beds. Some researchers believed that red sandstone formed in desert environments and was analogous to the red sand dunes of some modern tropical desert conditions, whereby the color was inherited from the fine-grained hematite film wrapped around the periphery of the sand particles [16], while other researchers suggested that red beds originated from lateritic weathering and formed in warm and humid environments where weathering of the nearby parent rocks produced the color originating from the residual red clay rich in iron oxides [17–21]. Thus, for a long time, the formation of continental red beds was assigned some paleoclimatic significance. However, geologists subsequently found that red beds formed in both arid and humid tropical climates [22–24]. The color of the red beds was formed during diagenesis, and the pigment hematite was derived from in situ alterations of iron-bearing mineral fragments or iron oxyhydroxide dihydroxylation [25]. Paleoclimatic conditions were not critical for the formation of the red color [7].

Geologists have confirmed that fine-grained hematite is the main cause of the red color through study of secondary reduction zones and the extraction of pigmentary minerals [21,26,27]. Confusion remains as to its source, existence, and formation process. In recent years, the genesis of red beds has once again become of interest to geoscientists, because the presence of hematite could be indicative of the iron cycle, water storage, and the existence of life on Mars [28–30]. Therefore, this paper attempts to review the progress of research on issues related to the continental red beds, summarize the mechanisms of its color formation, and explore future development directions.

2. Red Substance

2.1. Color-Rendering Characteristics

The color of red beds depends mainly on its inherited color and oxidized color, with the inherited color being the color exhibited by the clastic minerals themselves in the rock, while the oxidized color is usually caused by iron oxides or oxyhydroxides contained in its matrix and cement. It is generally accepted that the red color in the red beds originates from hematite, but there is some fluctuation in the hue of the red layer from different places. If the Munsell color system is used to describe the red beds, the color of most red beds usually varies between 5YR and 5 RP [16,31–33]. This is mainly because the red beds also include other iron oxides or oxyhydroxides such as goethite, ferrihydrite, and lepidocrocite. Table 1 shows the color variation values of eight synthetic samples measured by Scheinost and Schwertmann (1999) using the Munsell color system [34]. Hematite has the most reddish hue on average and goethite shows a yellowish-brown color, while the remaining iron oxides or oxyhydroxides are in between, which would explain the reddish-brown color of many red beds. However, different iron oxides differ greatly in tinting strength and hiding power. For example, the tinting strength of hematite is much greater than that of goethite at the same particle size and distribution, even though all iron oxides have strong tinting strength [33]. Even a small percentage of hematite mixed with goethite can mask the color of goethite [35]. Moreover, hematite has a greater covering power than goethite ($30–60 \ \mathrm{m^2 \cdot kg^{-1}}$ and $15–20 \ \mathrm{m^2 \cdot kg^{-1}}$, respectively) [36]. Therefore, it is necessary to study hematite in the red beds and, thus, its color genesis.

Table 1. Munsell colors of the Fe oxides (median and range) (Reprinted with permission from Ref. [34]. 1999, John Wiley and Sons).

	N	Hue	Value	Chroma
Hematite	59	1.2 YR 3.5 R–4.1 YR	3.6 2.4–4.4	5.2 1.5–7.9
Goethite	82	0.4 Y 7.3 YR–1.6 Y	6.0 4.0–6.8	6.9 6.0–7.9
Lepidocrocite	32	6.8 YR 4.9 YR–7.9 YR	5.5 4.6–5.9	8.2 7.1–9.9
Ferrihydrite	59	6.6 YR 2.8 YR–9.2 YR	4.9 2.3–6.3	6.3 1.9–7.3
Akaganéite	8	5.2 YR 1.2 YR–6.8 YR	3.8 2.8–4.3	5.8 4.4–7.3
Schwertmannite	16	8.5 YR 6.2 YR–0.3 Y	5.9 4.7–6.7	6.9 4.0–9.1
Feroxyhyte	10	4.2 YR 3.7 YR–5.4 YR	3.8 3.4–4.7	6.0 5.5–7.0
Maghemite	7	8.3 YR 6.2 YR–9.4 YR	3.1 2.5–3.6	3.2 2.5–4.1

Note: *N* = number of samples.

The shape and size, distribution, aggregation, and cation substitution of hematite particles can also lead to changes in color. Franz (1981) showed that acicular hematite has higher reflectance and scattering ability in the long-wavelength region of the spectrum than more symmetrical hematite particles and, therefore, tends to have a more yellowish hue [26]. The tinting strength of hematite powder increases with decreasing particle size, and the strongest coloring ability of hematite is achieved when the particle size decreases to the optimal particle size, i.e., the particle size with the largest scattering cross-section (about 1 μm) [33]. Therefore, even in very brightly colored red beds, hematite is often difficult to detect by XRD. In contrast, large crystals or dense aggregates of hematite are usually dark brown or black, due to much greater absorption than scattered throughout the visible region. The arrangement of hematite aggregates likewise affects the color variation. In the Devonian red beds of Scotland, Turner and Archer (1977) observed oriented aggregates of platelet-like hematite crystals epitaxially growing on altered biotite, and the combination of such small lamellar hematite crystals into oriented aggregates causes a color shift to purple [27,37]. In addition, other metal substitutions of some Fe in the crystals can also modify the color of hematite. For example, manganese- and titanium-substituted hematite is black, while, in the case of aluminum substitution, it shifts to a more reddish hue due to the reduction in particle size [38,39].

2.2. Distribution and Existence of Hematite

There are two main forms of authigenic hematite in red beds. One is in the form of single or polycrystalline aggregates coexisting with other authigenic minerals around the clastic particles, forming along the cleavage planes of mica (mainly biotite) or filling the pores between the particles; a small number of hematite particles are also present in fractures [40–43]. The crystal form of this hematite ranges from semi-hedral to euhedral, usually with relatively small quantities but large particles, mainly on the micrometer scale. There are also some submicron crystals. For example, Eren et al. (2013) discovered rhombohedral hematite crystal cross-sections with a grain size of about 0.3 mm in the red sandstones of the Early Cambrian Hüdai Formation in the Aydıncık (Mersin) Zone, central Taurides, southern Turkey (Figure 1a) [44]. Rasmussen and Muhling (2019) found radiolarian aggregates (Figure 1c) and slatted hematite filling granular voids (Figure 1d) in metamorphic red sandstones and shales in the Stirling Formation of Western Australia [45].

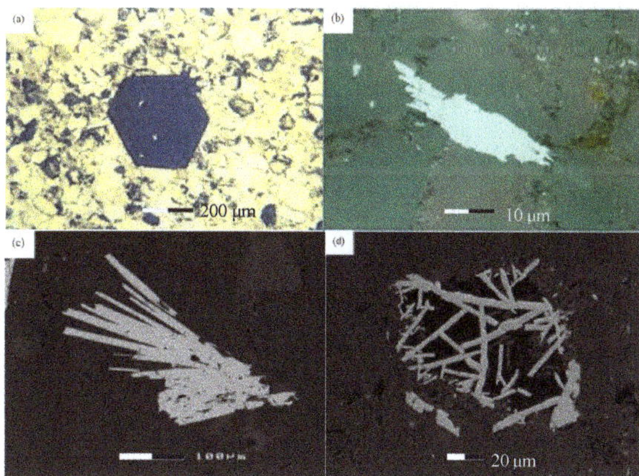

Figure 1. (**a**) Hematite in a thin section illustrating a euhedral hexagonal crystal form; (**b**) hematite outgrowths aligned in the fabric defined by strain fringes on quartz grains; (**c**) hematite plates partly filling a grain-shaped cavity; (**d**) hematite grain with large radiating blades. (**a**) is reprinted with permission from Ref. [44]. 2013, Turkish Journal of Earth Sciences; (**b–d**) are reprinted with permission from Ref. [45]. 2019, Elsevier.

Another type of hematite is distributed as microfine grained crystals mixed with clay minerals in clusters disseminated in the matrix and cement of the red beds [46], partially forming a thin film of iron-bearing clay around individual clastic particles; clay films are present on clastic particles in almost all red beds. Films are sometimes developed at the contacts with detrital grains and are covered by quartz secondary overgrowth or wrapped by carbonate cement; clay-containing films formed after quartz overgrowth also exist [42]. This hematite is distributed microscopically and diffusely on montmorillonite and illite/smectite wafers or filled among the micropores between the illite/smectite layers (Figure 2a,b). Most of the hematite is in tiny crystal agglomerates or in individual microcrystals form in the pores of the montmorillonite and exhibit a variety of shapes (Figure 2d) [47]. There are spherical, rod-shaped, leaf-shaped, and well-developed hexagonal flake hematite single crystals, as well as fibrous to lath-shaped, rosette clusters and spherical hematite/goethite aggregates (Figure 2c–f) [47,48]. These morphological features are closely related to the relative abundance of the iron-bearing mineral precursors in the rocks. Hematite particles dispersed in clay minerals are usually small but very abundant and are the main source of coloring of the red beds. The hematite mainly constitutes submicron or even nanoscale crystals; hence, it is difficult to identify it using conventional detection methods [49].

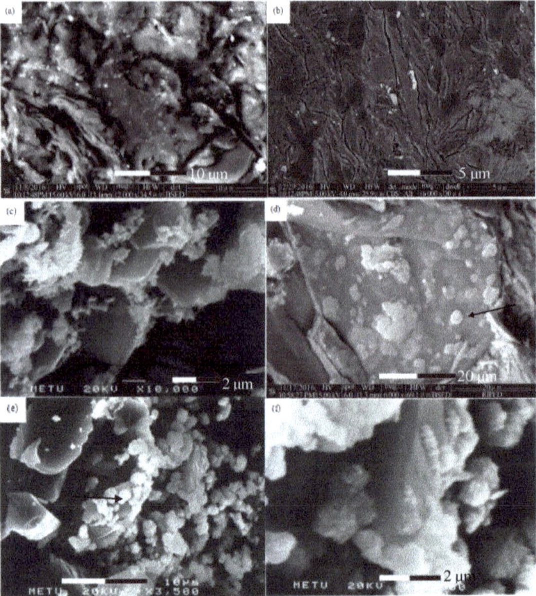

Figure 2. (**a**) Hematite scattered over illite/smectite layers; (**b**) fine and dispersed hematite distributed among clay minerals; (**c**) hematite pigment showing a patchy coating on detrital grains; (**d**) plated hexagonal hematite crystals distributed on the surface of a detrital mineral; (**e**) spherical hematite particles; (**f**) rod-shaped (or bacillus-shaped) hematite particles disseminated on the grain surface. (**a,b,d**) are reprinted with permission from Ref. [47]. 2020, Geoscience; (**c,e,f**) are reprinted with permission from Ref. [44]. 2013, Turkish Journal of Earth Sciences.

The form of authigenic hematite in the red beds seems to correspond to the age of the red beds. Compared with the hematite in the form of single or polycrystalline aggregates, the latter (mixed with clay minerals) is more widely distributed in time and space, appearing in red beds of all ages. In Mesozoic and Cenozoic red beds, hematite is almost only adsorbed on the surface and/or in the voids of clay minerals [41,48,50]. In older red beds, hematite can be depleted to varying degrees due to later tectonics, groundwater penetration, or the participation of hydrothermal fluids, and then replaced by euhedral to semi-hedral authigenic hematite grains formed in situ and secondary pores or aggregates [42,43]. Residues at the grain contacts or the edges of some reddish iron-bearing clays are surrounded by quartz secondary enlarged edges. Older red beds have a greater proportion of granular hematite particles and a darker overall color.

3. The Source and Formation of Hematite

3.1. Source

Almost all Fe-containing minerals are potential sources of Fe for red-bed chromogenic substances. During oxidative weathering, Fe is released from these minerals, mainly forming Fe-bearing clay minerals and Fe oxides/hydroxides. However, under reducing conditions, Fe carbonates, sulfides, and phosphates may also be formed [33,51]. During the formation of red beds, the alteration of unstable iron-rich particles is an important source of Fe for hematite [24,25,52–54].

The commonly altered mineral grains are iron–magnesium silicates (e.g., hornblende, pyroxene, and biotite), volcanic rock fragments and glass, magnetite, and ilmenite [45]. These grains formed at high temperature or high pressure during magmatism and metamorphism became unstable at surface and near-surface conditions in the sedimentary basin. During grain alteration, under reducing conditions, released iron goes into solution as Fe^{+2} and precipitates as iron oxide or oxyhydroxide under oxidizing conditions [55]. Walker et al. (1967, 1978) [22,56] elaborated on the formation mechanism of red beds during burial diagenesis and pointed out that the alteration of the interior of the iron–magnesium silicate layers by oxygen-containing groundwater during the burial process is the key. They showed that the hydrolysis of iron-bearing detrital minerals follows the Goldich dissolution reaction series and is governed by the Gibbs free energy of the specific reaction [22,56].

Mucke (1994) indicated that alteration after diagenesis can also occur through the oxidation of pyrite and siderite [57].

$$3O_2 + 4FeS_2 \rightarrow Fe_2O_3 \text{ (hematite)} + 8S; E = -789 \text{ kJ/mol.}$$

$$O_2 + 4FeCO_3 \rightarrow 2Fe_2O_3 \text{ (hematite)} + 4CO_2; E = -346 \text{ kJ/mol.}$$

Dissolution and displacement are the main processes leading to grain alteration in red beds. Partial dissolution of unstable minerals such as amphibole and pyroxene may proceed inward from the grain edges, while minerals such as plagioclase may preferentially disrupt the grain interior [32]. The different morphologies of partially dissolved minerals are an abiotic result of selective dissolution along lattice planes. Xiao et al. (2018), suggested that Fe and Ti-bearing silicate minerals such as biotite gradually release various elements during low-temperature alteration, with water-soluble elements such as K and Na being carried away with the transport of pore water, while water-insoluble iron–titanium oxides are deposited in situ around these altered minerals [49]. Clay mineral replacement of clastic particles is also a common alteration phenomenon in red beds, where the replaced clay forms in situ and shows random sheet crystal orientation [23].

Apart from the inheritance of parent minerals such as ferromagnesian silicates, clastic and clay minerals mechanically permeated by groundwater can also provide a direct source of iron for staining [7,54,58,59]. The mixing of hematite with clay minerals in the form of very fine-grained crystals is an important feature of red beds, and the relative proportions of hematite and clay vary considerably. In some cases, iron-bearing films may consist only of granular crystals of hematite that have precipitated in situ. In other cases, a thicker iron-containing clay film may be present. These Fe-bearing clays may have originated from surface soils or, more likely, from loose sediments formed in the adjacent floodplain [60]. An important feature of mechanical infiltration is that clay is not red when deposited, but reddens over time after contact with oxygenated groundwater [25]. In Cenozoic alluvium, the usual sediment color is yellow or red orange, due to iron oxyhydroxide, of which goethite and ferrihydrite are the main chromogenic minerals. Thus, the reddening of permeable clays represents an early stage of red material formation in red beds.

In addition, Blodgett et al. (1993) suggested that recirculation of red material may be the source of hematite in some red beds [32]. Such red beds would be produced by the erosion and redeposition of pre-existing red beds. These often have irregular color zoning, which is related to an unconformity in the weathering profile. Color boundaries cross lithological contacts and show more intense reddening near unconformities. Secondary red beds formed in this way are representative of late diagenesis, associated with previously deposited uplift, erosion, and surface weathering, requiring similar primary and diagenetic conditions [61].

3.2. Hematite Formation

There are two main potential mineralogical pathways for the formation of hematite during diagenesis: ferrihydrite transformation and ferric oxyhydroxide dehydroxylation. Ferrihydrite is the most common precursor for the occurrence of iron oxides in soils and sediments, and the transformation from ferrihydrite to hematite in red beds is an important process [62]. The conversion of ferrihydrite to hematite appears to start with the internal rearrangement and dehydroxylation of ferrihydrite aggregates, which are converted to hematite over time [63]. This transition may occur faster than the new formation of ferrihydrite and may occur too quickly to be detected in most geological settings [33]. Jiang et al. (2018), proposed a new model for the transformation of ferrihydrite to hematite in soils and sediments, which can be divided into five stages according to the chronological order of transformation: (I) magnetically ordered (or core-shell structured) superparamagnetic (SP) ferrihydrite formation, (II) rapid hematite formation from SP ferrihydrite, (III) relative hematite stabilization, (IV) maghemite nanoparticle neoformation, and (V) completion of the transformation from maghemite to hematite [64]. This model provides a new framework for interpreting the formation of iron oxides and their paleoenvironmental significance.

The transformation of goethite to hematite has long been considered the important pathway for red beds [65,66]. Weibel et al. (1999) observed pseudo-crystals of goethite transforming to hematite in red sandstone samples from the Triassic Skagerak Formation in Denmark. Withing increasing burial depth (550 m to 2500 m), the ground temperature increased from 47 °C to >105 °C [67]. Berner (1969) pointed out that goethite and iron hydroxides are generally unstable relative to hematite, and goethite is easily dehydroxylated under water shortage or high-temperature conditions (2FeOOH (goethite) $\rightarrow$ Fe$_2$O$_3$ (hematite) + H$_2$O). Experiments show that, at 250 °C, the Gibbs free energy (G) of goethite-hematite reaction is 2.76 kJ/mol [68]. Langmuir (1971) concluded that G becomes more and more negative as particle size decreases. The smaller the mineral particle, the easier it is for goethite to undergo dehydroxylation to form hematite [69]. Dehydroxylation of fine-grained goethite may occur quickly, especially during deposition [7]. Thus, it appears that clastic iron oxyhydroxides (such as goethite) spontaneously dehydrate into red hematite over time. This process helps to explain not only the gradual reddening of the alluvium but also why desert dunes of older geological ages are redder than the newer ones.

In addition, other processes leading to the formation of diagenetic hematite may include the pseudo-crystalline substitution of hematite for minerals such as magnetite [7,70] and microbial interactions [44,71]. It should be noted that usually the formation of hematite in red beds may not be caused by a single mechanism of action but is the result of a combination of mechanisms. On long geological time scales, the origin of hematite in some older red beds may even be more complex because of the possibility of dissolution and reprecipitation of hematite during one or more earlier tectonic or crustal fluid events [43,45].

4. Reduction and Leaching

The local development of light bands of white, gray, or green color in many red-dominated stratigraphic sequences is a typical feature of continental red beds. Many researchers refer to such light bands as secondary reduction zones [7]. Geologists have been observing secondary reduction zones for almost as long as the color has been discussed in the literature [32,72–75]. According to the mineralogical characteristics and geochemical properties of secondary reduction zones in red beds, two categories have been identified: leaching zones and reduction zones. Rocks in the leaching zone tend to be white, as the iron content is depleted relative to the surrounding red host rock, due to leaching of soluble iron [7,76]. In contrast, rocks in the reduction zone are mainly gray or green and show a slight deficit in iron content relative to the peripheral red host rock. There, Fe^{2+}/Fe^{3+} values can be significantly higher and can be attributed to the formation of sulfides, carbonates, or clay minerals (mainly chlorite) [74]. Some red beds are characterized by both leaching and reduction, which are manifested by the coexistence of multiple-colored sediments in such secondary reduction zones.

According to scale of development and morphological characteristics, secondary reduction zones can exhibit three common forms in field outcrops: reduction spots, reduction strips, and reduction areas.

4.1. Reduction Spots

Light-colored zones having a circular or elliptical shape can be referred to as reduction spots or reduction points. Spatially, these reduction spots are a series of irregularly shaped reduction spheres or ellipsoidal spheres, which are early reduction features formed by diffusion of reduced material from the core of reduction spots to the surrounding area. Most reduction spheres in red beds are near-perfect spheres showing clear boundaries with the outer red host rock (Figure 3a,d,e). Some of the ellipsoidal spheres have long axes extending parallel to the stratigraphy (Figure 3e,f), mostly between 1 and 3 cm in diameter [32,59]. Their colors range from white to off-white or light green to blue-green. Reduction spots can occur in both impermeable argillaceous limestone and pore-rich conglomerates; they are like the surrounding red host rock in terms of sediment composition, particle size, roundness, porosity, and sedimentary structure [9]. In terms of chemical composition, the red host rock has a higher total iron content and a lower Fe^{2+}/Fe^{3+} value than the reduction spot, but the Fe^{2+} content in the reduction spot is not necessarily higher than the Fe^{3+} value. Hematite is usually present in the matrix and cement of the red beds as microcrystalline particles or mixed with clay minerals, but is rarely present in reduction spots [7,59,77]. The formation of reduction spots is usually attributed to the in situ reduction of Fe^{3+} to Fe^{2+} from the dissolution of iron oxides. Some researchers have suggested that reduction spots form after the crystallization of pigment hematite, i.e., the sediment is red before the formation of reduction spots, and the hematite is converted to the corresponding reduction products by reduction [78,79]. Other researchers have suggested that the area where the reduction spots form never become red, i.e., the spots formed before the surrounding host rocks become pigmented, preventing the precipitation of hematite and preserving the original color of the sediment or reducing it further [7,25,55,59,77,80].

In some red beds, dark cores of black, brown, or red colors are also observed in the center of the reduced spots (Figure 3b). The dark core is often enriched with metallic elements such as Cu, V, U, and Ni and nonmetallic elements such as S and As [74,79,81,82]. Some researchers have suggested that the higher concentration of metallic elements such as copper in the reduction spots may be due to adsorption by organic matter [76–78]. More researchers have suggested that the formation of reduction spots containing dark cores is better attributed to the activity of bacteria or microorganisms that use organic matter as an energy source and reducing agent, allowing the dissolution of metallic elements and their precipitation in the cores of the reduction spots through bacterial-mediated reduction [74,75,81,83–85].

We note that carbonatite-bearing rock chip conglomerates also commonly appear in the center of the reduction spot in some conglomerate red beds. The formation of such reduction spots is closely related to the composition of the rock chips (Figure 3c,f).

Figure 3. (**a**) A large perfectly round white reduction spot in reddish-brown sandstone; (**b**) many reduction spots of different sizes in reddish-brown sandstone (note red core exists at center of a white reduction spot); (**c**) reduction ring around carbonate fragments in red sandstone; (**d**) sub-circular gray-white reduced spot in the red bed; (**e**) sub-elliptical light green reduction spot in red conglomerate (note its long axis is parallel to bedding); (**f**) gravelly debris in center of sub-elliptical reduction spot and crossed by late-forming fractures. Reduction spot photo from Danxia Mountain, Guangdong, China.

4.2. Reduction Strips

Secondary reduction zones distributed parallel to bedding planes, along with unconformity contacts and structural deformation zones or fractures, usually appear as strips or irregular bands in field outcrops. these are collectively referred to as reduction strips in this paper (Figure 4b,d). Compared with the surrounding red host rocks, the main composition of sediments in the reduction strip is similar, but with greater grain size, porosity, and permeability [59]. Therefore, the reduction strip is often developed on the sides of the relatively coarse-grained laminated units or tectonic fissures. The colors of the reduction strip are mainly white, gray, or off-white, followed by light pink, dark gray, green, etc. [42,54,86,87]. Reduction fluids use the pores or fissures in the reduction strips as channels and leach the free iron oxide from the sediments by reduction [88,89]. According to the chemical composition, reducing fluids can be divided into the following categories: (1) hydrocarbons (hydrocarbon organics associated with hydrocarbons) [90,91]; (2) organic acids (associated with surface facies formation) [54]; (3) acidic reduction fluids such as CO_2 and H_2S (associated with volcanism and hydrothermal action) [92,93]. In addition, reduction fluids from reservoirs occasionally contain high H_2S contents (>5 vol.%) as a result of thermochemical sulfate reduction of methane with capping anhydrite [94]. Most secondary reduction zones in red beds may not be a result of a single reduction, but include a variety of reducing fluids with multiple stages of reduction and leaching, and even the participation of bacteria or other microbial activities [54,87,95].

Figure 4. (**a**) Coal measures within the green facies; (**b**) green facies and relicts of red facies in outcrop showing bleaching of red sandstone to green; (**c**) gray sandstones and conglomerates separated by red mudstone interbeds; (**d**) reduction strips in the red bed are distributed along the bedding plane. (**a**,**b**) are reprinted with permission from Ref. [15]. 2019, Elsevier; (**c**) is reprinted with permission from Ref. [91]. 2019, Elsevier; (**d**) is from Danxia Mountain, Guangdong, China.

4.3. Reduction Areas

Reduction areas have a similar genesis mechanism to reduction strips but require an adequate source of reduction fluids to enable the reduction of larger-scale zones and even entire stratigraphic units (Figure 4a,c). Reduction areas must involve the transport and storage of large-scale reduction fluids. The formation of large reductive areas is primarily related to hydrocarbon reservoirs. The reductive fluids are mainly hydrocarbons, followed by CO_2. Most Jurassic and Cretaceous red beds in the Colorado Plateau of the United States developed abundant reducing areas with hydrocarbons as the main reducing fluid. These are closely related to nearby oil and gas reservoirs in space [89,96–98]. However, in other areas such as the red sandstone reduction zones near Green River, Utah, the reduction fluids were dominated by CO_2. Fe and other trace metal elements in the matrix and cement were dissolved in low-pH reduction fluids and eventually reprecipitated near the redox front with fluid transport [93,99,100]. In addition, because the creation of reduction areas usually consumes large amounts of fluids, when they carry dissolved metals such as Cu, U, V, Co, and Au, creating enrich mineralization near redox fronts, geologists often locate copper, uranium, and other sedimentary minerals near the reduction areas [7,15,42,82,101–104].

5. Conclusions and Future Work

Since the Mesoproterozoic Great Oxidation Event (GOE), the free oxygen content of the atmosphere has increased sharply, and most of the Earth's surface is generally an oxidizing environment. All iron oxides or oxyhydroxides have high tinting strength and hiding power. Even if their content is less than 1% iron oxides, it is enough to stain the sediment. Therefore, the color of most iron-bearing continental sediments is reddish or yellowish. It is certain that hematite has stronger coloring and covering power than other iron oxides and oxyhydroxides; hence, the color of the red beds is almost dependent on microfine hematite. Thus, the color evolution of the red beds is essentially the evolution of hematite. In the evolution of these red beds, hematite is mainly formed during diagenesis, and iron originates from iron-rich mineral particles such as iron-bearing silicates, sulfides, and oxides, which are released into pore solutions through alterations, and then precipitated under suitable geochemical conditions forming hematite or precursors. Hematite can also be formed by dehydroxylation of iron oxyhydroxides such as goethite, an important pathway for hematite formation during red-bed burial and diagenesis, although the intermediate product of the dehydration process remains controversial [39,105–107]. Even though the free energies of the different surfaces of hematite crystals are roughly similar and their relative stability is easily modified by environmental conditions, crystal morphology usually shows great variability [63,108]. Therefore, authigenic hematite in red beds exhibit various shapes such as spherical, rod, flake, and lath, likely caused by different diagenetic environments. The reductive leaching phenomenon of red beds is often accompanied by the interaction between reducing fluids and iron oxides. Thus, the secondary reduction zone not only indicates the redox state of iron ions but also allows the color evolution history of the red beds to be probed by the redistribution traces of iron phases.

According to the progress and existing problems of previous research on the color origin of continental red beds, the authors believe that more in-depth and detailed research related to red beds in the future needs to consider the following topics:

(1) Distinguishing the difference and connection between hematite formed by alteration of minerals such as iron silicates or oxides and hematite formed by iron-bearing clay to explore the influence of provenance on the color of red beds;

(2) Combining the thermodynamic behavior of iron oxide in the red beds, as well as the microscopic mechanism of formation of hematite and iron oxyhydroxide, to reveal the microscopic kinetic process of red sediments;

(3) Detailed quantifying of hematite and other iron oxyhydroxides in red beds to provide a more accurate scientific basis for the definition of red beds and to clarify the similarities and differences between red beds and other sedimentary strata;

(4) Exploring the relationships among the size, shape, and diagenetic environment of hematite crystals in the red beds;

(5) Studying the color fading phenomena to further deepen understanding of the history and kinetic process leading to red-bed color formation.

Author Contributions: Conceptualization, W.H. (Wang He) and Z.Y.; methodology, W.H. (Wang He); validation, W.H. (Wang He), Z.Y. and H.D.; investigation, W.H. (Wang He); resources, H.L.; writing—original draft preparation, W.H. (Wang He); writing—review and editing, W.H. (Weisheng Hou), Z.Y. and K.Z.; visualization, H.D. and J.H.; supervision, Z.Y., W.H. (Weisheng Hou) and K.Z. All authors have read and agreed to the published version of the manuscript.

Funding: This study was financially supported by the Guangdong provincial Natural Science Foundation (Grant 2021A1515011472) and the Guangdong Special Fund for National Park Construction (Grant 2021GJGY026).

Conflicts of Interest: The authors declare no conflict of interest.

References

1. Wang, C.; Hu, X. Cretaceous world and oceanic red beds. *Earth Sci. Front.* **2005**, *12*, 11–21. (In Chinese)
2. Hu, X. Distribution, Types and Origins of Phanerozoic Marine Red Beds. *Bull. Mineral. Petrol. Geochem.* **2013**, *32*, 335–342. (In Chinese)
3. Lyu, X.; Liu, Z. Distribution, compositions and significance of oceanic red beds. *Adv. Earth Sci.* **2017**, *32*, 1307–1318. (In Chinese)
4. Peng, H.; Pan, Z.; Yan, L.; Scott, S. A review of the research on red beds and Danxia landform. *Acta Geogr. Sin.* **2013**, *68*, 1170–1181. (In Chinese)
5. Russell, I.C. *Subaerial Decay of Rocks and Origin of the Red Color of Certain Formations*; US Government Printing Office: Washington, DC, USA, 1889; pp. 537–597.
6. Dorsey, G.E. The origin of the color of red beds. *J. Geol.* **1926**, *34*, 131–143. [CrossRef]
7. Turner, P. *Continental Red Beds*; Elsevier Scientific Publishing Company: Amsterdam, The Netherlands, 1980.
8. Migoń, P. Geomorphology of conglomerate terrains—Global overview. *Earth-Sci. Rev.* **2020**, *208*, 103302. [CrossRef]
9. Zhu, C.; Wu, L.; Zhu, T.; Hou, R.; Hu, Z.; Tan, Y.; Sun, W.; Jia, T.; Peng, H. Experimental studies on the Danxia landscape morphogenesis in Mt. Danxiashan, South China. *J. Geogr. Sci.* **2015**, *25*, 943–966. [CrossRef]

10. Dill, H.G.; Kadirov, O.; Tsoy, Y.; Usmanov, A. Palaeogeography of Neogene red bed sequences along the Aksa-Ata River in the Parkent-Nurekata intermontane basin (Tien Shan Mountains, Uzbekistan): With special reference to the magnetic susceptibility of siliciclastic rocks. *J. Asian Earth Sci.* **2007**, *29*, 960–977. [CrossRef]
11. Jiang, Z.; Liu, Q.; Dekkers, M.J.; Zhao, X.; Roberts, A.P.; Yang, Z.; Jin, C.; Liu, J. Remagnetization mechanisms in Triassic red beds from South China. *Earth Planet. Sci. Lett.* **2017**, *479*, 219–230. [CrossRef]
12. Thibal, J.; Etchecopar, A.; Pozzi, J.P.; Barthes, V.; Pocachard, J. Comparison of magnetic and gamma ray logging for correlations in chronology and lithology; example from the Aquitanian Basin (France). *Geophys. J. Int.* **1999**, *137*, 839–846. [CrossRef]
13. Sheldon, N.D. Do red beds indicate paleoclimatic conditions? A Permian case study. *Palaeogeogr. Palaeoclimatol. Palaeoecol.* **2005**, *228*, 305–319. [CrossRef]
14. Bai, H.; Kuang, H.; Liu, Y.; Peng, N.; Chen, X.; Wang, Y. Marinoan-aged red beds at Shennongjia, South China: Evidence against global-scale glaciation during the Cryogenian. *Palaeogeogr. Palaeoclimatol. Palaeoecol.* **2020**, *559*, 109967. [CrossRef]
15. Gao, R.; Xue, C.; Zhao, X.; Chen, X.; Li, Z.; Symons, D. Source and possible leaching process of ore metals in the Uragen sandstone-hosted Zn-Pb deposit, Xinjiang, China: Constraints from lead isotopes and rare earth elements geochemistry. *Ore Geol. Rev.* **2019**, *106*, 56–78. [CrossRef]
16. Folk, R.L. Reddening of desert sands: Simpson Desert, N. T., Australia. *J. Sediment. Petrol.* **1976**, *46*, 604–615. [CrossRef]
17. Krynine, P.D. The origin of red beds. *Trans. N. Y. Acad. Sci.* **1949**, *11*, 60–68. [CrossRef]
18. Krynine, P.D. *Petrology, Stratigraphy, and Origin of the Triassic Sedimentary Rocks of Connecticut*; State Geological and Natural History Survey: Berkeley, CA, USA, 1950; p. 247.
19. Van Houten, F.B. Climatic significance of red beds. *Descr. Paleoclimatology* **1961**, *21*, 89–139.
20. Van Houten, F.B. Origin of red beds—Some unsolved problems. In *Problems in palaeoclimatology, Proceedings of the NATO Paleoclimate Conference, Newcastle upon Tyne, UK, 7–12 January 1963*; Inter-Science Pub. Inc.: New York, NY, USA, 1964; pp. 647–660.
21. van Houten, F.B. Iron oxides in red beds. *Geol. Soc. Am. Bull.* **1968**, *79*, 399–416. [CrossRef]
22. Walker, T.R. Formation of red beds in modern and ancient deserts. *Geol. Soc. Am. Bull.* **1967**, *78*, 353–368. [CrossRef]
23. Walker, T.R. Color of recent sediments in tropical Mexico: A contribution to the origin of red beds. *Geol. Soc. Am. Bull.* **1967**, *78*, 917–920. [CrossRef]
24. Walker, T.R. Formation of red beds in moist tropical climates: A hypothesis. *Geol. Soc. Am. Bull.* **1974**, *85*, 633–638. [CrossRef]
25. Walker, T.R. Diagenetic origin of continental red beds. In *The Continental Permain in Central, West, and South Europe*; Springer: Berlin/Heidelberg, Germany, 1976; pp. 240–282.
26. Hund, F. Inorganic Pigments: Bases for Colored, Uncolored, and Transparent Products. *Angew. Chem. Int. Ed. Engl.* **1981**, *20*, 723–730. [CrossRef]
27. Torrent, J.; Schwertmann, U. Influence of hematite on the color of red beds. *J. Sediment. Petrol.* **1987**, *57*, 682–686. [CrossRef]
28. Chan, M.A.; Beitler, B.; Parry, W.T.; Ormo, J.; Komatsu, G. A possible terrestrial analogue for haematite concretions on Mars. *Nature* **2004**, *429*, 731–734. [CrossRef] [PubMed]
29. Chan, M.A.; Bowen, B.B.; Parry, W.; Ormö, J.; Komatsu, G. Red rock and red planet diagenesis. *GSA Today* **2005**, *15*, 4–10. [CrossRef]
30. Chen, S.A.; Heaney, P.J.; Post, J.E.; Fischer, T.B.; Eng, P.J.; Stubbs, J.E. Superhydrous hematite and goethite: A potential water reservoir in the red dust of Mars? *Geology* **2021**, *49*, 1343–1347. [CrossRef]
31. McBride, E.F. Significance of color in red, green, purple, olive, brown, and gray beds of Difunta Group, northeastern Mexico. *J. Sediment. Petrol.* **1974**, *44*, 760–773. [CrossRef]
32. Blodgett, R.H.; Crabaugh, J.P.; McBride, E.F. The color of red beds—A geologic perspective. *Soil Color* **1993**, *31*, 127–159.
33. Cornell, R.M.; Schwertmann, U. *The Iron Oxides: Structure, Properties, Reactions, Occurrences, and Uses*; Wiley-vch Weinheim: Weinheim, Germany, 2003; Volume 2.
34. Scheinost, A.C.; Schwertmann, U. Color identification of iron oxides and hydroxysulfates: Use and limitations. *Soil Sci. Soc. Am. J.* **1999**, *63*, 1463–1471. [CrossRef]
35. Barrón, V.; Torrent, J. Use of the Kubelka—Munk theory to study the influence of iron oxides on soil colour. *J. Soil Sci.* **1986**, *37*, 499–510. [CrossRef]
36. Fuller, C.W. Iron oxides, synthetics. In *Chemical and Process Technology Encyclopedia*; Considine, D.M., Ed.; McGraw-Hill Book Company: New York, NY, USA, 1974; p. 645.
37. Turner, P.; Archer, R. The role of biotite in the diagenesis of red beds from the Devonian of northern Scotland. *Sediment. Geol.* **1977**, *19*, 241–251. [CrossRef]
38. Scheinost, A.C.; Schulze, D.G.; Schwertmann, U. Diffuse reflectance spectra of Al substituted goethite: A ligand field approach. *Clays Clay Miner.* **1999**, *47*, 156–164. [CrossRef]
39. Jiang, Z.; Liu, Q.; Roberts, A.P.; Dekkers, M.J.; Barrón, V.; Torrent, J.; Li, S. The Magnetic and Color Reflectance Properties of Hematite: From Earth to Mars. *Rev. Geophys.* **2022**, *60*, e2020RG000698. [CrossRef]
40. Rasmussen, B.; Muhling, J.R. Monazite begets monazite: Evidence for dissolution of detrital monazite and reprecipitation of syntectonic monazite during low-grade regional metamorphism. *Contrib. Mineral. Petrol.* **2007**, *154*, 675–689. [CrossRef]
41. Eren, M.; Kadir, S.; Kapur, S.; Huggett, J.; Zucca, C. Colour origin of Tortonian red mudstones within the Mersin area, southern Turkey. *Sediment. Geol.* **2015**, *318*, 10–19. [CrossRef]

42. Bankole, O.M.; El Albani, A.; Meunier, A.; Rouxel, O.J.; Gauthier-Lafaye, F.; Bekker, A. Origin of red beds in the Paleoproterozoic Franceville Basin, Gabon, and implications for sandstone-hosted uranium mineralization. *Am. J. Sci.* **2016**, *316*, 839–872. [CrossRef]

43. Mahmud, S.A.; Hall, M.W.; Almalki, K.A. Mineralogy and spectroscopy of Owen Group sandstones, Australia: Implications for the provenance, diagenesis, and origin of coloration. *Geosci. J.* **2018**, *22*, 765–776. [CrossRef]

44. EREN, M. Colour origin of red sandstone beds within the Hüdai Formation (Early Cambrian), Aydıncık (Mersin), southern Turkey. *Turk. J. Earth Sci.* **2013**, *22*, 563–573. [CrossRef]

45. Rasmussen, B.; Muhling, J.R. Syn-tectonic hematite growth in Paleoproterozoic Stirling Range "red beds", Albany-Fraser Orogen, Australia: Evidence for oxidation during late-stage orogenic uplift. *Precambrian Res.* **2019**, *321*, 54–63. [CrossRef]

46. Luo, X.; Yang, Z.; Zhang, K.; Wang, H.; Du, H.; He, W. A Mineralogical study of red origin in Mount Danxia, Guangdong Province. *Acta Mineral. Sin.* **2021**, *41*, 1–10. (In Chinese)

47. Tan, C.; Yu, B.; Yuan, X.; Liu, C.; Wang, T.; Zhu, X. Color Origin of the Lower Triassic Liujiagou and Heshanggou Formations Red Beds in the Ordos Basin. *Geoscience* **2020**, *34*, 769–783. (In Chinese) [CrossRef]

48. Al Juboury, A.I.; Hussain, S.H.; McCann, T.; Aghwan, T.A. Clay mineral diagenesis and red bed colouration: A SEM study of the Gercus Formation (Middle Eocene), northern Iraq. *Geol. J.* **2020**, *55*, 7977–7997. [CrossRef]

49. Xiao, Y.; Li, Y.; Ding, H.; Li, Y.; Lu, A. The Fine Characterization and Potential Photocatalytic Effect of Semiconducting Metal Minerals in Danxia Landforms. *Minerals-Basel* **2018**, *8*, 554. [CrossRef]

50. Singh, S.; Awasthi, A.K.; Khanna, Y.; Kumari, A.; Singh, B.; Kumar, A.; Popli, C. Sediment colour as recorder of climate and tectonics: Cenozoic continental red beds of the Himalayan foreland basin in NW India. *Catena* **2021**, *203*, 105298. [CrossRef]

51. Schindler, M.; Michel, S.; Batcheldor, D.; Hochella, M.F. A nanoscale study of the formation of Fe-(hydr)oxides in a volcanic regolith; implications for the understanding of soil forming processes on Earth and Mars. *Geochim. Cosmochim. Acta* **2019**, *264*, 43–66. [CrossRef]

52. Walker, T.R. Red beds in the western interior of the United States. *US Geol. Surv. Prof. Pap.* **1975**, *853*, 49–56.

53. Miki, T.; Matsueda, H.; Yim, W.W.S. Petrography and geochemistry of Cretaceous (?) red beds in Hong Kong. *J. Southeast Asian Earth Sci.* **1990**, *4*, 99–106. [CrossRef]

54. Aehnelt, M.; Hilse, U.; Pudlo, D.; Heide, K.; Gaupp, R. On the origin of bleaching phenomena in red bed sediments of Triassic Buntsandstein deposits in Central Germany. *Geochemistry* **2021**, *81*, 125736. [CrossRef]

55. Weibel, R. Diagenesis in oxidising and locally reducing conditions—An example from the Triassic Skagerrak Formation, Denmark. *Sediment. Geol.* **1998**, *121*, 259–276. [CrossRef]

56. Walker, T.R.; Waugh, B.; Grone, A.J. Diagenesis in first-cycle desert alluvium of Cenozoic age, southwestern United States and northwestern Mexico. *Geol. Soc. Am. Bull.* **1978**, *89*, 19–32. [CrossRef]

57. MüCke, A. Part i. postdiagenetic ferruginization of sedimentary rocks (sandstones, oolitic ironstones, kaolins and bauxites)—Including a comparative study of the reddening of red beds. In *Developments in Sedimentology*; Elsevier: Amsterdam, The Netherlands, 1994; Volume 51, pp. 361–395.

58. Muchez, P.; Viaene, W.; Dusar, M. Diagenetic control on secondary porosity in flood plain deposits: An example of the Lower Triassic of northeastern Belgium. *Sediment. Geol.* **1992**, *78*, 285–298. [CrossRef]

59. Bensing, J.P.; Mozley, P.S.; Dunbar, N.W. Importance of Clay in Iron Transport and Sediment Reddening: Evidence from Reduction Features of the Abo Formation, New Mexico, U.S.A. *J. Sediment. Res.* **2005**, *75*, 562–571. [CrossRef]

60. Matlack, K.S.; Houseknecht, D.W.; Applin, K.R. Emplacement of clay into sand by infiltration. *J. Sediment. Petrol.* **1989**, *59*, 77–87. [CrossRef]

61. Johnson, S.A.; Glover, B.W.; Turner, P. Multiphase reddening and weathering events in Upper Carboniferous red beds from the English West Midlands. *J. Geol. Soc. Lond.* **1997**, *154*, 735–745. [CrossRef]

62. Chukhrov, F.V. On mineralogical and geochemical criteria in the genesis of red beds. *Chem. Geol.* **1973**, *12*, 67–75. [CrossRef]

63. Guo, H.; Barnard, A.S. Thermodynamic modelling of nanomorphologies of hematite and goethite. *J. Mater. Chem.* **2011**, *21*, 11566–11577. [CrossRef]

64. Jiang, Z.; Liu, Q.; Roberts, A.P.; Barrón, V.; Torrent, J.; Zhang, Q. A new model for transformation of ferrihydrite to hematite in soils and sediments. *Geology* **2018**, *46*, 987–990. [CrossRef]

65. Al-Rawi, Y. Origin of red color in the gercus formation (eocene), Northeastern Iraq. *Sediment. Geol.* **1983**, *35*, 177–192. [CrossRef]

66. Stel, H. Diagenetic crystallization and oxidation of siderite in red bed (Buntsandstein) sediments from the Central Iberian Chain, Spain. *Sediment. Geol.* **2009**, *213*, 89–96. [CrossRef]

67. Weibel, R.; Grobety, B. Pseudomorphous transformation of goethite needles into hematite in sediments of the Triassic Skagerrak Formation, Denmark. *Clay Miner.* **1999**, *34*, 657. [CrossRef]

68. Berner, R.A. Goethite stability and the origin of red beds. *Geochim. Cosmochim. Acta* **1969**, *33*, 267–273. [CrossRef]

69. Langmuir, D. Particle size effect on the reaction goethite = hematite + water. *Am. J. Sci.* **1971**, *271*, 147–156. [CrossRef]

70. Zhao, J.; Brugger, J.; Pring, A. Mechanism and kinetics of hydrothermal replacement of magnetite by hematite. *Geosci. Front.* **2019**, *10*, 29–41. [CrossRef]

71. García-Rivas, J.; Suárez, M.; Torres, T.; Sánchez-Palencia, Y.; García-Romero, E.; Ortiz, J. Geochemistry and Biomarker Analysis of the Bentonites from Esquivias (Toledo, Spain). *Minerals-Basel* **2018**, *8*, 291. [CrossRef]

72. Chen, G. A solution to the origin of white spots in red rock formations. *Geol. Rev.* **1941**, *Z3*, 393–396. (In Chinese)

73. Hofmann, R.A. Mineralogy and geochemistry of reduction spheroids in red beds. *Mineral. Petrol.* **1991**, *44*, 107–124. [CrossRef]
74. Spinks, S.C.; Parnell, J.; Bowden, S.A. Reduction spots in the Mesoproterozoic age: Implications for life in the early terrestrial record. *Int. J. Astrobiol.* **2010**, *9*, 209–216. [CrossRef]
75. Fox, D.C.M.; Spinks, S.C.; Thorne, R.L.; Barham, M.; Aspandiar, M.; Armstrong, J.G.T.; Uysal, T.; Timms, N.E.; Pearce, M.A.; Verrall, M.; et al. Mineralogy and geochemistry of atypical reduction spheroids from the Tumblagooda Sandstone, Western Australia. *Sedimentology* **2019**, *67*, 677–698. [CrossRef]
76. Hartmann, M. Einige geochemische Untersuchungen an Sandsteinen aus Perm und Trias. *Geochim. Cosmochim. Acta* **1963**, *27*, 459–499. [CrossRef]
77. Durrance, E.M.; Meads, R.E.; Ballard, R.; Walsh, J.N. Oxidation state of iron in the Littleham Mudstone Formation of the New Red Sandstone Series (Permian-Triassic) of southeast Devon, England. *Geol. Soc. Am. Bull.* **1978**, *89*, 1231–1240. [CrossRef]
78. Mykura, H.; Hampton, B.P. On the mechanism of formation of reduction spots in the Carboniferous/Permian red beds of Warwickshire. *Geol. Mag.* **1984**, *121*, 71–74. [CrossRef]
79. Hofmann, B.A. Reduction spheroids from northern Switzerland: Mineralogy, geochemistry and genetic models. *Chem. Geol.* **1990**, *81*, 55–81. [CrossRef]
80. Picard, M.D. Iron oxides and fine-grained rocks of Red Peak and Crow Mountain Members, Chugwater (Triassic) Formation, Wyoming. *J. Sediment. Res.* **1965**, *35*, 464–479. [CrossRef]
81. Parnell, J.; Brolly, C.; Spinks, S.; Bowden, S. Metalliferous biosignatures for deep subsurface microbial activity. *Orig. Life Evol. Biosph.* **2016**, *46*, 107–118. [CrossRef] [PubMed]
82. Parnell, J.; Still, J.; Spinks, S.; Bellis, D. Gold in Devono-Carboniferous red beds of northern Britain. *J. Geol. Soc. Lond.* **2016**, *173*, 245–248. [CrossRef]
83. Spinks, S.C.; Parnell, J.; Still, J.W. Redox-controlled selenide mineralization in the Upper Old Red Sandstone. *Scott. J. Geol.* **2014**, *50*, 173–182. [CrossRef]
84. Taylor, B.; Beaudoin, G. *Sulphur Isotope Stratigraphy of the Sullivan Pb-Zn-Ag Deposit, B. C.: Evidence for Hydrothermal Sulphur, and Bacterial and Thermochemical Sulphate Reduction*; Mineral Deposits Division of the Geological Association of Canada: Ottawa, ON, Canada, 2000; Volume 1, pp. 696–719.
85. Dill, H.G.; Berner, Z.A. Sedimentological and structural processes operative along a metalliferous catena from sandstone-hosted to unconformity-related Pb-Cu-Zn deposits in an epicontinental basin, SE Germany. *Ore Geol. Rev.* **2014**, *63*, 91–114. [CrossRef]
86. Parry, W.T.; Chan, M.A.; Beitler, B. Chemical bleaching indicates episodes of fluid flow in deformation bands in sandstone. *AAPG Bull.* **2004**, *88*, 175–191. [CrossRef]
87. Xie, D.; Yao, S.; Cao, J.; Hu, W.; Wang, X.; Zhu, N. Diagenetic alteration and geochemical evolution during sandstones bleaching of deep red-bed induced by methane migration in petroliferous basins. *Mar. Pet. Geol.* **2021**, *127*, 104940. [CrossRef]
88. Chan, M.A.; Parry, W.T.; Bowman, J.R. Diagenetic hematite and manganese oxides and fault-related fluid flow in Jurassic sandstones, southeastern Utah. *AAPG Bull.* **2000**, *84*, 1281–1310.
89. Beitler, B.; Chan, M.A.; Parry, W.T. Bleaching of Jurassic Navajo Sandstone on Colorado Plateau Laramide highs; evidence of exhumed hydrocarbon supergiants? *Geology (Boulder)* **2003**, *31*, 1041–1044. [CrossRef]
90. Rainoldi, A.L.; Franchini, M.; Beaufort, D.; Patrier, P.; Giusiano, A.; Impiccini, A.; Pons, J. Large-Scale Bleaching of Red Beds Related To Upward Migration of Hydrocarbons: Los Chihuidos High, Neuquen Basin, Argentina. *J. Sediment. Res.* **2014**, *84*, 373–393. [CrossRef]
91. Zhang, L.; Liu, C.; Lei, K. Green altered sandstone related to hydrocarbon migration from the uranium deposits in the northern Ordos Basin, China. *Ore Geol. Rev.* **2019**, *109*, 482–493. [CrossRef]
92. Rushton, J.C.; Wagner, D.; Pearce, J.M.; Rochelle, C.A.; Purser, G. Red-bed bleaching in a CO2 storage analogue: Insights from Entrada Sandstone fracture-hosted mineralization. *J. Sediment. Res.* **2020**, *90*, 48–66. [CrossRef]
93. Mortezaei, K.; Amirlatifi, A.; Ghazanfari, E.; Vahedifard, F. Potential CO2 leakage from geological storage sites: Advances and challenges. *Environ. Geotech.* **2021**, *8*, 3–27. [CrossRef]
94. Dill, H.G.; Wehner, H.; Blum, N. The origin of sulfide mineralization in arenaceous rocks beneath carbonaceous horizons in fluvial depositions of late Paleozoic through Cenozoic age (SE Germany). *Chem. Geol.* **1993**, *104*, 159–173. [CrossRef]
95. Konhauser, K.O.; Kappler, A.; Roden, E.E. Iron in microbial metabolisms. *Elements* **2011**, *7*, 89–93. [CrossRef]
96. Garden, I.R.; Guscott, S.C.; Burley, S.D.; Foxford, K.A.; Walsh, J.J.; Marshall, J. An exhumed palaeo-hydrocarbon migration fairway in a faulted carrier system, Entrada Sandstone of SE Utah, USA. *Geofluids* **2001**, *1*, 195–213. [CrossRef]
97. Petrovic, A.; Khan, S.D.; Chafetz, H.S. Remote detection and geochemical studies for finding hydrocarbon-induced alterations in Lisbon Valley, Utah. *Mar. Pet. Geol.* **2008**, *25*, 696–705. [CrossRef]
98. Eichhubl, P.; Davatz, N.C.; Becker, S.P. Structural and diagenetic control of fluid migration and cementation along the Moab fault, Utah. *AAPG Bull.* **2009**, *93*, 653–681. [CrossRef]
99. Wigley, M.; Kampman, N.; Dubacq, B.; Bickle, M. Fluid-mineral reactions and trace metal mobilization in an exhumed natural CO2 reservoir, Green River, Utah. *Geology (Boulder)* **2012**, *40*, 555–558. [CrossRef]
100. Jung, N.; Han, W.S.; Watson, Z.T.; Graham, J.P.; Kim, K. Fault-controlled CO2 leakage from natural reservoirs in the Colorado Plateau, East-Central Utah. *Earth Planet. Sci. Lett.* **2014**, *403*, 358–367. [CrossRef]
101. Parnell, J.; Spinks, S.; Brolly, C. Tellurium and selenium in Mesoproterozoic red beds. *Precambrian Res.* **2018**, *305*, 145–150. [CrossRef]

102. Parnell, J.; Wang, X.; Raab, A.; Feldmann, J.; Brolly, C.; Michie, R.; Armstrong, J. Metal Flux from Dissolution of Iron Oxide Grain Coatings in Sandstones. *Geofluids* **2021**, *2021*, 1–14. [CrossRef]
103. Dill, H.G. The "chessboard" classification scheme of mineral deposits: Mineralogy and geology from aluminum to zirconium. *Earth-Sci. Rev.* **2010**, *100*, 1–420. [CrossRef]
104. Castillo-Oliver, M.; Melgarejo, J.C.; Torró, L.; Villanova-de-Benavent, C.; Campeny, M.; Díaz-Acha, Y.; Amores-Casals, S.; Xu, J.; Proenza, J.; Tauler, E. Sandstone-Hosted Uranium Deposits as a Possible Source for Critical Elements: The Eureka Mine Case, Castell-Estaó, Catalonia. *Minerals* **2020**, *10*, 34. [CrossRef]
105. Ozdemir, O.; Dunlop, D.J. Intermediate magnetite formation during dehydration of goethite. *Earth Planet. Sci. Lett.* **2000**, *177*, 59–67. [CrossRef]
106. Prasad, P.S.R.; Shiva Prasad, K.; Krishna Chaitanya, V.; Babu, E.V.S.S.; Sreedhar, B.; Ramana Murthy, S. In situ FTIR study on the dehydration of natural goethite. *J. Asian Earth Sci.* **2006**, *27*, 503–511. [CrossRef]
107. Zhang, W.; Huo, C.; Feng, G.; Li, Y.; Wang, J.; Jiao, H. Dehydration of goethite to hematite from molecular dynamics simulation. *J. Mol. Struct. THEOCHEM* **2010**, *950*, 20–26. [CrossRef]
108. Guo, H.; Barnard, A.S. Naturally occurring iron oxide nanoparticles: Morphology, surface chemistry and environmental stability. *J. Mater. Chem. A* **2013**, *1*, 27–42. [CrossRef]

minerals

Article

A 3D Predictive Method for Deep-Seated Gold Deposits in the Northwest Jiaodong Peninsula and Predicted Results of Main Metallogenic Belts

Mingchun Song [1,*], Shiyong Li [2,3], Jifei Zheng [4,5], Bin Wang [1,2], Jiameng Fan [1], Zhenliang Yang [1], Guijun Wen [1], Hongbo Liu [3], Chunyan He [3], Liangliang Zhang [1] and Xiangdong Liu [1]

[1] Shandong Provincial No. 6 Exploration Institute of Geology and Mineral Resources, Weihai 264209, China; wangbinjlu@163.com (B.W.); a_afanfan@163.com (J.F.); yzl198849@126.com (Z.Y.); lywgj@126.com (G.W.); 15634333042@163.com (L.Z.); sddk807@163.com (X.L.)
[2] College of Earth Sciences, Jilin University, Changchun 130061, China; dikechulsy@126.com
[3] Shandong Institute of Geophysical and Geochemical Exploration, Jinan 250013, China; wtylhb@163.com (H.L.); hcy_gg@163.com (C.H.)
[4] College of Earth Sciences and Resources, China University of Geoscience, Beijing 100083, China; zhengjf@sd-gold.com
[5] Shandong Gold Group Co., Ltd., Jinan 250014, China
* Correspondence: mingchuns@163.com

Citation: Song, M.; Li, S.; Zheng, J.; Wang, B.; Fan, J.; Yang, Z.; Wen, G.; Liu, H.; He, C.; Zhang, L.; et al. A 3D Predictive Method for Deep-Seated Gold Deposits in the Northwest Jiaodong Peninsula and Predicted Results of Main Metallogenic Belts. *Minerals* 2022, 12, 935. https://doi.org/10.3390/min12080935

Academic Editor: Stanisław Mazur

Received: 24 May 2022
Accepted: 21 July 2022
Published: 25 July 2022

Publisher's Note: MDPI stays neutral with regard to jurisdictional claims in published maps and institutional affiliations.

Abstract: With the rapid depletion of mineral resources, deep prospecting is becoming a frontier field in international geological exploration. The prediction of deep mineral resources is the premise and foundation of deep prospecting. However, conventional metallogenic predictive methods, which are mainly based on surface geophysical, geochemical, and remote sensing data and geological information, are no longer suitable for deep metallogenic prediction due to the large burial depth of deep-seated deposits. Consequently, 3D metallogenic prediction becomes a critical method for delineating deep prospecting target areas. As a world-class giant gold metallogenic province, the Jiaodong Peninsula is at the forefront in China in terms of deep prospecting achievements and exploration depth. Therefore, it has unique conditions for 3D metallogenic prediction and plays an important exemplary role in promoting the development of global deep prospecting. This study briefly introduced the method, bases, and results of the 3D metallogenic prediction in the northwest Jiaodong Peninsula and then established 3D geological models of gold concentration areas in the northwest Jiaodong Peninsula using drilling combined with geophysics. Since gold deposits in the northwest Jiaodong Peninsula are often controlled by faulting in the 3D space, this study proposed a method for predicting deep prospecting target areas based on a stepped metallogenic model and a method for predicting the deep resource potential of gold deposits based on the shallow resources of ore-controlling faults. Multiple characteristic variables were extracted from the 3D geological models of the gold concentration areas, including the buffer zone and dip angle of faults, the changing rate of fault dip angle, and the equidistant distribution of orebodies. Using these characteristic variables, five deep prospecting target areas in the Jiaojia and Sanshandao faults were predicted. Moreover, based on the proven gold resources at an elevation of −2000 m and above, the total gold resources of the Sanshandao, Jiaojia, and Zhaoping ore-controlling faults at an elevation of −5000−−2000 m were predicted to be approximately 3377–6490 t of Au. Therefore, it is believed that the total gold resources in the Jiaodong Peninsula are expected to exceed 10,000 t. These new predicted results suggest that the northwest Jiaodong Peninsula has huge potential for the resources of deep gold deposits, laying the foundation for further deep prospecting.

Keywords: deep prospecting; 3D metallogenic prediction; characteristic variables; stepped metallogenic model; resource potential; Jiaodong Peninsula

1. Introduction

The supply of gold in China does not meet the present demand. In 2021, the gold production in China was 330 t, while the gold consumption was 1121 t. As the shallow-surface mineral resources in East China are increasingly low, deep prospecting has inevitably become a method for resolving the resource crisis. The prediction of deep mineral resources is the premise and foundation of deep prospecting. However, metallogenic prediction is a complex engineering task [1,2]. As relevant methods, technologies progress and geological understandings deepen, the predicted and prospecting results will change significantly. As the most important base of gold in China, the Jiaodong Peninsula has witnessed the complexity and uncertainty of metallogenic prediction in its prospecting and metallogenic prediction history. At the beginning of the 21st century, the gold resources in the northwest Jiaodong Peninsula were predicted to be 2492 t of Au in total, based on previous regional metallogenic predictions [3,4]. During 2007–2012, the gold resources at a depth of 0–2000 m in the Jiaodong Peninsula were predicted to be 3963 t of Au using the "three-in-one" prospecting prediction theory that integrates metallogenic geological bodies, metallogenic structural plane, and metallogenic characteristics [5]. At present, the accumulative proven gold resources in the Jiaodong Peninsula total more than 5000 t of Au [6], far exceeding previously predicted results. Since the Jiaodong Peninsula is a world-class giant gold metallogenic province, the scientific and accurate assessment of gold resource potential in this area plays an important exemplary role in promoting the development of deep prospecting.

Deep-seated gold orebodies are covered by rock layers with a thickness of more than 1000 m or even thousands of meters. The mineralization information is strongly suppressed, thus, metallogenic predictive methods based on surface geophysical, geochemical, remote-sensing data and geological information are ineffective. High-precision deep geophysical exploration (gravity, electromag, seismic) and 3D visualization analysis are effective for predicting and exploring deep resources [7–11]. Furthermore, 3D metallogenic prediction based on 3D geological modeling and fault ore-controlling law has achieved important research achievements in many areas of the world [12–19]. Based on the deep geophysical exploration and 3D modeling of gold concentration areas in the northwest Jiaodong Peninsula, this study proposed two new methods: a method for predicting deep prospecting target areas based on a stepped metallogenic model and a method for predicting the deep resource potential based on gold resources in shallow parts. Applying both methods, this study predicted the deep prospecting target areas and resource potential of major gold metallogenic belts in the northwest Jiaodong Peninsula, revealing an exciting deep prospecting prospect. This study lays the foundation for further deep prospecting in the Jiaodong Peninsula.

2. Geological Background and Overview of Gold Deposits

The Jiaodong Peninsula lies at the southeastern margin of the North China Craton and at the northeastern end of the Dabie-Sulu ultrahigh-pressure (UHP) metamorphic belt [20]. The Jiaobei terrane in the western Jiaodong Peninsula and the Weihai terrane in the eastern Jiaodong Peninsula fall in the North China Craton and the Sulu UHP metamorphic belt, respectively. Moreover, the Jiaolai Basin is superimposed on the Jiaobei terrane and the southern part of the Weihai terrane (Figure 1). The Jiaobei terrane mainly consists of Neoarchean granite-greenstone belts and Paleoproterozoic-Neoproterozoic metamorphic strata, the Weihai terrane is mainly composed of Neoproterozoic granitic gneiss bearing UHP eclogites, and the Jiaolai Basin mainly includes Cretaceous volcanic-sedimentary rock series [21–24]. Jurassic-Cretaceous granitic intrusive rocks are widely emplaced in the Jiaobei and Weihai terranes [25–32], whereas only a small number of Triassic granitoids are exposed in the Weihai terrane.

Faults are well-developed in the Jiaodong Peninsula. Among them, NE-NNE-trending faults with dip of SE or NW are the most developed. They are followed by nearly EW-NEE-trending faults. Additionally, EW-trending faults are sporadically exposed, showing poor continuity. The gold deposits in the Jiaodong Peninsula are mainly controlled by NE-NNE faults including the Sanshandao, Jiaojia, Zhaoping, Xilin-Douya, and Jinniushan faults [32–35].

There are more than 200 gold deposits with proven resources in the Jiaodong Peninsula (Figure 1), with gold resources greater than 5000 t [6]. The gold deposits in this peninsula are intensively distributed and are divided into three metallogenic sub-regions, namely Jiaoxibei (Laizhou-Zhaoyuan), Qipengfu (Qixia-Penglai-Fushan), and Muru (Muping-Rushan). These three regions consist of six metallogenic belts, namely Sanshandao, Jiaojia, Zhaoping, Qixia-Daliuhang, Taocun, and Muru. These metallogenic belts are composed of 13 gold ore-fields, namely Sanshandao, Jiaojia, Lingbei, Anshi, Dazhuangzi, Linglong, Dayingezhuang, Jiudian, Qixia, Daliuhang, Laishan, Pengjiakuang, and Denggezhuang. The gold mineralization in the Jiaodong Peninsula primarily include altered rock in fractured zones and quartz vein, followed by a small quantity of altered breccia, altered conglomerate, interlayer decollement-detachment zone, and pyrite-carbonate vein types. The characteristics, ore-controlling regularity and genesis of Jiaodong gold deposits have been studied extensively by predecessors [36–41]. The deep prospecting carried out in the peninsula since the beginning of this century has discovered more than 3000 t of proven gold resources at a depth of 600–2000 m, exceeding the previously proven gold resources at a depth of 500 m and less [6,42].

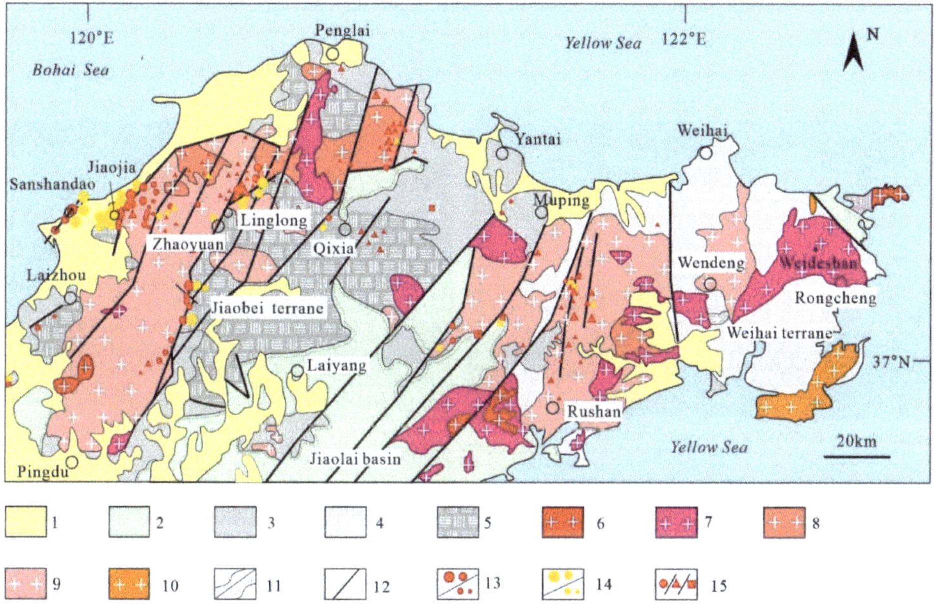

Figure 1. Map showing the regional geology and gold deposit distribution in the Jiaodong Peninsula. 1—Quaternary; 2—Cretaceous; 3—Paleoproterozoic and Neoproterozoic; 4—Neoproterozoic bearing eclogite granitic gneiss; 5—Archean granite-greenstone belt; 6—Cretaceous Laoshan granite; 7—Cretaceous Weideshan granite; 8—Cretaceous Guojialing granite; 9—Jurassic Linglong granite; 10—Triassic granitoid; 11—Geological boundary of conformity/unconformity; 12—Fault; 13—Shallow gold deposits (very large and large/medium-scale and small); 14—Deep-seated gold deposits (very large and large/medium-scale and small); 15—Gold deposit of altered-rock-type/quartz-vein-type/altered-breccia-type.

3. 3D Predictive Method of Deep-Seated Gold Deposits

3.1. 3D Modeling of Main Gold Concentration Areas in the Northwest Jiaodong Peninsula

This study established the geological models of the Sanshandao supergiant gold deposit in the Sanshandao fault zone, the Jiaojia supergiant gold deposit in the Jiaojia fault zone, and the Lingnan-Shuiwangzhuang and Dayingezhuang gold deposits in the Zhaoping fault zone.

The basic data used for modeling included regional geological data on scales of 1:50,000 and 1:10,000, exploration line sections, borehole histograms, digital elevation data, and high-precision geophysical profiles. The modeling parameters included the scale of planar geological maps of 1:10,000, the scale of exploration line sections of 1:2000, and the grid density in the horizontal direction of 60 m × 60 m. Meanwhile, the minimum thickness was set to be 0.1 m.

The 3D geological models of the Sanshandao, Jiaojia, Lingnan-Lijiazhuang, and Dayingezhuang gold deposits were established using data of 311, 500, 680, and 291 boreholes, respectively, as well as the basic data used for modeling. For example, Figure 2 shows the 3D geological model of the Sanshandao gold deposit (Figure 2a) and the superposition of gold orebodies on the plane of the Sanshandao fault (Figure 2b). Each of the 3D geological models was composed of two parts: the known model of the shallow part and the inferred model of the deep part. The former was controlled by systematic drilling engineering. As a known part, it was used to summarize metallogenic rules and extract favorable metallogenic information through spatial comprehensive analysis. The latter, which was constructed based on the former and relevant geophysical interpretation and inference, was used to provide intuitive attributes for deep metallogenic prediction.

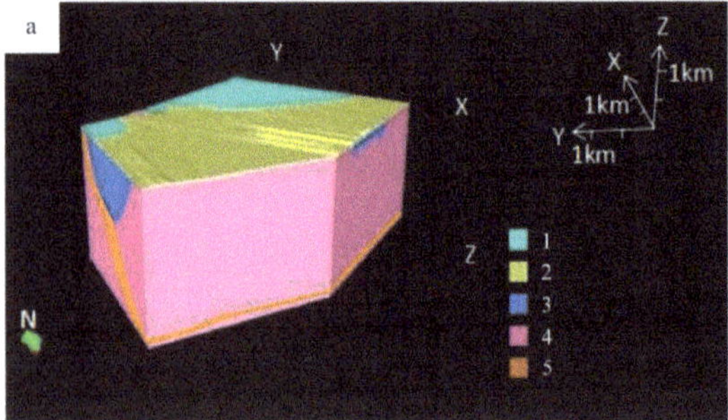

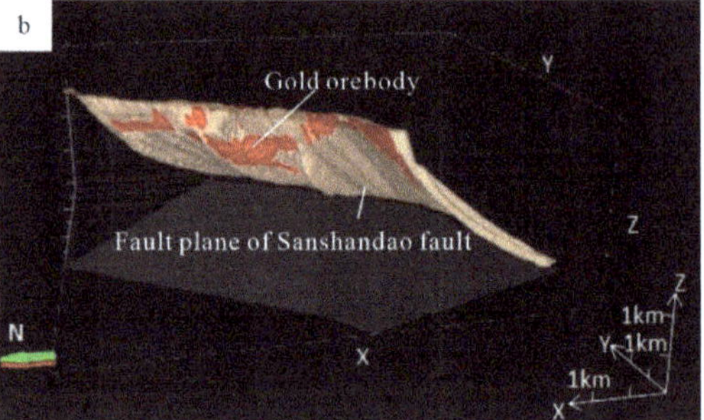

Figure 2. 3D geological model of the Sanshandao supergiant gold deposit (**a**) and superposition map of fault plane on gold orebodies (**b**). 1—Sea area; 2—Quaternary; 3—Early Precambrian metamorphic rock series; 4—Linglong-type granite; 5—Alteration zone of Sanshandao fault.

3.2. Method for Predicting Deep Metallogenic Target Areas Based on a Stepped Metallogenic Model

3.2.1. Overview of Predictive Method

Breakthroughs have been made in deep prospecting in the Jiaodong Peninsula since the beginning of the 21st century, leading to the discovery of large-scale gold resources mainly at a depth of 700–2000 m. Based on the study of the characteristics and ore-hosting regularity of deep faults, Song MC et al. [41] proposed a stepped metallogenic model of deep-seated gold deposits. Specifically, ore-controlling faults show a stepped pattern due to alternate steep-gentle dip angles, and gold orebodies are mainly rich in stepped fault parts with gentle dip angles and the fault parts with a transition between steep-gentle dip angles. Given that it is difficult to capture the mineralization-related predictive information on the ground surface due to the large burial depth of deep-seated gold deposits, this study proposed a method based on the stepped metallogenic model according to the following three characteristics [43]. First, ore-controlling faults of gold deposits have large scales. Second, the hanging walls and footwalls of the faults are composed of Early

Precambrian metamorphic rock series and Jurassic-Cretaceous granitoids, which greatly differ in physical properties. Third, the deep locations and morphological characteristics of faults can be detected using high-precision geophysical methods (gravity, electromag, seismic). This predictive method is designed to extract the predictive factors for identifying ore-controlling faults and the characteristics of metallogenic plane based on deep exploration and 3D geological modeling and to delineate the favorable ore-hosting locations of faults as prospecting target areas.

The 3D geological models of known gold concentration areas and their deep parts were divided into large numbers of 3D cubic blocks (120 m × 120 m × 10 m or 120 m × 120 m × 15 m). Each block was regarded as a homogeneous body with consistent properties, and thus the regularity of change in the properties of all blocks approximately reflected the regularity of internal change in geological bodies. Such cube blocks are called cell blocks. The storage address in a computer of each cell block corresponds to its location in the natural deposit. The prospecting target areas were determined as follows. Firstly, various quantitative information for deep prospecting assessment was comprehensively analyzed and processed according to the ore-controlling geological conditions of known deposits and the regularity of spatial (especially deep) change in prospecting indicators, and accordingly, 3D prospecting predictive models were established. Subsequently, the quantitative information described in Section 3.2.2. was assigned to each cell block, i.e., each favorable indicator is coded. Once the coding of the many prospective indicators is finished, a global favorable indicator is computed by combining the favorable coded indicator on each cell. Finally, the areas with higher prospective global scores were identified as prospecting target areas.

3.2.2. Main Predictive Bases

a. Favorable ore-hosting parts of faults. ① The altered-rock-type gold deposits in fractured zones in the Jiaodong Peninsula are mainly controlled by large-scale regional detachment faults in the NNE trending [33,44–46]. The fissure space in the fractured zones is favorable for the enrichment of gold-rich fluids [26]. The gold orebodies in these gold deposits mainly occur in the fractured alteration zones on the footwall of the main fault plane, which has fault gouges as the roof; ② The junctions of EW- and NE-trending faults; ③ The bending and turning parts of faults along fault strikes. ④ The development parts of the secondary faults on the footwall of main fault plane, fault junctions, and parts where fault branches combine. By comparison, the parts closer to the major faults are more favorable ore-hosting parts. ⑤ Gold orebodies occur in fault parts where the dip directions and dip angles of faults change, and fault sections with gentle dip angles are favorable ore-hosting parts [41,46–48]. For example, the gold orebodies in the Sanshandao gold concentration area mainly occur in fault sections with a steep-to-gentle transition of fault dip angle and fault concave sections with a gentle fault dip (Figure 2b), i.e., open space in faulting/striking faults allowing mineralized fluid circulation;

b. Distribution patterns of orebodies. ① The equidistant distribution pattern of orebodies. That is, deposits, orebodies, and mineralization enrichment zones in the Jiaodong Peninsula are roughly equidistantly distributed at about 500 m intervals. They are some similitudes with the Abitibi zone [13,14]; ② Pitching characteristics of orebodies [49,50]. Orebodies occurring in the NE- and NNE-trending faults pitch southwestward when the fault dip direction is NW and pitch northeastward when the fault dip direction is SE. Therefore, the orebodies roughly pitch in an SW-NE-trending straight line;

c. Ore-hosting geological bodies. ① The gold deposits in the Jiaodong Peninsula mainly occur in Jurassic Linglong-type granites, followed by Cretaceous Guojialing-type granites and Neoarchean-Paleoproterozoic metamorphic rock series. In addition, a few gold deposits occur at the bottom of the Early Cretaceous Laiyang Group. ② The contact zones between different geological bodies, especially those between the Early

Precambrian metamorphic rock series and the Mesozoic granitoids, are favorable ore-hosting parts;

d. Geophysical bases. The efficiency of geophysical methods on gold prospecting is obvious [51–60]. The main geophysical indicators of gold prospecting in the Jiaodong Peninsula include the following aspects. ① Linear gradient zones of gravity anomalies. Favorable metallogenic parts mainly include the edges of zones with low gravity (i.e., the transition zones between gravity highs and gravity lows could indicate fossil hydrothermal zone with dissolved rocks) and the contact zones between large-scale gravity lows and gravity highs, especially the turning parts of the gradient zones. The favorable metallogenic parts of deep-seated gold deposits include moniliform and elongate zones with high-amplitude magnetic anomalies, especially the bending parts of magnetic anomaly contours (protruding and concave parts), the edges of blocky gravity and magnetic anomaly zones, and the edges of small-scale blocky and moniliform positive magnetic anomaly zones. The magnetic anomaly would be associated to magnetite/pyrrhotite minerals often associated to gold mineralization. ② The boundaries between high- and low-resistance electric fields. On the apparent resistivity sections, the sparse and wide contour lines that synchronously bend downward and show U- or V-shaped low resistance mark the parts favorable for the occurrence of gold deposits. Boundaries between low/high resistivity zones could be linked to the presence of conductive minerals such pyrite often pathfinder minerals for gold mineralization.

3.2.3. Prediction Process

a. Analysis and processing of the modeling data on typical ore deposits and their surrounding mining areas

Comprehensively collect the geological, geophysical, geochemical, and mineral data of prediction areas and their surrounding areas, and determine the basic characteristics of typical deposits. Then, study metallogenic geological bodies, metallogenic structures, metallogenic characteristics, and prospective indicators, as well as the vertical and horizontal mineralization zoning characteristics of deposits. Based on this, summarize metallogenic rules and establish metallogenic models;

b. Deep geophysical exploration

Study the physical properties of ores and minerals and the geophysical fields of prediction areas; carry out measurements of deep geophysical profiles; extract key information for prospecting, and establish geophysical models. Classify different geological bodies and identify faults according to the main physical properties and geophysical models;

c. 3D modeling and 3D analysis of ore-hosting structures

Construct 3D geological models of deposits; emphatically analyze the coupling relationships between the changing characteristics of fault surfaces and the distribution and enrichment of orebodies; extract and assess predictive factors; identify areas with high prospecting information amounts as prospecting favorable areas and prospecting target areas.

3.3. Method for Predicting Deep Gold Resource Potential in the Ore-Controlling Fault Zones Based on Shallow Resources

3.3.1. Methodological Overview

According to the trend extrapolation principle, this predictive method was designed to extrapolate the resources in the deep prediction areas with metallogenic conditions similar to the shallow parts from the statistical analysis of the proven resources in the shallow parts.

3.3.2. Predictive Factors

The main predictive factors include ore-controlling faults, the cumulative proven gold resources at an elevation of approximately -2000 m and above, the average gold content per unit area at an elevation of approximately -2000 m and above, and the area of ore-controlling faults at an elevation of -5000–-2000 m on the vertical longitudinal projection map.

The deep resources were estimated using this equation: deep resources = ore-bearing rate × the area of the deep prediction areas. In this equation, the ore-bearing rate is equal to the average gold content per unit area in the shallow part, and the area of the deep prediction areas is equal to the area of ore-controlling faults at an elevation of -5000–-2000 m on the vertical longitudinal projection map. In other words, the vertical longitudinal projection map was obtained by cutting a 3D geological model. Then, the average gold content per unit area (i.e., ore-bearing rate) was determined based on the distribution of orebodies at different depths in the vertical longitudinal projection map. Subsequently, the boundaries of the ore-controlling faults in the vertical longitudinal projection map were reasonably determined based on the prediction depth, and then the area of the faults was accordingly calculated. Finally, the potential gold resources in the deep part were estimated by making an analogy with those in the known shallow area.

3.3.3. Prediction Process

The resource potential of deep-seated gold deposits of a fault was predicted as follows: 3D geological modeling → obtaining the statistics of accumulative proven resources in the shallow part using the 3D geological model → obtaining the vertical longitudinal projection map by cutting the 3D geological model → obtaining the statistics of the ore-bearing rate → determining the area of the deep prediction areas → estimating the resources in the deep part.

4. Prediction of Deep Prospecting Target Areas

4.1. Deep Prospecting Target Areas in the Jiaojia Fault

4.1.1. Determining Cell Blocks and Extracting Favorable Mineralization Information

The range and basic parameters of the modeling of the Jiaojia fault were determined based on the existing geological data of orebodies in the fault (especially the distribution of exploration lines), as well as the morphology, attitude, and spatial distribution of known orebodies. The established 3D geological model was divided into 660,800 cell blocks based on the line spacing × column spacing × layer spacing of 120 m × 120 m × 10 m. Among these cell blocks, 7263 were occupied by orebodies (also referred to as ore-hosting cell blocks). Ore-hosting cell blocks were assigned 1, while other cell blocks were assigned 0. Subsequently, the ore-controlling factors of ore-hosting cell blocks were extracted and their statistics were obtained, and accordingly, a quantitative predictive model was determined. Then, the predictive parameters were assigned to each cell block as per its attributes.

a.   Cell blocks with favorable information of ore-hosting faults.

① Structural buffer zones. Gold deposits are strictly controlled by faults, and gold orebodies mainly occur in the fractured alteration zones on the footwall of the main fault plane with fault gouges as the roof, with a small number of gold orebodies occurring on the hanging walls of the main fault plane. Therefore, the zone between 300 m above the footwall and 100 m below the hanging wall of the main fault plane was defined as a structural buffer zone favorable for mineralization. In this study, the structural buffer zones covered 108,572 cell blocks (Figure 3), which accounted for 16.43% of the total cell blocks in the model (660,800). Moreover, the structural buffer zones included 7124 ore-hosting cell blocks, accounting for 98.09% of the total ore-hosting cell blocks (7263) in the model. ② The turning parts of the fault dip angles. Orebodies are mainly rich in stepped fault parts with gentle dip angles. In this study, the fault parts with a steep-to-gentle transition of fault dip angle covered 53,285 cell blocks, which accounted for 8.06% of the total cell blocks in the model. These parts included 2866 ore-hosting cell blocks, which accounted for 39.46% of the total ore-hosting cell blocks in the model. ③ The changing rate of fault dip angle. The fault surfaces were divided into a number of square blocks, to each of which a slope attribute was assigned. The variance of the slope values of square blocks within a fixed range was calculated to determine the changes in the fault surface. The changing rate of fault dip angle involved 99,776 cell blocks, which accounted for 15.10% of the total cell blocks in the model. Furthermore, the changing rate of fault dip angle involved 5670 ore-hosting cell blocks, which accounted for 78.07% of the total ore-hosting cell blocks in the model;

b.   Cell blocks with favorable information of orebody distribution.

By extracting the distribution range of ore bodies and mineralization enrichment areas, it can be seen that the ore body enrichment areas are nearly equally spaced along the strike and dip at about 500 m intervals (Figure 4). Accordingly, 273,240 cell blocks were divided along the strike with equal spacing, which accounted for 41.35% of the total cell blocks in the model. These cell blocks included 5932 ore-hosting cell blocks, which accounted for 81.67% of the total ore-hosting cell blocks. The orebody concentration areas showing equidistant distribution along fault dip directions included 251,856 cell blocks, which accounted for 38.11% of the total cell blocks in the model. These cell blocks included 5884 ore-hosting cell blocks, which accounted for 81.01% of the total ore-hosting cell blocks in the model.

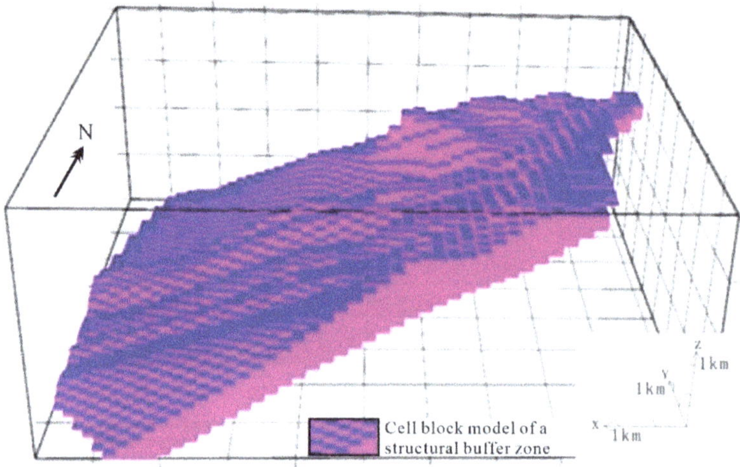

Figure 3. Distribution of cell blocks covered by structural buffer zones in the Jiaojia fault.

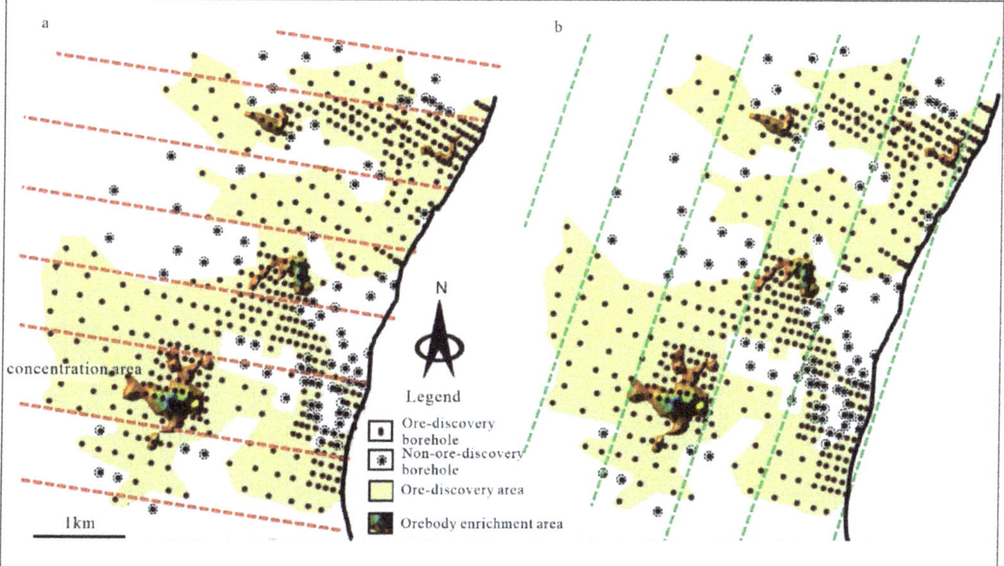

Figure 4. Horizontal projection of orebodies along fault strikes (**a**) and dip directions (**b**) in the Jiaojia deposit concentration area. The red dotted line in figure (**a**) and the green dotted line in figure (**b**) show the equally spaced distribution of orebody enrich areas.

4.1.2. 3D Predictive Model and Statistics of Information Amounts

The deep prospecting predictive model of the Jiaojia fault (Table 1) was established based on the 3D geological model and the favorable metallogenic information extracted from the geological model. This predictive model involved eight characteristic variables for statistical analysis. They include Early Precambrian metamorphic rock series, Linglong granite, contact zone of Early Precambrian metamorphic rock series and Linglong granite, fault buffer zone, gentle part of a fault section with a steep-to-gentle transition of fault dip angle, changing rate of fault dip angle, equidistant distribution along fault strike and equidistant distribution along fault dip direction, respectively. Each characteristic variable was assigned 1 if it was true for a cell block and 0 otherwise. Based on the statistics of these characteristic variables of each cell block, the sum of the values of these characteristic variables of each cell block was calculated as the information amount for metallogenic prediction).

4.1.3. Predicted Results

All the eight characteristic variables were assessed using the information amount method, obtaining metallogenic information amounts, which were then assigned to block models. Cell blocks with high information amounts have a high mineralization probability. Table 2 shows the statistics of the proportion of various information amount intervals of cell blocks. The information amount scopes where the information amounts tend to stabilize and converge are favorable for mineralization. According to the stability analysis of the proportion statistics, the information amounts tended to stabilize and converge when they were $\geq$3.222 (Figure 5). Therefore, the information amount interval of $\geq$3.222 was the favorable interval for mineralization (Figure 6a). This information amount interval was divided into three grades: 3.222–4.184, 4.184–5.018, and $\geq$5.018. The information amount range of $\geq$5.018 was regarded as the range of predicted target areas based on the comparison with known mineralized zones which are generally greater than 4.5 (Figure 7). As a result, two predicted target areas were delineated in the Jiaojia fault (Figure 6b).

Table 1. Deep predictive model of the Jiaojia fault.

Ore-Controlling Condition	Predictive Factor	Characteristic Variable	Characteristic Value
Geologic condition	Favorable geological body for mineralization	Early Precambrian metamorphic rock series	Direct screening and utilization
		Linglong granite	Direct screening and utilization
		Contact zone of Early Precambrian metamorphic rock series and Linglong granite	500 m width
Structural conditions	Ore-controlling faults	Fault buffer zone	500 m
	Fault plane	Gentle part of a fault section with a steep-to-gentle transition of fault dip angle	Direct screening and utilization
		Changing rate of fault dip angle	Characteristic value of the changing rate of fault dip angle
Ore-hosting conditions	Orebody distribution pattern	Equidistant distribution along fault strike	About 1.5 km along fault strike
		Equidistant distribution along fault dip direction	About 1.8 km along fault dip direction

Table 2. Proportion of metallogenic information amount intervals of cell blocks of the Jiaojia fault.

S.N.	Information Amount	Number of Cell Blocks	Proportion (%)	S.N.	Information Amount	Number of Cell Blocks	Proportion (%)
1	$\geq$0.000	660,800	100.00	17	$\geq$3.231	69,024	10.45
2	$\geq$0.681	429,566	65.01	18	$\geq$3.375	68,023	10.29
3	$\geq$0.754	365,257	55.27	19	$\geq$3.430	66,996	10.14
4	$\geq$1.435	298,945	45.24	20	$\geq$3.912	64,178	9.71
5	$\geq$1.588	181,999	27.54	21	$\geq$3.985	59,866	9.06
6	$\geq$1.643	180,484	27.31	22	$\geq$4.056	57,437	8.69
7	$\geq$1.787	175,404	26.54	23	$\geq$4.111	57,117	8.64
8	$\geq$2.269	121,332	18.36	24	$\geq$4.129	53,276	8.06
9	$\geq$2.324	120,369	18.22	25	$\geq$4.184	51,412	7.78
10	$\geq$2.342	112,905	17.09	26	$\geq$4.666	48,310	7.31
11	$\geq$2.397	110,896	16.78	27	$\geq$4.810	37,228	5.63
12	$\geq$2.468	106,281	16.08	28	$\geq$4.865	35,688	5.40
13	$\geq$2.541	99,034	14.99	29	$\geq$5.018	23,389	3.54
14	$\geq$3.023	94,558	14.31	30	$\geq$5.699	22,072	3.34
15	$\geq$3.078	92,210	13.95	31	$\geq$5.772	17,072	2.58
16	$\geq$3.222	73,592	11.14	32	$\geq$6.453	12,779	1.93

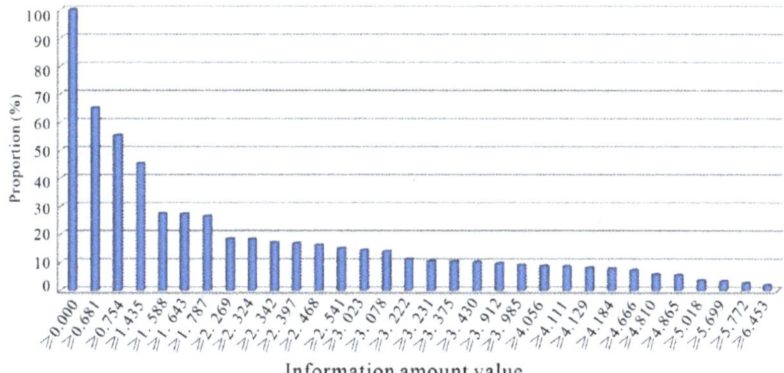

Figure 5. Histogram showing the proportion of metallogenic information amount intervals of cell blocks of the Jiaojia fault.

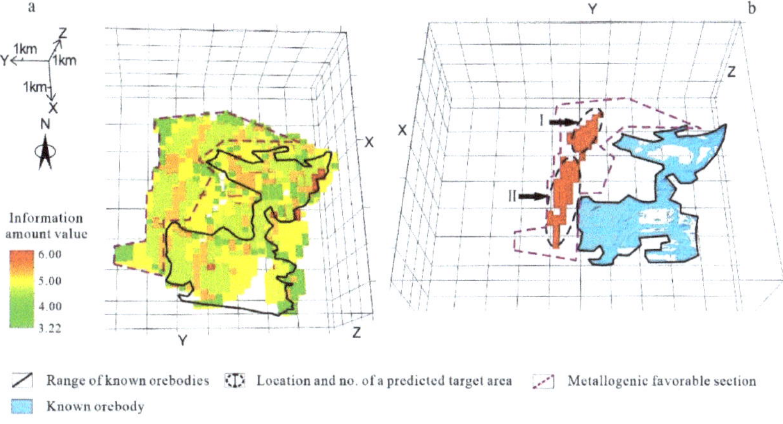

Figure 6. Advantageous metallogenic target (**a**) and predicted target areas (**b**) of deep metallogenic in the Jiaojia fault

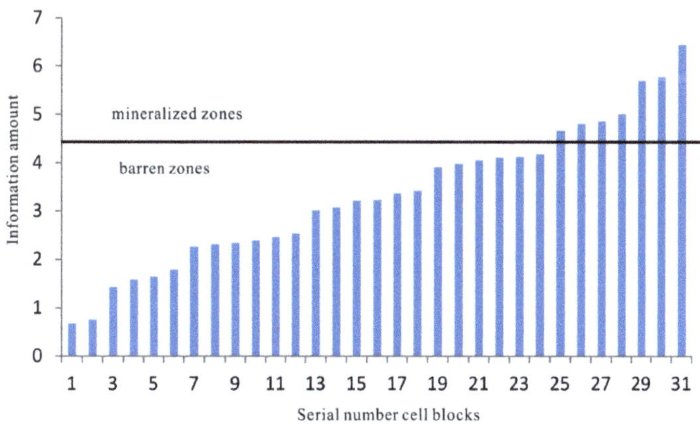

Figure 7. Histogram of metallogenic information amount showing the mineralized zones and the barren zones in the Jiaojia fault.

Target area I (Beijueyujia) is located in Beijueyujia Village, Jincheng Town, Laizhou City. It has an elevation of -2349–-1154 m and included 1201 predicted ore-hosting cell blocks. Target area II (Xiji) lies in Xiji Village, Jincheng Town, Laizhou City. It has an elevation range of -2490–-2000 m and includes 1588 predicted ore-hosting cell blocks.

4.1.4. Probability and Verification of Prediction Results

The information amount used in this paper to delineate the predicted target area is determined according to the information amount of the known ore body. Statistics show that the information amount of the known ore body unit is generally ≥ 4.5. The above information content ≥ 3.222 is taken as the range of favorable metallogenic area, and the probability of ore occurrence is estimated to be more than 70%. Taking the information amount ≥ 5.018 as metallogenic target area, the theoretical probability of ore occurrence should be 100%. At present, the two predicted metallogenic target areas have been verified by deep drilling. For example, the 3266.06 m deep drilling in target area 2 revealed multi-layer industrial ore bodies at 2810–2854 m [61].

4.2. Deep Prospecting Target Areas in the Sanshandao Fault Zone

4.2.1. Determining Cell Blocks and Extracting Favorable Mineralization Information

The 3D geological model of the Sanshandao fault zone was divided into 1,747,886 cell blocks based on the line spacing × column spacing × layer spacing of 120 m × 120 m × 15 m. These cell blocks included 5323 ore-hosting cell blocks. Ore-hosting cell blocks were assigned 1, while other cell blocks were assigned 0. The favorable metallogenic information of this fault was extracted in the same manner as in the Jiaojia fault.

4.2.2. 3D Predictive Model and Statistics of Information Amounts

The deep prospecting predictive model of the Sanshandao fault (Table 3) was established based on the 3D geological model of deposits and the favorable metallogenic information extracted from the geological model. This predictive model involved eight characteristic variables for statistical analysis. Each characteristic variable was assigned 1 if it was true for a cell block and 0 otherwise. The statistics of these characteristic variables of each cell block were obtained (Table 4).

4.2.3. Predicted Results

The proportion of metallogenic information amount intervals of cell blocks (Table 5) in the 3D predictive model showed that the information amounts tended to stabilize and converge when they were ≥ 9.597 (Figure 8). Therefore, the information amount interval of ≥ 9.597 was favorable for mineralization (Figure 9a). This interval was divided into three grades: 9.597–12.303, 12.303–13.080, and ≥ 13.080. The information amount range of ≥ 13.080 was considered the range of predicted target areas based on the comparison with known mineralized zones which are generally greater than 12 (Figure 10). As a result, three predicted target areas were delineated (Figure 9b).

Table 3. Deep predictive model of the Sanshandao fault.

Ore-Controlling Condition	Predictive Factor	Characteristic Variable	Characteristic Value
		Early Precambrian metamorphic rock series	Direct screening and utilization
Geologic condition	Favorable geological body for mineralization	Linglong granite	Direct screening and utilization
		Contact zone of Early Precambrian metamorphic rock series and Linglong granite	500 m width

Table 3. *Cont.*

Ore-Controlling Condition	Predictive Factor	Characteristic Variable	Characteristic Value
Structural conditions	Ore-controlling faults	Fault buffer zone	500 m
	Fault plane	Gentle part of a fault section with a steep-to-gentle transition of fault dip angle	Direct screening and utilization
		Changing rate of fault dip angle	Characteristic value of the changing rate of fault dip angle
Ore-hosting conditions	Orebody distribution pattern	Equidistant distribution along fault strike	About 1.5 km along fault strike
		Equidistant distribution along fault dip direction	About 1.0 km along fault dip direction

Table 4. Calculated results of prospecting information amounts of the Sanshandao fault.

Favorable Factor	Number Ore-Hosting Cell Blocks	Number of Cell Blocks with Favorable Factor	Number of Ore-Hosting Cell Blocks with Favorable Factor	Total Cell Block Number of the Geological Model	Information Amount
Fault buffer zones	5323	25,781	2556	1,747,886	3.483
Fault parts with a steep-to-gentle transition of fault dip angle	5323	33,935	3215	1,747,886	3.438
Changing rate of fault dip angle	5323	36,466	2344	1,747,886	3.050
Equidistant distribution along fault strike	5323	29,964	2044	1,747,886	3.109
Equidistant distribution along fault dip direction	5323	120,981	5273	1,747,886	2.661

Table 5. Proportion of metallogenic information amount intervals of cell blocks of the Sanshandao fault.

S.N.	Information Amount	Number of Cell Blocks	Proportion (%)	S.N.	Information Amount	Number of Cell Blocks	Proportion (%)
1	$\geq$2.661	18,150	100.00	17	$\geq$9.149	2379	29.77
2	$\geq$3.050	2007	72.30	18	$\geq$9.194	712	26.14
3	$\geq$3.109	196	69.24	19	$\geq$9.208	1495	25.06
4	$\geq$3.438	570	68.94	20	$\geq$9.253	412	22.78
5	$\geq$3.483	4171	68.07	21	$\geq$9.582	5946	22.15
6	$\geq$5.711	2505	61.71	22	$\geq$9.597	159	13.07
7	$\geq$5.777	1490	57.89	23	$\geq$9.642	226	12.83
8	$\geq$6.099	7279	55.61	24	$\geq$9.971	38	12.49
9	$\geq$6.144	3343	44.51	25	$\geq$10.030	9	12.43
10	$\geq$6.159	210	39.40	26	$\geq$12.258	2971	12.41
11	$\geq$6.488	189	39.08	27	$\geq$12.303	1003	7.88
12	$\geq$6.533	320	38.80	28	$\geq$12.632	1511	6.35
13	$\geq$6.547	23	38.31	29	$\geq$12.691	1162	4.05
14	$\geq$6.592	117	38.27	30	$\geq$13.080	19	2.27
15	$\geq$6.921	201	38.09	31	$\geq$15.741	1470	2.24
16	$\geq$8.820	5252	37.79				

Target area I is located 3.8 km northeast (in the direction of 85°) of Sanshandao Town, Laizhou City (Figure 9b). It has an elevation of −3122−−2349 m and included 440 predicted ore-hosting cell blocks. Target area II is located 3.5 km southeast (in the direction of 100°) of Sanshandao Town, Laizhou City. It has an elevation range of −3121 to −2306 m and included 368 predicted ore-hosting cell blocks. Target area III is located 3.0 km southeast (in the direction of 120°) of Sanshandao Town, Laizhou City. It has an elevation range of −2701 to −2291 m and includes 181 predicted ore-hosting cell blocks.

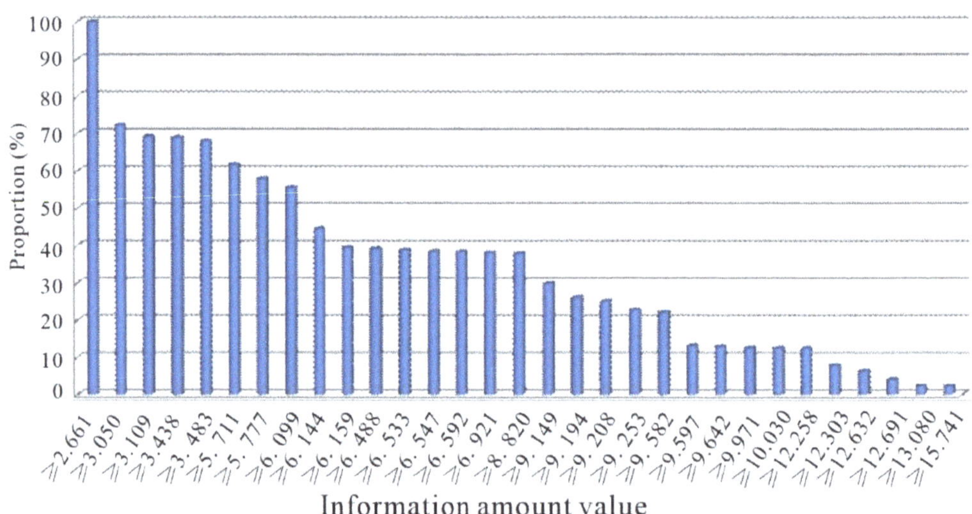

Figure 8. Histogram showing the proportion of metallogenic information amount intervals of cell blocks of the Sanshandao fault.

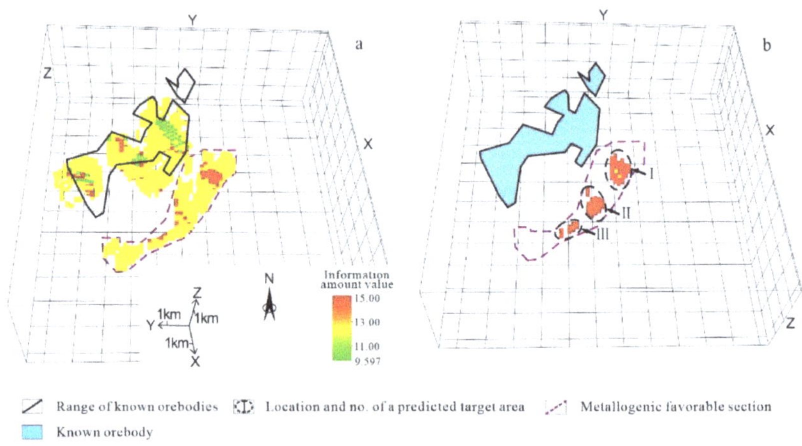

Figure 9. Deep metallogenic favorable zone (**a**) and the predicted deep target areas (**b**) in the Sanshandao fault.

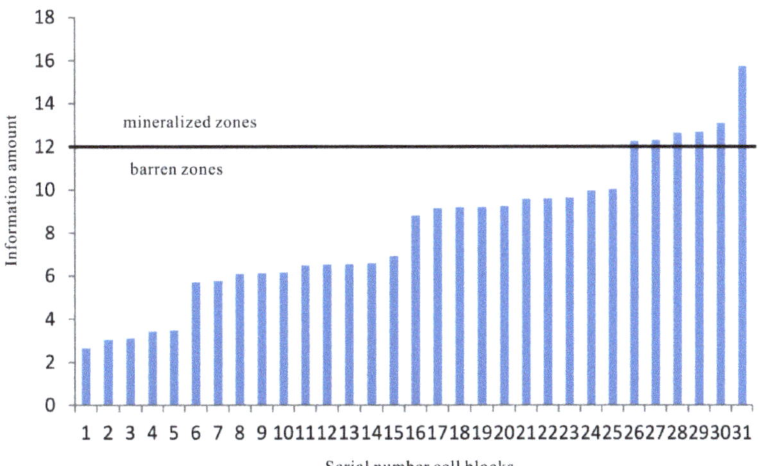

Figure 10. Histogram of metallogenic information amount showing the mineralized zones and the barren zone in the Sanshandao fault.

5. Prediction of Deep Resource Potential

5.1. Resource Potential of the Sanshandao Fault

5.1.1. Proved Resources

Based on the previous exploration results [6], the proved gold resources in the Sanshandao fault are analyzed in this paper. The Sanshandao fault has cumulative proved gold ore reserves of 333.65 Mt, gold metal reserves of 1,240,679 kg, an average orebody thickness of 7.93 m, and an average grade of 3.72 g/t. The first mineralization enrichment zone (elevation: −500–0 m) has gold ore reserves of 81.50 Mt, gold resources of 238,503 kg, an average orebody thickness of 8.10 m, and an average grade of 2.93 g/t. The second enrichment zone (elevation: −2000−−500 m) has gold ore reserves of 252.14 Mt, gold metal reserves of 1,002,176 kg, an average orebody thickness of 7.90 m, and an average grade of 3.98 g/t. The ratio of the resources of the second enrichment zone to those of the first enrichment zone is 4.20.

5.1.2. Predictive Parameters of Resources

Prediction range: The prediction range was determined by extrapolating the distribution range of identified deposits to deep parts. The distribution range of identified deposits in the Sanshandao fault was determined as follows. Its southern boundary was exploration line No. 171 on the southernmost side of the Xinli ore block and its northern boundary was exploration line No. 70 on the northernmost side of the sea-area ore block in the northern Sanshandao fault. Moreover, its western part was the outcrops of the Sanshandao fault, and its eastern part was the projection on the ground surface of the junction of the −2000 m elevation line and the deep part of the Sanshandao fault (Figure 11). The prediction range had the same southern, northern, and western boundaries as the distribution range of identified deposits, and its elevation was −5000−−2000 m.

Ore-bearing rates: the ore-bearing rates corresponding to the elevations of −500 to 0 m and −2000 to −500 m were calculated using the ratio of the proven resources to the projection area of the distribution range of identified deposits on the vertical longitudinal projection map (Figure 12, Table 6).

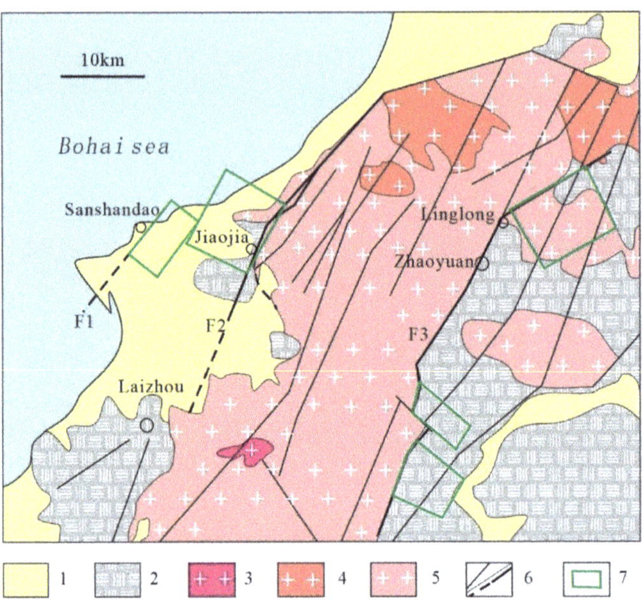

Figure 11. Planar range of the prediction areas of deep resource potential. 1—Quaternary; 2—Early Precambrian metamorphic rock; 3—Cretaceous Weideshan granite; 4—Cretaceous Guojialing granite; 5—Jurassic Linglong granite; 6—Fault; 7—Range of ore-forming prediction areas.

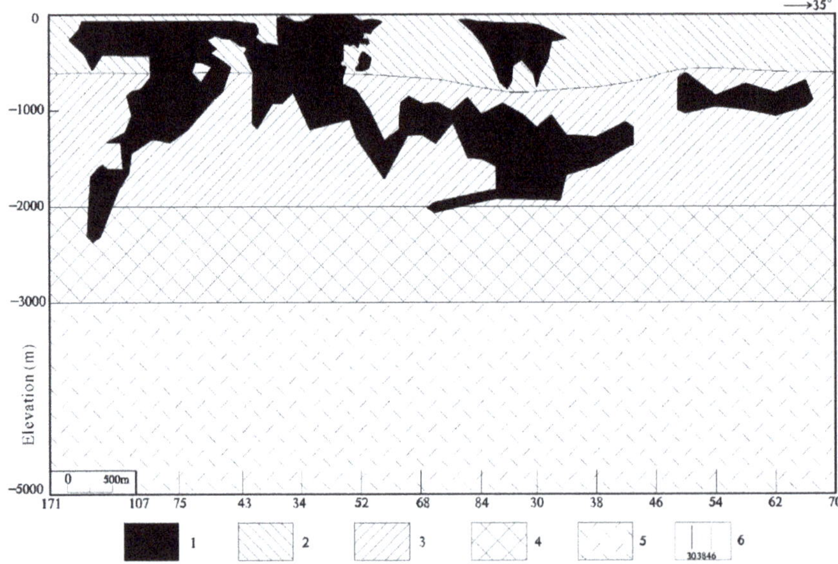

Figure 12. Vertical longitudinal projection map for the deep resource prediction of the Sanshandao fault. 1—Range of controlled gold orebodies; 2—Scope for the calculation of the shallow ore-bearing rate; 3—Scope for the calculation of the deep ore-bearing rate; 4—Prediction range at an elevation of −3000–−2000 m; 5—Prediction range at an elevation range of −5000–−3000 m; 6—Exploration lines and their numbers.

Table 6. Ore-hosting rates of the Sanshandao fault.

Elevation (m)	Proven Resource Reserves		Area (m^2)	Grade (g/t)	Thickness (m)	Ore-Bearing Rate
	Ore Reserves (Mt)	Metal Reserves (kg)				
−500–0	81.5	238,503	5,516,844	2.93	8.10	0.0432
−2000–−500	252.14	1,002,176	11,443,156	3.98	7.90	0.0876

5.1.3. Resources Prediction

The area of the deep prediction areas at an elevation of −5000 to −2000 m was calculated based on the vertical longitudinal projection map. Subsequently, the gold resources in the deep prediction areas were estimated according to the ore-bearing rates at elevations of −500 to 0 m and −2000 to −500 m. The predicted gold resources of the deep prediction areas at an elevation of −5000 to −2000 m in the Sanshandao fault included approximately (375–560) Mt of ore reserves and 1100–2228 t of gold metal reserves (Figure 12, Table 7).

Table 7. Statistics of predicted resources in the Sanshandao fault.

Elevation (m)	Area (km^2)	Ore-Bearing Rate		Predicted Resources			
				Ore Reserves (Mt)		Metal Reserves (t)	
		Shallow Part	Deep Part	Minimum	Maximum	Minimum	Maximum
−3000–−2000	8.48	0.0432	0.0876	125	187	367	743
−5000–−3000	16.96	0.0432	0.0876	250	373	733	1485
Total				375	560	1100	2228

5.2. Resource Potential of the Jiaojia and Zhaoping Faults

The prediction range of the Jiaojia fault is as follows. Its southern boundary was the exploration line No. 408 on the southernmost side of the Qianchen ore block and its northern boundary was the exploration line No. 201 on the northernmost side of the Xincheng ore block. Its eastern and eastern parts were the outcrops of the Jiaojia fault and the production of the Jiaojia fault on the ground surface, respectively. Moreover, it had an elevation of −5000 to −2000 m.

For the purposes of this study, the Zhaoping fault was divided into two prediction sections: the northern section and the central-southern section. The prediction range of the Lingnan-Shuiwangzhuang gold concentration area in the northern Zhaoping fault is as follows. Its southern boundary was exploration line No. 9 of the Lingnan ore block and its northern boundary was exploration line No. 29 on the northernmost side of the Shuiwangzhuang ore block. Moreover, its western and eastern parts were the outcrops of the Zhaoping fault and the projection of the Zhaoping fault on the ground surface, respectively. The prediction range of the Dayin'gezhuang gold concentration area in the central-southern section of the Zhaoping fault is as follows. Its southern and northern boundaries were exploration lines No. 54 and No. 120 of the Dayin'gezhuang gold orefield, and its western and eastern parts were the outcrops of the Zhaoping fault and the projection of the Zhaoping fault on the ground surface, respectively. The prediction range of the Xiadian ore block concentration area in the central-southern section of the Zhaoping fault is as follows. Its southern boundary was exploration line No. 441 of the Xiadian ore block, and its northern boundary was exploration line No. 35 of the Daobeizhuangzi ore block. Moreover, its western and eastern parts were the outcrops of the Zhaoping fault and the projection of the outcrops on the ground surface, respectively. The elevation range of the prediction areas of the Zhaoping fault was −5000 to −2000 m. The proven resources and the parameters for resources prediction of Jiaojia and Zhaoping faults were determined using the same procedure as the Sanshandao fault zone. The predicted results are listed in Table 8.

Table 8. Predicted gold resources of major ore-controlling faults in the northwest Jiaodong Peninsula.

| Name | Proved Resources | | Predicted Resources of Deep Parts at an Elevation of −5000−−2000 m | | | | Total Resources at an Elevation of −5000 m and Above | | | |
| | | | Ore Reserves (Mt) | | Metal Reserves (kg) | | Ore Reserves (Mt) | | Metal Reserves (t) | |
	Ore Reserves (Mt)	Metal Reserves (kg)	Minimum	Maximum	Minimum	Maximum	Minimum	Maximum	Minimum	Maximum
Sanshandao fault	334	1241	375	560	1100	2228	709	894	2341	3469
Jiaojia fault	429	1334	495	912	1433	2379	924	1341	2767	3713
Northern section of the Zhaoping fault	202	700	200	319	619	1146	402	521	1319	1846
Central-southern section of the Zhaoping fault	128	385	147	446	255	737	275	383	831	1122
Total	1093	3660	1217	2237	3377	6490	2310	3330	7073	10,150

In sum, for the metallogenetic belts of the Sanshandao, Jiaojia, and Zhaoping faults in the Jiaodong Peninsula, their predicted gold ore reserves and predicted gold metal reserves at an elevation of −5000 to −2000 m were approximately 1217–2237 × Mt and 3377–6490 t, respectively, and their total gold ore reserves and gold metal reserves at an elevation of −5000 m and above were approximately 2310–3330 × Mt and 7073–10,150 t, respectively. As the metallogenic conditions in the deep prediction area are different from those in the shallow area, the accuracy of the geophysical method will decrease with the increase in the depth, so the error of the prediction results is uncertain.

The above forecast data are only the predicted results of the deep parts of the main known gold deposits in the northwest Jiaodong Peninsula. In combination with the future prediction and prospecting of other areas in the Jiaodong Peninsula with great potential for deep prospecting, the total resources of gold deposits in the Jiaodong Peninsula will possibly double in quantity and increase from the current 5000 t to 10,000 t. Therefore, the Jiaodong gold concentration area is expected to become the world's second largest gold metallogenic area.

At present, the exploration depths of identified gold deposits in the Jiaodong Peninsula are mostly less than 2000 m, with only a few exploratory boreholes reaching a depth of more than 3000 m [61,62]. For example, exploratory borehole ZK96-5 in the Xiling mining area in the northern Sanshandao fault has a final hole depth of 4006.17 m, making it the deepest borehole for gold exploration in China [62]. This borehole revealed a fractured alteration zone with a thickness of tens of meters at a depth of approximately 3500 m. Borehole K01 drilled in the deep part of the Jiaojia supergiant gold deposit in the middle part of the Jiaojia fault by the Shandong Institute of Geological Sciences has a depth of 3266.06 m. It revealed multi-layer industrial orebodies at a depth of 2810–2854 m [61]. Borehole K3401 drilled in the deep part of the Shuiwangzhuang mining area in the northern Zhaoping fault by the Shandong Provincial No. 6 Exploration Institute of Geology and Mineral Resources has a final hole depth of 3000.58 m. It revealed a fractured zone in the Jiuqujiangjia 208 fault at an elevation of approximately −2700 m. The sampling analysis indicated that the fractured zone has a maximum gold grade of nearly 7 g/t. The metallogenic information obtained from these deep exploratory boreholes is consistent with the predicted results of this study, indicating that the deep parts at an elevation of −2000 m and below have great potential for deep prospecting. In addition, geological and geophysical explorations show that the Jiaojia and Sanshandao faults' dip angles gradually decrease towards their deep parts along their dip directions and that the two faults intersect at the elevation of approximately −4500 m [42,43,59]. These findings, combined with the comprehensive analysis of the evidence of denudation degree, metallogenic depth, deep drilling verification, and the extension of faults towards deep parts, have led to the conclusion that the Jiaojia and

Sanshandao faults have favorable gold metallogenic conditions and enormous prospecting potential at an elevation of −4500 m and below.

6. Conclusions

a. Since deep-seated deposits cannot be predicted using conventional metallogenic predictive methods, it has become an important means of deep metallogenic prediction to establish a 3D geological model based on high-precision geophysical exploration and the drilling engineering in known areas. Through the dissection of the ore-controlling regularities of gold deposits in the Jiaodong Peninsula, this study proposed a method for predicting deep prospecting target areas based on a stepped metallogenic model and a method for predicting the deep resource potential of gold deposits based on the shallow resources of ore-controlling faults;

b. The information amounts of eight characteristic variables, including Early Precambrian metamorphic rock series, Linglong granite, contact zone of Early Precambrian metamorphic rock series and Linglong granite, fault buffer zone, gentle part of a fault section with a steep-to-gentle transition of fault dip angle, changing rate of fault dip angle, equidistant distribution along fault strike and equidistant distribution along fault dip direction, were extracted from the 3D geological models of main gold concentration areas. Using the statistics of the information amounts, a total of five deep prospecting target areas in the Jiaojia and Sanshandao faults were predicted;

c. Based on the proven gold resources at an elevation of −2000 m and above, the total gold resources at an elevation of −5000 to −2000 m in the Sanshandao, Jiaojia, and Zhaoping ore-controlling faults were predicted to be approximately 3377–6490 t of Au. Therefore, it is very likely that the total gold resources in the Jiaodong Peninsula are expected to exceed 10,000 t. However, due to the high ground temperature and pressure environment of deep resources, future mining will face great challenges.

Author Contributions: M.S. conceived and designed the research ideas; S.L., H.L. and C.H. participated in the field investigation; J.F., Z.Y., L.Z. and X.L. performed the data processing, and M.S., J.Z., S.L., B.W. and G.W. reviewed and edited the draft. All the data were obtained from previous work performed by the project team. All authors have read and agreed to the published version of the manuscript.

Funding: This study was financially supported by the National Natural Science Foundation of China-Shandong Joint Fund Project (U2006201).

Institutional Review Board Statement: Not applicable.

Informed Consent Statement: Not applicable.

Data Availability Statement: Not applicable.

Acknowledgments: The authors are grateful for the constructive comments by the anonymous reviewers.

Conflicts of Interest: The authors declare no conflict of interest.

References

1. Zhao, P.D.; Chi, S.D.; Li, Z.D.; Cao, X.Z. *Theory and Methods of Mineral Exploration*; China University of Geosciences Press: Wuhan, China, 2006; pp. 1–334. (In Chinese)
2. Zhu, Y.S. *Introduction to the Methodology of Mineral Resource Evaluation*; Geological Publishing House: Beijing, China, 1984. (In Chinese)
3. Wang, S.C.; Liu, Y.Q.; YI, P.H.; Zhang, Y.Q. *Multipurpose Information Metallogenic Prognosis of Gold Deposits and Gold Mineralization Area in Shandong Province*; Geological Publishing House: Beijing, China, 2003; pp. 1–245. (In Chinese)
4. Li, S.X.; Liu, C.C.; AN, Y.H.; Wang, W.C.; Huang, T.L.; Yang, C.H. *Geology of Gold Deposits in Shandong Peninsula*; Geological Publishing House: Beijing, China, 2007; pp. 1–423. (In Chinese)
5. Ye, T.Z.; Lv, Z.C.; Pang, Z.S. *Prospecting Prognosis Theory and Method of Exploration Area*; Geological Publishing House: Beijing, China, 2015; pp. 1–703. (In Chinese)

6. Song, M.C.; Ding, Z.J.; Zhang, J.J.; Song, Y.X.; Bo, J.W.; Wang, Y.Q.; Liu, H.B.; Li, S.Y.; Li, J.; Li, R.X.; et al. Geology and mineralization of the Sanshandao supergiant gold deposit (1200 t) in the Jiaodong Peninsula, China: A review. *China Geol.* **2021**, *4*, 686–719. [CrossRef]

7. Lü, Q.T.; Qi, G.; Yan, J.Y. 3D geological model of Shizishan ore field constrained by gravity and magmatic interactive modeling: A case history. *Geophysics* **2013**, *78*, B25–B35. [CrossRef]

8. Malehmir, A.; Tryggvason, A.; Juhlin, C.; Rodriguez-Tablante, J.; Weihed, P. Seismic imaging and potential field modeling to delineate structures hosting VHMS deposits in the Skellefte Ore District, Northern Sweden. *Tectonophysics* **2006**, *426*, 319–334. [CrossRef]

9. Malehmir, A.; Tryggvason, A.; Lickorish, H.; Weihed, P. Regional structural profiles in the western part of the Palaeoproterozoic Skellefte Ore District, northern Sweden. *Precambrian Res.* **2007**, *159*, 1–18. [CrossRef]

10. Wang, G.W.; Huang, L. 3D geological modeling for mineral resource assessment of the Tongshan Cu deposit, Heilongjiang Province, China. *Geosci. Front.* **2012**, *3*, 483–491. [CrossRef]

11. Wang, G.W.; Li, R.X.; Carranza, J.M.E.; Zhang, S.T.; Yan, C.H.; Zhu, Y.Y.; Qu, J.N.; Hong, D.M.; Song, Y.W.; Han, J.W.; et al. 3D geological modeling for prediction of subsurface Mo targets in the Luanchuan district, China. *Ore Geol. Rev.* **2015**, *71*, 592–610. [CrossRef]

12. Rabeaut, O.; Legault, M.; Cheilletz, A.; Jébrak, M.; Royer, J.J.; Cheng, L.Z. Gold potential of a Hidden Archean fault zone: The case of the Cadillac–Larder lake fault. *Explor. Min. Geol.* **2010**, *19*, 99–116. [CrossRef]

13. Royer, J.J.; Cheilletz, A.; Rabeau, O.; Jébrak, L.Z. Is deposit location predictable? Example of the orogenic gold deposits in the Abitibi Province. In Proceedings of the 11th Biennial SGA Meeting, Antofagasta, Chile, 26–29 September 2011. 3p.

14. Rabeau, O.; Royer, J.J.; Jébrak, M.; Cheilletz, A. Log-uniform distribution of gold deposits along major Archean fault zones. *Miner. Depos.* **2013**, *48*, 817–824. [CrossRef]

15. Farahbakhsh, E.; Hezarkhani, A.; Eslamkish, T.; Bahroudi, A.; Chandra, R. Three-dimensional weights of evidence modeling of a deep-seated porphyry Cu deposit. *Geochem. Explor. Environ. Anal.* **2020**, *20*, 480–495. [CrossRef]

16. Goldfarb, R.J.; Groves, D.I.; Gardoll, S. Orogenic gold and geological time: A global synthesis. *Ore Geol. Rev.* **2001**, *18*, 1–75. [CrossRef]

17. Sibson, R.H.; Scott, J. Stress/fault controls on the containment and release of overpressured fluids: Examples from gold-quartz vein systems in Juneau, Alaska; Victoria, Australia and Otago, New Zealand. *Ore Geol. Rev.* **1998**, *13*, 293–306. [CrossRef]

18. Tripp, G.I.; Vearncombe, J.R. Fault/fracture density and mineralization: A contouring method for targeting in gold exploration. *J. Struct. Geol.* **2004**, *26*, 1087–1108. [CrossRef]

19. Mejia-Herrera, P.; Royer, J.J.; Caumon, G.; Cheilletz, A. Curvature attribute from surface-restoration as predictor variable in Kupferschiefer copper potentials: An example from the Fore-Sudetic region. *Nat. Resour. Res.* **2014**, *24*, 275–290. [CrossRef]

20. Kusky, T.M.; Polat, A.; Windley, B.F.; Burke, K.C.; Dewey, J.F.; Kidd, W.S.F.; Maruyama, S.; Wang, J.P.; Deng, H.; Wang, Z.S.; et al. Insights into the tectonic evolution of the North China Craton through comparative tectonic analysis: A record of outward growth of Precambrian continents. *Earth Sci. Rev.* **2016**, *162*, 387–432. [CrossRef]

21. Li, S.Z.; Zhao, G.C.; Santosh, M.; Liu, X.; Dai, L.M.; Suo, Y.H.; Tam, P.Y.; Song, M.C.; Wang, P.C. Paleoproterozoic structural evolution of the southern segment of the Jiao-Liao-Ji Belt, North China Craton. *Precambrian Res.* **2012**, *200–203*, 59–73. [CrossRef]

22. Wang, L.; Kusky, T.M.; Polat, A.; Wang, S.; Jiang, X.; Zong, K.; Wang, J.; Deng, H.; Fu, J. Partial melting of deeply subducted eclogite from the Sulu orogen in China. *Nat. Commun.* **2014**, *5*, 5604. [CrossRef]

23. Zhou, J.B.; Han, W.; Song, M.C.; Li, L. Zircon U-Pb ages of the cetaceous sedimentary rocks in the Laiyang Basin, eastern China and their tectonic implications. *J. Asian Earth Sci.* **2020**, *194*, 103965. [CrossRef]

24. Yang, K.F.; Fan, H.R.; Santosh, M. Reactivation of the Archean Lower Crust: Implications for Zircon Geochronology, Elemental and Sr–Nd–Hf Isotopic Geochemistry of Late Mesozoic Granitoids from Northwestern Jiaodong Terrane, the North China Craton. *Lithos* **2012**, *146–147*, 112–127. [CrossRef]

25. Zhao, R.; Wang, Q.F.; Deng, J.; Santosh, M.; Liu, X.F.; Cheng, H.Y. Late Mesozoic magmatism and sedimentation in the Jiaodong Peninsula: New constraints on lithospheric thinning of the North China Craton. *Lithos* **2018**, *322*, 312–324. [CrossRef]

26. Charles, N.; Augier, R.; Gumiaux, C.; Monié, P.; Chen, Y.; Faure, M.; Zhu, R.X. Timing, duration and role of magmatism in wide rift systems: Insights from the Jiaodong Peninsula (China, East Asia). *Gondwana Res.* **2013**, *24*, 412–428. [CrossRef]

27. Goss, C.S.; Wilde, S.A.; Wu, F.Y.; Yang, J. The age, isotopic signature and significance of the youngest Mesozoic granitoids in the Jiaodong Terrene, Shandong Province, North China Craton. *Lithos* **2010**, *120*, 309–326. [CrossRef]

28. Huang, X.L.; He, P.L.; Wang, X.; Zhong, J.W.; Xu, Y.G. Lateral variation in oxygen fugacity and halogen contents in early Cretaceous magmas in Jiaodong area, East China: Implication for triggers of the destruction of the North China Craton. *Lithos* **2016**, *248–251*, 478–492. [CrossRef]

29. Tang, H.Y.; Zheng, J.P.; Yu, C.M.; Ping, X.Q.; Ren, H.W. Multistage crust-mantle interactions during the destruction of the North China Craton: Age and composition of the Early Cretaceous intrusions in the Jiaodong Peninsula. *Lithos* **2014**, *190–191*, 52–70. [CrossRef]

30. Wang, L.G.; Qiu, Y.M.; McNaugmton, N.J. Constraints on crust evolution and gold metallogeny in the northwestern Jiaodong Peninsula, China, from SHRIMP U-Pb zircon studies of granitoids. *Ore Geol. Rev.* **1998**, *13*, 275–291. [CrossRef]

31. Yan, Q.S.; Metcalfe, I.; Shi, X.F.; Zhang, P.; Li, F. Early Cretaceous granitic rocks from the southern Jiaodong Peninsula, eastern China: Implications for lithospheric extension. *Int. Geol. Rev.* **2019**, *61*, 821–838. [CrossRef]

32. Xia, Z.M.; Liu, J.L.; Ni, J.L.; Zhang, T.T.; Yun, W. Structure, evolution and regional tectonic implications of the Queshan metamorphic core complex in eastern Jiaodong Peninsula of China. *Sci. China Earth Sci.* **2016**, *59*, 997–1013. [CrossRef]
33. Deng, J.; Wang, C.M.; Bagas, L.; Carranza, E.J.M.; Lu, Y.J. Cretaceous–Cenozoic tectonic history of the Jiaojia Fault and gold mineralization in the Jiaodong Peninsula, China: Constraints from zircon U–Pb, illite K–Ar, and apatite fission track thermo chronometry. *Miner. Depos.* **2015**, *50*, 987–1006. [CrossRef]
34. Deng, J.; Yang, L.Q.; Li, H.R.; Groves, D.I.; Santosh, M.; Wang, Z.L.; Sai, S.X.; Wang, S.R. Regional structural control on the distribution of world-class gold deposits: An overview from the Giant Jiaodong Gold Province, China. *Geol. J.* **2019**, *54*, 378–391. [CrossRef]
35. Yang, L.; Zhao, R.; Wang, Q.F.; Liu, X.; Carranza, E.J.M. Fault geometry and fluid-rock reaction: Combined controls on mineralization in the Xinli gold deposit, Jiaodong Peninsula, China. *J. Struct. Geol.* **2018**, *111*, 14–26. [CrossRef]
36. Deng, J.; Yang, L.Q.; Groves, D.I.; Zhang, L.; Qiu, K.F.; Wang, Q.F. An integrated mineral system model for the gold deposits of the giant Jiaodong province, eastern China. *Earth Sci. Rev.* **2020**, *208*, 103274. [CrossRef]
37. Zhang, L.; Weinberg, R.F.; Yang, L.Q.; Groves, D.I.; Deng, J. Mesozoic orogenic gold mineralization in the Jiaodong Peninsula, China: A focused event at 120 ± 2 Ma during cooling of pregold granite intrusions. *Econ. Geol.* **2020**, *115*, 415–441. [CrossRef]
38. Li, L.; Santosh, M.; Li, S.R. The "Jiaodong type" gold deposits: Characteristics, origin and prospecting. *Ore Geol. Rev.* **2015**, *65*, 589–611. [CrossRef]
39. Wen, B.J.; Fan, H.R.; Santosh, M.; Hu, F.F.; Pirajno, F.; Yang, K.F. Genesis of two different types of gold mineralization in the Linglong gold field, China: Constrains from geology, fluid inclusions and stable isotope. *Ore Geol. Rev.* **2015**, *65*, 643–658. [CrossRef]
40. Li, J.J.; Zhang, P.P.; Li, H. Formation of the Liaoshang gold deposit, Jiaodong Peninsula, eastern China: Evidence from geochronology and geochemistry. *Geol. J.* **2020**, *55*, 5903–5913. [CrossRef]
41. Song, M.C.; YI, P.H.; Xu, J.X.; Cui, S.X.; Shen, K.; Jiang, H.L.; Yuan, W.H.; Wang, H.J. A step metallogenetic model for gold deposits in the northwestern Shandong Peninsula, China. *Sci. China Earth Sci.* **2012**, *42*, 992–1000. [CrossRef]
42. Song, Y.X.; Song, M.C.; Ding, Z.J.; Wei, X.F.; Xu, S.H.; Li, J.; Tan, X.F.; Li, S.Y.; Zhang, Z.L.; Jiao, X.M.; et al. Major advances on deep prospecting in Jiaodong gold ore cluster and its metallogenic characteristics. *Gold Sci. Technol.* **2017**, *25*, 4–18, (In Chinese with English abstract).
43. Song, M.C.; Xue, G.Q.; Liu, H.B.; Li, Y.X.; He, C.Y.; Wang, H.J.; Wang, B.; Song, Y.X.; Li, S.Y. A Geological-Geophysical Prospecting Model for Deep-Seated Gold Deposits in the Jiaodong Peninsula, China. *Minerals* **2021**, *11*, 1393. [CrossRef]
44. Yang, X.A.; Zhao, G.C.; Song, Y.B.; Tian, F.; Dong, H.W.; Gao, J.W. Characteristics of ore-controlling detachment fault and future prospecting in the Muping-Rushan metallogenic belt, Eastern Shandong Province. *Geotecton. Metallog.* **2011**, *35*, 339–347.
45. Yang, L.Q.; Deng, J.; Wang, Z.L.; Zhang, L.; Guo, L.N.; Song, M.C.; Zheng, X.L. Mesozoic gold metallogenic system of the Jiaodong Gold Province, eastern China. *Acta Petrol. Sin.* **2014**, *30*, 2447–2467. (In Chinese with English abstract).
46. Zhang, L.; Yang, L.Q.; Wang, Y.; Weinberg, R.F.; An, P.; Chen, B.Y. Thermochronologic constrains on the processes of formation and exhumation of the Xinli orogenic gold deposit, Jiaodong Peninsula, eastern China. *Ore Geol. Rev.* **2017**, *81*, 140–153. [CrossRef]
47. Mao, X.C.; Ren, J.; Liu, Z.K.; Chen, J.; Tang, L.; Deng, H.; Bayless, R.C.; Yang, B.; Wang, M.J.; Liu, C.M. Three-dimensional prospectivity modeling of the Jiaojia-type gold deposit, Jiaodong Peninsula, Eastern China: A case study of the Dayingezhuang deposit. *J. Geochem. Explor.* **2019**, *203*, 27–44. [CrossRef]
48. Zhang, L.; Groves, D.I.; Yang, L.Q.; Wang, G.W.; Liu, X.D.; Li, D.P.; Song, Y.X.; Shan, W.; Sun, S.C.; Wang, Z.K. Relative roles of formation and preservation on gold endowment along the Sanshandao gold belt in the Jiaodong gold province, China: Importance for province- to district-scale gold exploration. *Miner. Depos.* **2020**, *55*, 325–344. [CrossRef]
49. Song, M.C.; Deng, J.; Yi, P.H.; Yang, L.Q.; Cui, S.X.; Xu, J.X.; Zhou, M.L.; Huang, T.L.; Song, G.Z.; Song, Y.X. The kiloton class Jiaojia gold deposit in eastern Shandong Province and its genesis. *Acta Geol. Sin. Engl. Ed.* **2014**, *88*, 801–824. [CrossRef]
50. Wang, S.R.; Yang, L.Q.; Wang, J.G.; Wang, E.J.; Xu, Y.L. Geostatistical determination of ore shoot plunge and structural control of the Sizhuang world-class epizonal orogenic gold deposit, Jiaodong Peninsula, China. *Minerals* **2019**, *9*, 214. [CrossRef]
51. Drummond, B.J.; Goleby, B.R. Seismic reflection images of the major ore-controlling structures in the eastern goldfields province Western Australia. *Explor. Geophys.* **1993**, *24*, 473–478. [CrossRef]
52. West, D.; Witherly, K. Geophysical exploration for gold in deeply weathered terrains, two tropical cases. *Explor. Geophys.* **1995**, *26*, 124–130. [CrossRef]
53. Takakura, S. CSAMT and MT investigations of an active gold depositing environment in the Osorezan geothermal area, Japan. *Explor. Geophys.* **1995**, *26*, 172–178. [CrossRef]
54. Di, Q.Y.; Xue, G.Q.; Yin, C.C.; Li, X. New methods of controlled-source electromagnetic detection in China. *Sci. China Earth Sci.* **2020**, *50*, 1219–1227. [CrossRef]
55. Di, Q.Y.; Xue, G.Q.; Zeng, Q.D.; Wang, Z.X.; An, Z.G.; Lei, D. Magnetotelluric exploration of deep-seated gold deposits in the Qingchengzi orefield, eastern Liaoning (China), using a SEP system. *Ore Geol. Rev.* **2020**, *122*, 103501. [CrossRef]
56. Guo, Z.W.; Xue, G.Q. Electromagnetic methods for mineral exploration in China: A review. *Ore Geol. Rev.* **2020**, *118*, 103357. [CrossRef]
57. Xue, G.Q.; Li, Z.Y.; Guo, W.B.; Fan, J.S. The exploration of sedimentary bauxite deposits using the reflection seismic method: A case study from the Henan Province, China. *Ore Geol. Rev.* **2020**, *127*, 103832. [CrossRef]

58. Xue, G.Q.; Zhang, L.B.; Hou, D.Y.; Liu, H.T.; Luo, X.N. Integrated geological and geophysical investigations for the discovery of deeply buried gold–polymetallic deposits in China. *Geol. J.* **2020**, *55*, 1771–1780. [CrossRef]
59. Song, M.C.; Wan, G.P.; Cao, C.G.; He, C.Y. Geophysical–geological interpretation and deep-seated gold deposit prospecting in Sanshandong–Jiaojia area, eastern Shandong Province, China. *Acta Geol. Sin. Engl. Ed.* **2012**, *86*, 640–652.
60. Duan, L.A.; Guo, Y.C.; Han, X.M.; Wang, J.T.; Zhao, P.F.; Wang, L.P.; Wei, Y.F. Discovery of a medium-scale tectonic altered rock type gold deposit (13.5 t) on the northeastern margin of Jiaolai Basin, Shandong Province, China and its new application of exploration direction. *China Geol.* **2022**. [CrossRef]
61. Yu, X.F.; Li, D.P.; Tian, J.X.; Yang, D.P.; Shan, W.; Geng, K.; Xiong, Y.X.; Chi, N.J.; Wei, P.F.; Liu, P.R. Deep gold mineralization features of Jiaojia metallogenic belt, Jiaodong gold Province: Based on the breakthrough of 3000 m exploration drilling. *China Geol.* **2020**, *3*, 385–401.
62. Chen, S.X.; Zhang, Y.C.; Liu, Z.D. Drilling Technology of ZK96-5 Well in Xiling Gold Deposit, Laizhou, Shandong Province. In Proceedings of the 17th National Exploration Engineering (Geotechnical Drilling Engineering) Academic Exchange Conference, Nanchang, China, 11–13 October 2013; pp. 121–125. (In Chinese).

Article

Orebody Modeling Method Based on the Coons Surface Interpolation

Zhaohao Wu [1,2], Lin Bi [1,2,3], Deyun Zhong [1,2,3,*], Ju Zhang [1,2], Qiwang Tang [3] and Mingtao Jia [1,2,3]

[1] School of Resources and Safety Engineering, Central South University, Changsha 410083, China
[2] Research Center of Digital Mine, Central South University, Changsha 410083, China
[3] Changsha Digital Mine Co., Ltd., Changsha 410221, China
* Correspondence: deyun_zhong@csu.edu.cn

Abstract: This paper presents a surface modeling method for interpolating orebody models based on a set of cross-contour polylines (geological polylines interpreted from the raw geological sampling data) using the bi-Coons surface interpolation method. The method is particularly applicable to geological data with cross-contour polylines acquired during the geological and exploration processes. The innovation of this paper is that the proposed method can automatically divide the closed loops and automatically combine the sub-meshes. The method solves the problem that it is difficult to divide closed loops from the cross-contour polylines with complex shapes, and it greatly improves the efficiency of modeling based on complex cross-contour polylines. It consists of three stages: (1) Divide closed loops using approximate planes of contour polylines; each loop is viewed as a polygon combined with several polylines, that is the n-sided region. (2) After processing the formed n-sided regions, Coons surface interpolation is improved to complete the modeling of every single loop (3) Combine all sub-meshes to form a complete orebody model. The corresponding algorithm was implemented using the C++ programing language on 3D modeling software. Experimental results show that the proposed orebody modeling method is useful for efficiently recovering complex orebody models from a set of cross-contour polylines.

Keywords: geological modeling; orebody modeling; coons surface interpolation; contour interpolation; polygon triangulation

Citation: Wu, Z.; Bi, L.; Zhong, D.; Zhang, J.; Tang, Q.; Jia, M. Orebody Modeling Method Based on the Coons Surface Interpolation. *Minerals* **2022**, *12*, 997. https://doi.org/10.3390/min12080997

Academic Editor: Behnam Sadeghi

Received: 1 July 2022
Accepted: 3 August 2022
Published: 6 August 2022

Publisher's Note: MDPI stays neutral with regard to jurisdictional claims in published maps and institutional affiliations.

1. Introduction

Three-dimensional geological modeling and visualization is one of the hot topics in the field of geosciences [1]. As a geological body with high economic value, orebody modeling is also one of the current research hot topics and is of great significance for promoting mine digitization and improving mining efficiency [2]. In 2002, Mallet [3] proposed a new method called discrete smooth interpolation (DSI), based on which a number of orebody modeling works have been published [4–7]. Additionally, some commercial software [8] has been developed using different technologies including triangulate surfaces, radial functions, and parametric surfaces [9]. However, due to the complex shape of the orebody, much research work still needs to be done in orebody modeling. Our research is dedicated to developing the software for modeling orebody using small triangles based on the Coons surface [10].

We can often obtain the cross-contour polylines of planes and sections of orebodies through geological logging in the geological exploration and production exploration of mines, especially of some precious metal mines with flat orebody shapes. These contour polylines are interlaced with complex shapes. Therefore, if the explicit modeling method based on the contour stitching method is used, the model will be constructed with poor quality and low efficiency [11]. Modeling orebodies with high speed and great quality is of great significance to improving the production efficiency of mines [12]. The orebody modeling method we propose based on Coons surface interpolation can automatically divide the contour polylines into closed loops and interpolate and model the formed n-sided regions; then, the constructed sub-meshes will be combined without manual intervention. Based on the above steps, the complete orebody model can be quickly constructed.

1.1. Related Research Work

Based on the above analysis, we transform the 3D modeling of the orebody into the Coons surface interpolation of the interpreted contour polylines. Therefore, we will briefly summarize the research work on free-form surface reconstruction, contour interpolation modeling, and orebody modeling.

1.1.1. Free-Form Surface Reconstruction

Free-form surface reconstruction is a key technology of orebody 3D modeling [13–15]. In practical application, the commonly used surface reconstruction methods include implicit function surface, triangular Bezier surface, B-spline surface, NURBS surface, Coons surface, polygon model, and so on.

The implicit function surface [16–18] refers to the surface represented by equation $F(x, y, z) = 0$, which can represent quadric surfaces such as spherical surfaces, cylindrical surfaces, and more complex surfaces. When reconstructing the surface, the implicit function equations are constructed using the existing data, and then the surface model can be generated by obtaining isosurfaces through the Marching Cube method or other methods. Though using the implicit function to reconstruct a surface has the advantages of small calculation and easy solution, there are also some problems such as unclear geometric meaning. In the future, a great amount of research work needs to be carried out to make extensive use of implicit modeling.

In 1982, Farin [19] proposed the method of constructing the triangular Bezier surface based on the triangulation of scattered data. This surface is explicit and C^1 continuous with flexible construction. Using the triangular Bezier surface for surface reconstruction requires less manual intervention and is easy to automate. However, it also has some disadvantages, such as the insufficient ability for surface modification and poor controllability.

The B-spline curve and surface [20–22] have many excellent properties, such as geometric invariance, convex hull, convexity preservation, and local support, which lead to the powerful function of representing and designing free-form curves and surfaces. It is a commonly used free-form surface suitable for engineering shape design, but it cannot be well applied to represent and design elementary curves and surfaces. Therefore, researchers proposed the NURBS surface method [23–25] for surface reconstruction, which both has the powerful function of representing and designing free-form curves and surfaces like B-spline and can accurately represent quadric arcs and quadric surfaces. The unification of analytical geometry and free-form curves and surfaces is well realized by this method. Due to the introduction of a weight factor and a manipulation control vertex, it provides sufficient flexibility for various shape designs. However, the parameterization effect will become very poor with an inappropriate weight factor. In addition, there are still many problems to be further solved in the calculation of NURBS surface intersection.

In 1964, Professor S. A. Coons of the Massachusetts Institute of Technology (MIT) proposed a general method to describe surfaces, the Coons surfaces method [10], which has unique advantages in engineering applications. Given the boundary conditions, Coons patches of the corresponding degree can be constructed. Therefore, in theory, the Coons method can be used to construct arbitrary order patches through boundary constraints [26,27], which makes Coons surfaces show strong advantages in representing and designing surfaces. However, in the Coons surface method, it is difficult to control the surface shape. When the boundary curve is determined, the shape can only be modified by changing the torsion vector [28].

The most commonly used polygon model method is the triangle model method based on triangulation. The triangulation technology in the plane has matured [29], and great research progress has been made in the triangulation of surface data and tetrahedrons of spatial scattered data [30,31]. The characteristics of the polygon model are: (1) adjacent surface patches can only be C^0 continuous and (2) data amounts are large. When there needs high accuracy of a model, a large number of polygons need to be used to represent the surface approximately, which has a huge amount of data. As a result, this method is not suitable for constructing complex models with high accuracy. Therefore, in this paper, the polygon model method is used only in the modeling of simple single-sided regions.

1.1.2. Contour Interpolation Modeling

Visualization based on contour interpolation refers to the modeling of three-dimensional entities using a sequence of two-dimensional contours, which is an important research direction in scientific computing visualization. According to the number of contour polylines of two adjacent layers, it can be divided into single contour modeling and multi-contour modeling.

Single contour modeling refers to the three-dimensional model reconstruction of the contour polylines when there is only one contour polyline in each layer on two adjacent planes, which is relatively simple. Researchers have proposed many optimization methods for single contour modeling, such as the minimum surface product method [32], the maximum volume method [33] based on the global search strategy, the shortest diagonal method [34], and the synchronous advance method of adjacent contours [35] based on local calculation and decision.

Multi-contour modeling refers to 3D model reconstruction when there are multiple contour polylines on two adjacent planes, which difficulty is the correspondence and branch processing between contour polylines. In terms of the correspondence of contour polylines, Meyers et al. [36] proposed the minimum spanning tree method, but this method still has some limitations. In some special cases, the contour polylines cannot correctly correspond, and subsequent manual processing is required. For the branching problem between contour lines, Ekoule et al. [37] proposed a method for constructing the middle contour polyline in the middle of the two adjacent planes of contour polylines. This method transforms the branching problem into the connection problem between a series of single contour polylines. However, in some cases, this method will leave holes on the constructed model surface.

Jones et al. [38] proposed an isosurface construction method using volume data, which requires that contour polylines have simple shapes and be completely closed. There is no need to judge the corresponding relationship and branch relationship when using this method for surface reconstruction, and it is suitable for both convex and non-convex contour polylines. However, it also has the disadvantage of a large amount of calculation. Additionally, Zhong et al. [39] proposed a method of reconstructing the 3D orebody by cross-section contour polylines, which allows for adding geometric constraints and can easily control the shape of the model, but many discontinuous parts may be generated after reconstruction if the data are relatively sparse. Moreover, Wu et al. [40] proposed a 3D orebody modeling method based on the normal estimation of cross-contour polylines. First, the normals of the cross-contour polylines will be estimated, based on which the radial basis function will be used for interpolating and modeling. This method improves the modeling automation of contour polylines, but it requires strict intersection between contour polylines, and they can be completely broken at intersections.

1.1.3. Orebody Modeling

The 3D modeling of the orebody is to interpret the topological information and determine the geometric shape of the orebody using mine geological exploration data and production exploration data [41]. According to the different principles of model construction, it can be divided into explicit modeling and implicit modeling [42]. In explicit modeling, first, the contour polylines of the orebody should be delineated manually, and then the 3D reconstruction of two-dimensional contour polylines can be realized by the contour stitching method [37]. Because explicit modeling requires a great deal of manual interaction, the limitation of low efficiency of this method is gradually highlighted for orebodies with a large number of sections and complex shapes. In addition, when the modeling data change locally, the modelers need to reinterpret the data, delineate the orebody, and model through contour stitching. This process is complex, which is not conducive to the dynamic updating of the model [43]. In contrast with explicit modeling, implicit modeling is based on the implicit function. The relationships of the spatial sampling data are used to construct an implicit function, and the 3D model can be constructed by the spatial interpolation algorithm [18]. This method not only greatly reduces the manual interaction but also realizes the dynamic editing of the model, which also improves the automation and efficiency of modeling [44].

1.2. Solution Strategy

In this paper, we consider transforming the problem of orebody modeling into a modeling problem based on the Coons surface interpolation of n-sided regions. The proposed method requires the contour polylines to be in an approximate plane within the tolerance range. Firstly, we creatively propose a closed-loop division method for contour polylines interpreted from data obtained in mine production and exploration. In this method, the approximate plane of each contour polyline is used to cut all polylines, and the polylines in the final subspace are connected and grouped within the tolerance range to form closed loops. Secondly, we preprocess the formed n-sided region. For the simple single-sided region, the method based on constrained Delaunay triangulation is used for modeling. In terms of complex single-sided regions and non-quadrilateral regions, they are transformed into four-sided regions by different methods. Finally, the Coons surface is used for interpolation and modeling based on the processed four-sided regions to realize the 3D visualization of the orebody. In summary, we creatively propose a closed-loop division and modeling method of orebody contour polylines.

2. Overview of the Method

The basic idea of this modeling method using Coons surface interpolation is as follows. Through the steps of n-sided region processing and function solving, the Coons surface interpolation is used to construct sub-meshes through the formed closed loops, which are obtained by interpreted cross-contour polylines. Finally, the sub-meshes will be combined to realize the modeling of the complete orebody.

The proposed method mainly consists of the following three steps. Figure 1 shows the overall process of the method.

1. Use approximate planes of each contour polyline to cut all polylines for the closed-loop division.
2. Preprocess the formed loops and separately model the processed four-sided regions through Coons surface interpolation.
3. Combine all the sub-meshes to construct a complete orebody model.

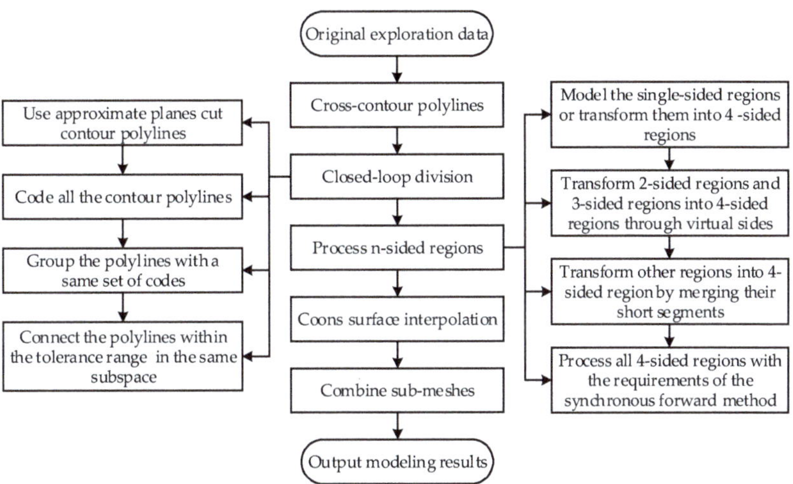

Figure 1. Overall flow chart.

3. Method

3.1. Coons Surface

With the continuous development of computer technology and modern industry, the application scope of Coons surfaces has gradually expanded from the shape design of cars, ships, and aircraft to various fields such as architectural design, medical research, and geological modeling. Coons surfaces can be divided into three categories according to boundary control conditions. The boundary control conditions of the first kind of Coons surface only contain four boundary curves or part of them. The boundary control conditions of the second kind of Coons surface include both boundary curves and boundary tangent vectors. For the third kind of Coons surface, the boundary control conditions include boundary curves and boundary tangent vectors, as well as boundary second-order derivative vectors. The bicubic Coons surface used in interpolation in this paper belongs to the second kind of Coons surface, which can ensure the interpolated patches are continuous both in position and slope.

To satisfy the slope continuity and the position continuity at the four boundaries, the four boundaries $P(u, j)$ and $P(i, v)$ and their cross-border tangent vectors $P_v(u, j)$ and $P_u(i, v)$ must be given at the same time, as shown in Figure 2.

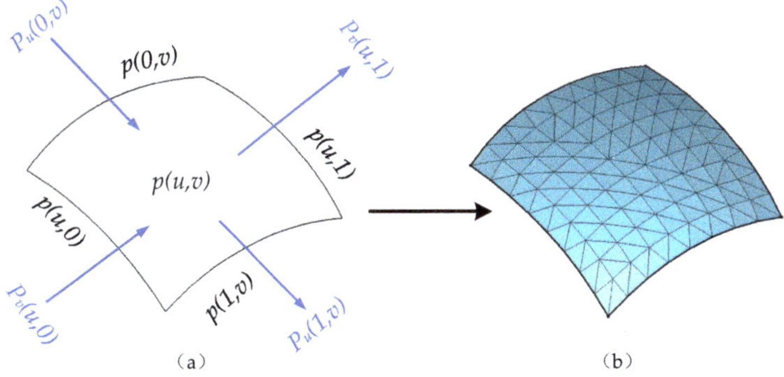

Figure 2. Schematic diagram of the interpolation of the second kind Coons surface: (**a**) boundaries and their cross-border tangent vectors of the surface patch; (**b**) the interpolated surface.

Under these constraints, the second kind of Coons surface equation [10] can be derived:

$$Q(u,v) = -\begin{bmatrix} -1 & F_0(u) & F_1(u) & G_0(u) & G_1(u) \end{bmatrix} \times \begin{bmatrix} 0 & P(u,0) & P(u,1) & P_v(u,0) & P_v(u,1) \\ P(0,v) & P(0,0) & P(0,1) & P_v(0,0) & P_v(0,1) \\ P(1,v) & P(1,0) & P(1,1) & P_v(1,0) & P_v(1,1) \\ P_u(0,v) & P_u(0,0) & P_u(0,1) & P_{uv}(0,0) & P_{uv}(0,1) \\ P_u(1,v) & P_u(1,0) & P_u(1,1) & P_{uv}(1,0) & P_{uv}(1,1) \end{bmatrix} \begin{bmatrix} -1 \\ F_0(v) \\ F_1(v) \\ G_0(v) \\ G_1(v) \end{bmatrix}, u,v \in [0,1] \qquad (1)$$

This fifth-order matrix is the boundary information matrix of the second kind of Coons patches (Coons patches with given boundaries and cross-border tangent vectors). F_0, F_1, G_0 and G_1 are mixed functions, which meet the following conditions [10]:

$$F_i(j) = G_i'(j) = \begin{cases} 1, i = j \\ 0, i \neq j \end{cases}$$
$$F_i'(j) = G_i(j) = 0, i,j = 0,1 \qquad (2)$$

Based on Equation (2), we can solve that [10]:

$$\begin{cases} F_0(u) = 2u^3 - 3u^2 + 1 \\ F_1(u) = -2u^3 + 3u^2 \\ G_0(u) = u^3 - 2u^2 + u \\ G_1(u) = u^3 - u^2 \end{cases} \qquad (3)$$

The above methods and formulas are extremely rigorous in theory, but too much boundary information is needed, which is not convenient for calculation and application. To meet its application requirements and simplify the calculation process as much as possible, Professor Coons proposed using corner information and mixed functions to define the boundary curves and their cross-border tangent vectors.

If the point vectors, tangent vectors of u direction, tangent vectors of v direction, and mixed partial derivative vectors of the four corners are known, as shown in Figure 3, by applying the mixed function F_0, F_1, G_0 and G_1, we can express the four boundaries of the surface patch and their cross-border tangent vectors as follows [10].

$$\begin{cases} P(0,v) = F_0(v)P(0,0) + F_1(v)P(0,1) + G_0(v)P_v(0,0) + G_1(v)P_v(0,1) \\ P(1,v) = F_0(v)P(1,0) + F_1(v)P(1,1) + G_0(v)P_v(1,0) + G_1(v)P_v(1,1) \\ P(u,0) = F_0(u)P(0,0) + F_1(u)P(0,1) + G_0(u)P_u(0,0) + G_1(u)P_u(0,1) \\ P(u,1) = F_0(u)P(1,0) + F_1(u)P(1,1) + G_0(u)P_u(1,0) + G_1(u)P_u(1,1) \\ P_u(u,0) = F_0(u)P_u(0,0) + F_1(u)P_u(0,1) + G_0(u)P_{uv}(0,0) + G_1(u)P_{uv}(0,1) \\ P_u(u,1) = F_0(u)P_u(1,0) + F_1(u)P_u(1,1) + G_0(u)P_{uv}(1,0) + G_1(u)P_{uv}(1,1) \\ P_v(u,0) = F_0(u)P_v(0,0) + F_1(u)P_v(0,1) + G_0(u)P_{uv}(0,0) + G_1(u)P_{uv}(0,1) \\ P_v(u,1) = F_0(u)P_v(1,0) + F_1(u)P_v(1,1) + G_0(u)P_{uv}(1,0) + G_1(u)P_{uv}(1,1) \end{cases} \qquad (4)$$

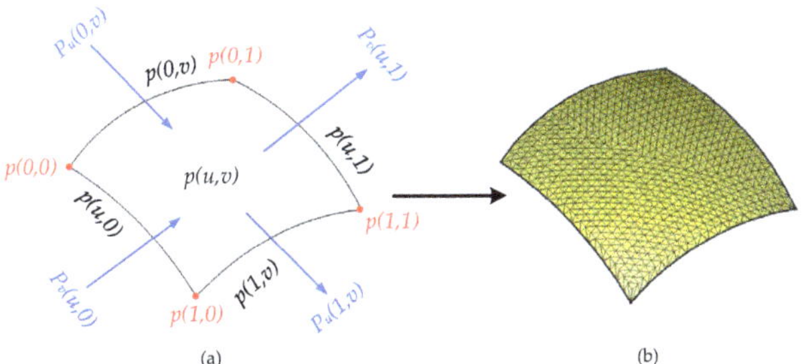

Figure 3. Schematic diagram of the interpolation of the bicubic Coons surface: (**a**) corner information, boundaries, and their cross-border tangent vectors of the surface patch; (**b**) the interpolated surface.

After bringing Equation (2) into Equation (1) and simplifying it, we can obtain the simplified surface equation [10]:

$$\left\{ Q(u,v) = [F_0(u)\,F_1(u)\,G_0(u)\,G_1(u)] \times \begin{bmatrix} P(0,0) & P(0,1) & P_v(0,0) & P_v(0,1) \\ P(1,0) & P(1,1) & P_v(1,0) & P_v(1,1) \\ P_u(0,0) & P_u(0,1) & P_{uv}(0,0) & P_{uv}(0,1) \\ P_u(1,0) & P_u(1,1) & P_{uv}(1,0) & P_{uv}(1,1) \end{bmatrix} \begin{bmatrix} F_0(v) \\ F_1(v) \\ G_0(v) \\ G_1(v) \end{bmatrix}, u,v \in [0,1] \right. \tag{5}$$

where F_0, F_1, G_0 and G_1 are the same as Equation (3).

The cubic mixed function and Equation (3) can be written as [10]:

$$Q(u,v) = UMBM^T V^T \tag{6}$$

where:

$$U = \begin{bmatrix} u^3 & u^2 & u & 1 \end{bmatrix},\ V = \begin{bmatrix} v^3 & v^2 & v & 1 \end{bmatrix}, u,v \in [0,1] \tag{7}$$

$$M = \begin{bmatrix} 2 & -2 & 1 & 1 \\ -3 & 3 & -2 & -1 \\ 0 & 0 & 1 & 0 \\ 1 & 0 & 0 & 0 \end{bmatrix} \tag{8}$$

$$B = \begin{bmatrix} P(0,0) & P(0,1) & P_v(0,0) & P_v(0,1) \\ P(1,0) & P(1,1) & P_v(1,0) & P_v(1,1) \\ P_u(0,0) & P_u(0,1) & P_{uv}(0,0) & P_{uv}(0,1) \\ P_u(1,0) & P_u(1,1) & P_{uv}(1,0) & P_{uv}(1,1) \end{bmatrix} \tag{9}$$

The parametric equation of the Coons surface is [10]:

$$\begin{cases} x(u,v) = UMB_x M^T V^T \\ y(u,v) = UMB_y M^T V^T \\ z(u,v) = UMB_z M^T V^T \end{cases} \tag{10}$$

where B_x, B_y, and B_z are the coordinate components of boundary information matrix B.

3.2. Closed-Loop Division

Cut all contour polylines using the approximate planes where each contour polyline is located. The contour polyline on the positive side (the positive and negative sides can be specified arbitrarily without affecting the final dividing result) is coded as 1 in this plane bit. The contour polyline on the negative side is coded as 2 in this plane bit. The contour polyline on the approximate plane is coded as 0 because it belongs to both positive and negative sides, and it will be expanded in the subsequent steps. Each time an approximate plane is added for cutting, the length of each polyline mark string is increased by 1, whose value is 0, 1, or 2.

As shown in Figure 4, there are six contour polylines in the figure, of which the bottom two are coplanar. Cut the contour polylines with the five approximate planes $\alpha, \beta, \gamma, \delta, \varepsilon$ where the contour polylines are located and divide them into four subspaces ①, ②, ③, ④. The direction indicated by the arrow is the positive direction of the plane, and the tag array length of each contour polyline is 5.

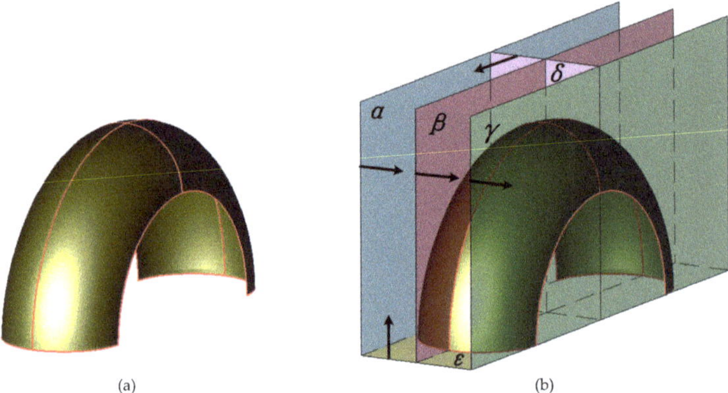

(a) (b)

Figure 4. Schematic diagram of the closed-loop division: (**a**) the example model and the cross-contour polylines of it; (**b**) the formed subspace after using approximate planes to cut polylines.

As shown in Figures 4 and 5, the contour polylines *a*, *b*, *c*, *d* are cut by the approximate planes and coded according to the above rules. The initial coding result is shown in Table 1.

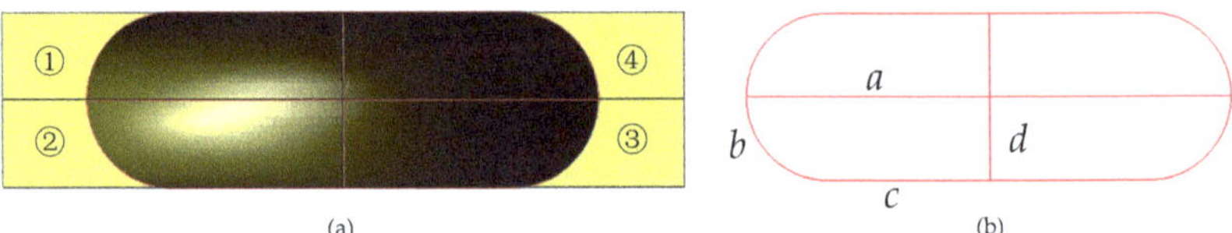

(a) (b)

Figure 5. The top view of formed subspaces: (**a**) the top view of the complete subspaces; (**b**) the top view of the divided contour polylines.

Table 1. Initial coding table after cutting polylines with the approximate planes.

Segments	Plane Bit α	β	γ	δ	ε
a	1	0	2	1	1
b	1	1	2	1	0
c	1	1	0	1	1
d	1	1	2	0	1

Copy the contour polyline containing 0 in the code array. On the plane bit where 0 is located, two contour polylines are coded as 1 and 2, and the other codes remain unchanged. For example, expand *a*: 10,211 to a_1: 11,211 and a_2: 12,211. After processing all the contour polylines, the final coding result is obtained as shown in Table 2.

The contour polylines with the same code array are the contour polylines in the same subspace. As shown in Table 2, a_1, b_1, c_2, d_1 marked by red have the same code arrays, which means that they are in the same subspace. Then, in the same subspace, the contour polylines within the specified tolerance are connected to form the closed loops.

Table 2. Final coding table after extending the side whose code array contains 0.

Segments \ Plane Bit	α	β	γ	δ	ε
a_1	1	1	2	1	1
a_2	1	2	2	1	1
b_1	1	1	2	1	1
b_2	1	1	2	1	2
c_1	1	1	1	1	1
c_2	1	1	2	1	1
d_1	1	1	2	1	1
d_2	1	1	2	2	1

It is worth noting that if the topological adjacency relationship between some contour polylines is too complex, the above automatic closed-loop division algorithm may not be able to obtain the required results. At this time, it is necessary to manually process and group the contour polylines to complete the closed-loop division.

3.3. Process n-Sided Regions

Using Coons surface for interpolation calculation needs the boundary information of four sides. However, many of the closed loops divided before are non-quadrilateral. Therefore, we need to preprocess the n-sided regions before interpolation calculation.

3.3.1. Process the Single-Sided Region

For the simple single-sided region, the polygon triangulation method based on constrained Delaunay triangulation is used to directly model it, as shown in Figure 6. For the complex single-sided region, we consider transforming it into a four-sided region by analyzing its shape characteristics and then performing the Coons surface interpolation.

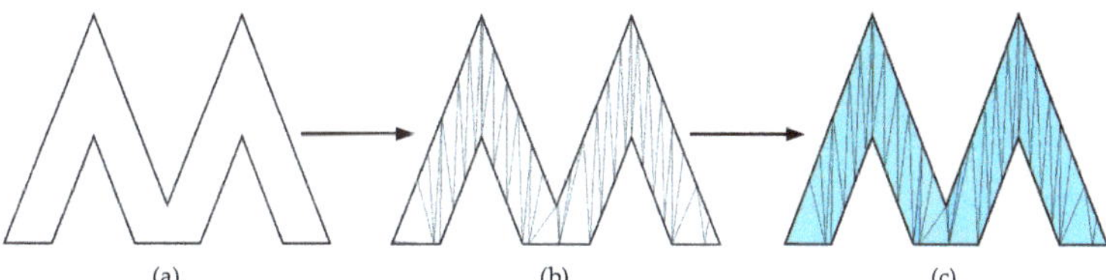

Figure 6. Schematic diagram of polygon triangulation: (**a**) the original polygon; (**b,c**) the process and the result of polygon triangulation.

3.3.2. Process the Two-Sided and the Three-Sided Regions

For the two-sided and the three-sided regions, we consider generating the virtual sides to transform them into four-sided regions.

As shown in Figure 7, for the two-sided region, two virtual edges are generated at two common intersections A and B on both sides. Then, n transition lines are generated between A and B, which means n virtual points are inserted on the two virtual sides. When performing interpolation calculation, the coordinate values of n points on the same virtual side are the same. The coordinates of $A_i, i = 1, 2, \ldots, n$ are the same as A. The coordinates of $B_i, i = 1, 2, \ldots, n$ are the same as B.

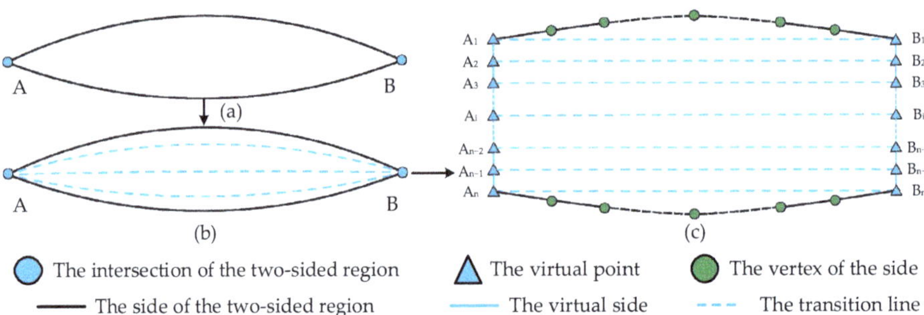

Figure 7. Schematic diagram of the two-sided region processing: (**a**) the original two-sided region; (**b,c**) the process and result of adding virtual sides and virtual points. The black dotted lines are omitted lines.

For the three-sided region, a virtual side is constructed at the intersection of any two sides. The transition lines are generated, whose number is the same as the number of vertices on the third side. Then, the same number of virtual points are inserted on the virtual side, and the coordinates of the virtual points are the same as that of the intersection where the virtual side is inserted.

3.3.3. Process Other Regions

For the five-sided region and regions that have more than five sides, first, search for the shortest side among all sides, and then search for the shorter side adjacent to this side for merging and connection. Repeat the above operation until this region is merged into a four-sided region.

3.3.4. Process All the Four-Sided Regions

After the above preprocessing, all the regions are transformed into four-sided regions. Since the synchronous forward method is used for interpolation, the number of vertices on the opposite side is required to be the same. However, the vertices cannot be directly added to the sides because of the requirement of forming manifold graphics. In this paper, we consider translating the sides of the four-sided region to the middle according to the specified distance, which will generate four new sides. As shown in Figure 8b, the same numbers of vertices are inserted on the opposite sides, which makes the same t of the vertices at the same position on the opposite sides. The t is the ratio of the distance from the vertice to the starting point to the side length. At last, the polygon triangulation method is used to model the newly formed sides and old sides, as shown in Figure 8c.

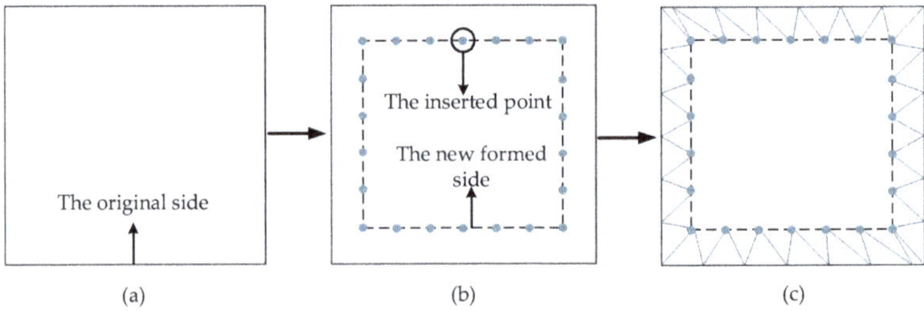

Figure 8. Schematic diagram of the side processing of the four-sided region: (**a**) the original four-sided region; (**b**) the formed sides and the points inserted on them; (**c**) the result of the polygon triangulation of the new and old sides.

3.4. Sided Region Modeling

After preprocessing, the four-sided regions can be modeled using bicubic Coons surface interpolation. Since the synchronous forward method is used to perform the Coons surface interpolation, the numbers of the vertices on the opposite sides are the same, which are m and n. This means that the number of the formed interpolation points in the grid will be $m \times n$. As shown in Figure 9, set the first group of opposite sides respectively as u_0 and u_1, the second group of opposite sides respectively as v_0 and v_1.

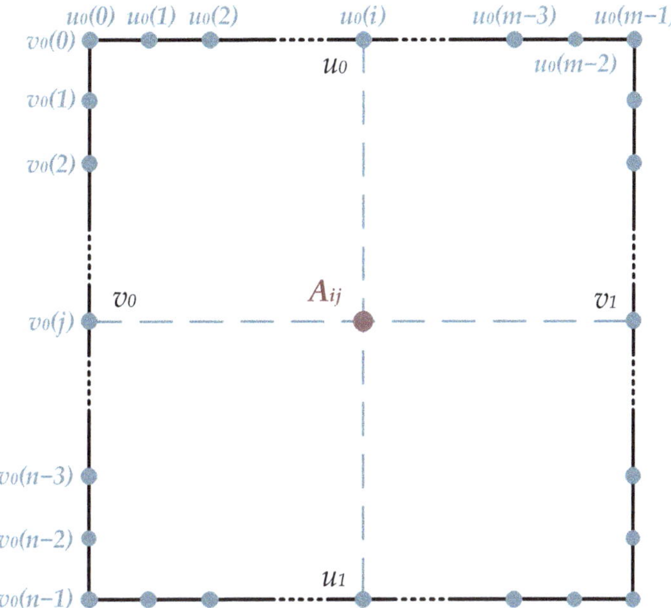

Figure 9. Schematic diagram of the Coons surface interpolation.

Based on Equation (9), the x, y, and z of each vertex can be calculated respectively by harmonic function, according to the coordinates of four corners and four relevant points, as well as the t of the four relevant points. Because the synchronous forward method is adopted, the t values of the opposite sides are the same.

Set the t of the first group sides u_0 and u_1 as $\{T_i | 0 \leq i < m,\ 0 \leq T_i \leq 1\}$, and the t of the second group sides v_0 and v_1 as $\{t_i | 0 \leq i < n,\ 0 \leq t_i \leq 1\}$. Then, set the point interpolated by the i point on the u_0 and the j point on the v_0 as A_{ij}, whose calculation formulas of coordinates are:

$$
\begin{aligned}
x = u_0(i)_x \times F_0(t_j) \quad &+ v_0(j)_x \times F_0(T_i) + u_1(i)_x \times F_1(t_j) + v_1(j)_x \times F_1(T_i) \\
&- (u_0(0)_x \times F_0(T_i) + u_0(m-1)_x \times F_1(T_i)) \times F_0(t_j) \\
&+ (u_1(0)_x \times F_0(T_i) + u_1(m-1)_x \times F_1(T_i)) \times F_1(t_j)
\end{aligned} \tag{11}
$$

$$
\begin{aligned}
y = u_0(i)_y \times F_0(t_j) \quad &+ v_0(j)_y \times F_0(T_i) + u_1(i)_y \times F_1(t_j) + v_1(j)_y \times F_1(T_i) \\
&- (u_0(0)_y \times F_0(T_i) + u_0(m-1)_y \times F_1(T_i)) \times F_0(t_j) \\
&+ (u_1(0)_y \times F_0(T_i) + u_1(m-1)_y \times F_1(T_i)) \times F_1(t_j)
\end{aligned} \tag{12}
$$

$$
\begin{aligned}
z = u_0(i)_z \times F_0(t_j) \quad &+ v_0(j)_z \times F_0(T_i) + u_1(i)_z \times F_1(t_j) + v_1(j)_z \times F_1(T_i) \\
&- (u_0(0)_z \times F_0(T_i) + u_0(m-1)_z \times F_1(T_i)) \times F_0(t_j) \\
&+ (u_1(0)_z \times F_0(T_i) + u_1(m-1)_z \times F_1(T_i)) \times F_1(t_j)
\end{aligned} \tag{13}
$$

where the harmonic function is:

$$\begin{cases} F_0(t) = 2t^3 - 3t^2 + 1 \\ F_1(t) = -2t^3 + 3t^2 \end{cases} \tag{14}$$

$u_0(0)_x, u_0(0)_y, u_0(0)_z$ refers to respectively x, y, z coordinates of the first point on the u_0; $u_0(m-1)_x, u_0(m-1)_y, u_0(m-1)_z$ refer to, respectively, x, y, z coordinates of the last point on the u_0; $u_1(0)_x, u_1(0)_y, u_1(0)_z$ refers to respectively x, y, z coordinates of the first point on the u_1; $u_1(m-1)_x, u_1(m-1)_y, u_1(m-1)_z$ refer to, respectively, x, y, z coordinates of the last point on the u_1; $u_0(i)_x, u_0(i)_y, u_0(i)_z$ refer to, respectively, x, y, z coordinates of the point on the u_0, whose t is T_i; $v_0(j)_x, v_0(j)_y, v_0(j)_z$ refer to, respectively, x, y, z coordinates of the point on the v_0, whose t is t_j; $u_1(i)_x, u_1(i)_y, u_1(i)_z$ refer to, respectively, x, y, z coordinates of the point on the u_1, whose t is T_i; $v_1(j)_x, v_1(j)_y, v_1(j)_z$ refer to, respectively, x, y, z coordinates of the point on the v_1, whose t is t_j.

Performing the interpolation calculation according to Equations (10)–(13), the modeling of this four-sided region can be completed, as shown in Figure 10.

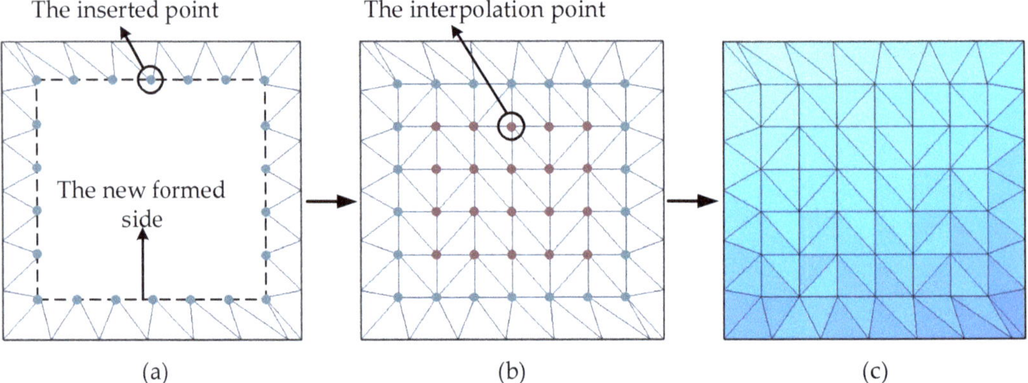

Figure 10. Schematic diagram of the four-sided region modeling:(**a**) the preprocessed four-sided region; (**b**,**c**) the process and the result of the Coons surface interpolation and modeling.

3.5. Combine Sub-Meshes

Multiple polygon data sets are set into one complete data set using the mature polygon data merging technology, which means all the sub-meshes are merged into a complete triangular mesh model. In the process of merging, all the polygon data will be extracted, but the extraction and expansion of point and cell attributes (scalar, vector, and normal) can only be carried out when multiple data sets contain them. As shown in Figure 11, we can obtain a complete triangular mesh model.

We use C++ to implement the relevant algorithms of the modeling method based on the Coons surface interpolation and test them on a 64-bit computer with Windows 10 professional edition. This computer has 3.70 Ghz Intel (R) Core (TM) i7-8700u, 32GB ram, and NVIDIA GeForce RTX 2080 GPU.

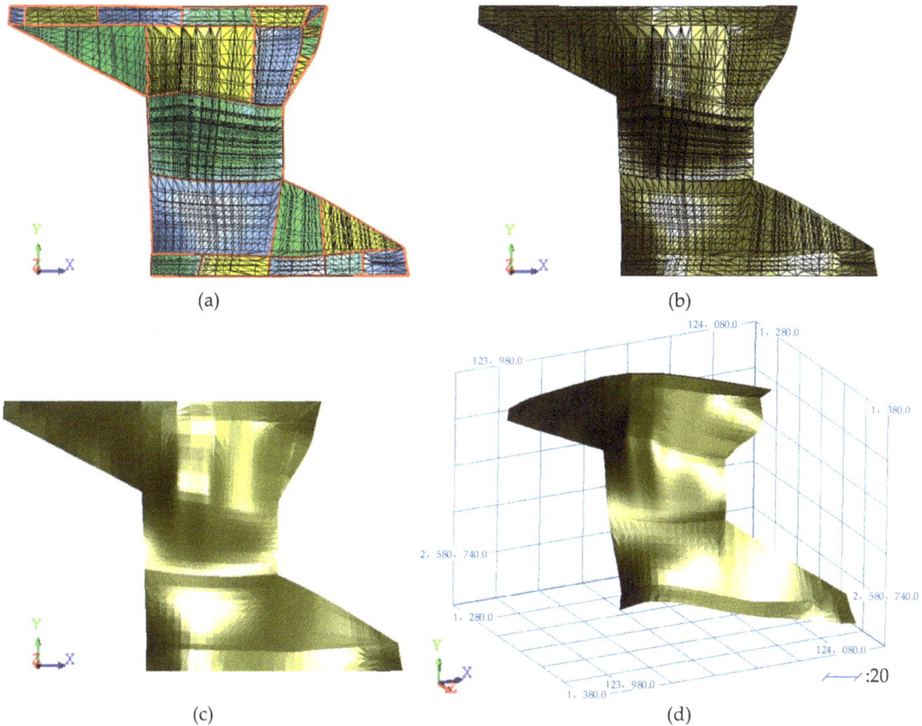

Figure 11. Schematic diagram of the sub-mesh combination: (**a**) the sub-meshes; (**b–d**) the complete model.

4. Results and Discussion

In this paper, five groups of different data are used to test the modeling method proposed in this work. These data are from the orebody of a tin mine in Yunnan, China, and are obtained in production and exploration. Based on these data, the original contour polylines can be produced through geological logging. The following experimental figures clearly show the process and the result of single closed-loop modeling and sub-mesh combination.

4.1. Examples

Figure 12 shows the processes of closed-loop division, single closed-loop modeling, and sub-mesh combination of the first group of cross-contour polylines. Figure 12a shows the original cross contour polylines of the orebody, each of which is in the same plane within the tolerance range. Then, the contour polylines are cut with the approximate planes. As shown in Figure 12b, each contour polyline is divided and grouped. After that, single closed-loop modeling is carried out on the contour polylines through the Coons surface interpolation, and the results are shown in Figure 12c. It can be seen that every closed loop has been modeled, but the sub-meshes are separate from each other. Finally, the sub-meshes are combined into a complete orebody model using polygon data merging technology, as shown in Figure 12d.

Figures 13 and 14 show the modeling process on contour polylines with different shapes. It can be seen that the algorithm implemented here can still carry out closed-loop division, single closed-loop modeling, and sub-mesh combination on more complex contour lines with good effects.

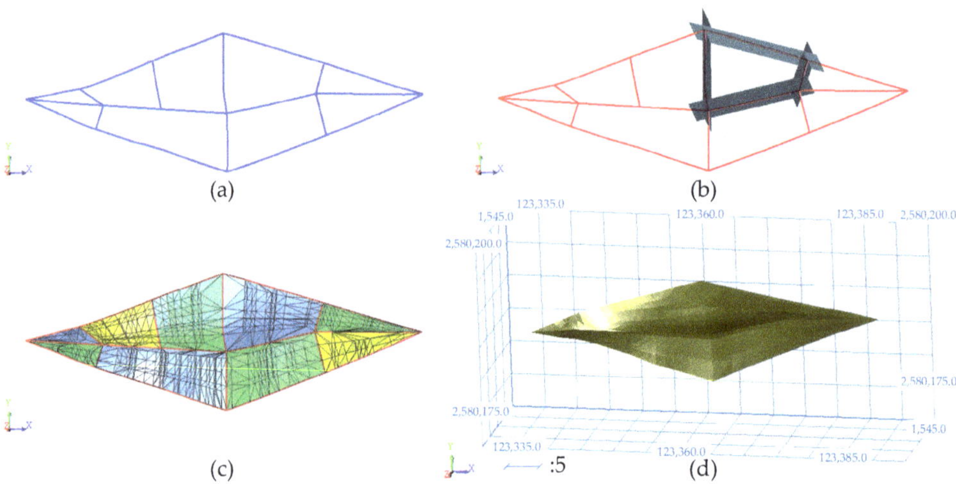

Figure 12. Experimental results of the first set of contour polylines: (**a**) the original contour polylines; (**b**) closed-loop division; (**c**) single closed-loop modeling; (**d**) sub-mesh combination.

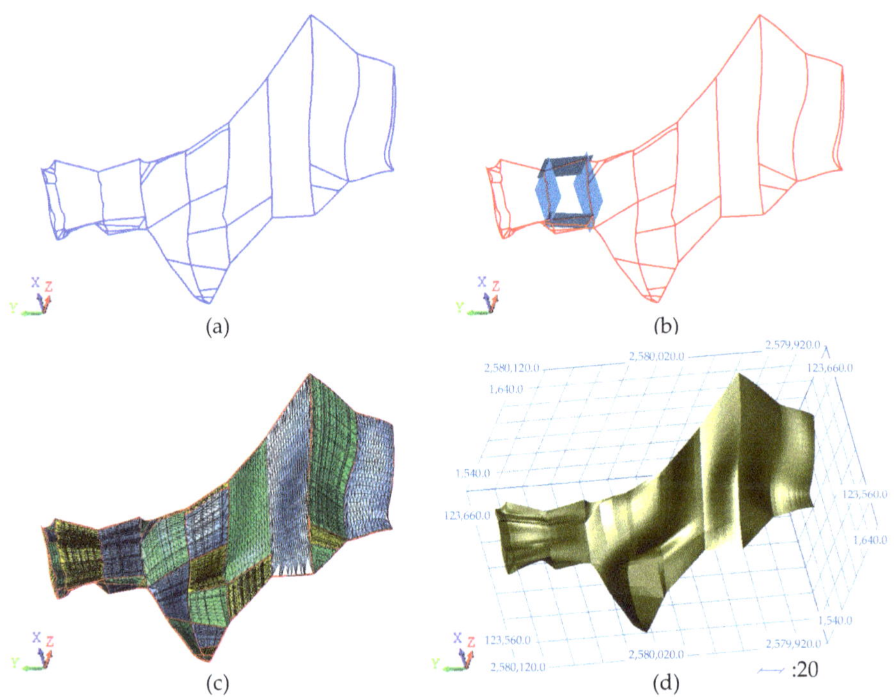

Figure 13. Experimental results of the second set of contour polylines: (**a**) the original contour polylines; (**b**) closed-loop division; (**c**) single closed-loop modeling; (**d**) sub-mesh combination.

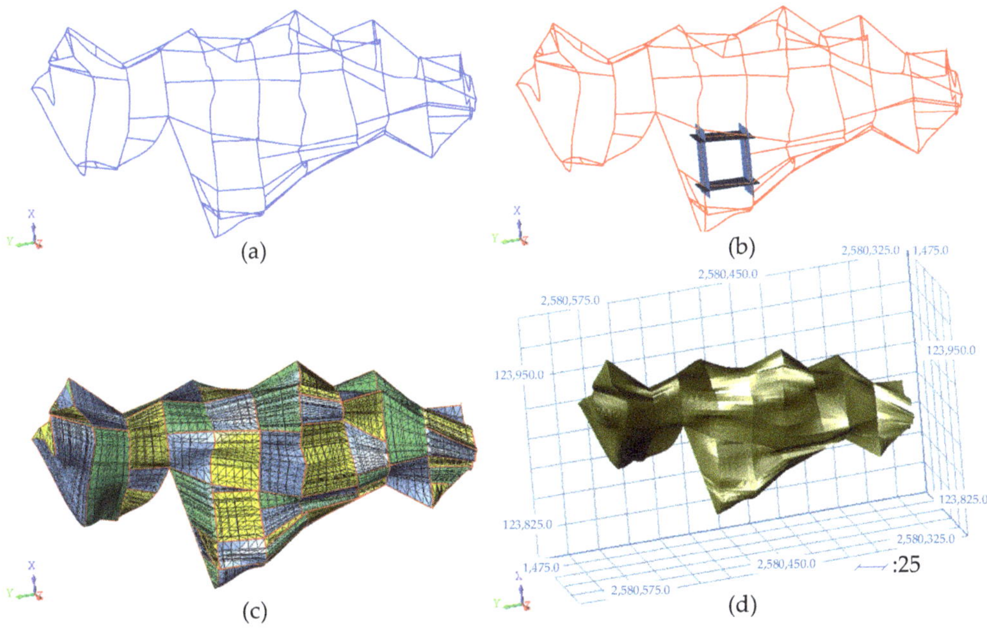

Figure 14. Experimental results of the third set of contour polylines: (**a**) the original contour polylines; (**b**) closed-loop division; (**c**) single closed-loop modeling; (**d**) sub-mesh combination.

Figure 15 shows the modeling process of more complex contour polylines, in which most of the contour polylines do not strictly intersect. Figure 15a shows the original contour polylines. First, search the original intersections of the contour polylines, and the result is shown in Figure 15b. The green points are the original intersections that need no further processing. It can be seen that only a few polylines possess the original intersections. Red points are isolated points that have no other contour polylines within the specified scope. Then, the contour polylines are moved to intersect within the tolerance range. Figure 15c shows the intersections after processing, and most of the contour polylines have intersected with each other. After that, the closed-loop division is carried out based on the intersected contour polylines, and the result is shown in Figure 15d. Based on the closed loops, sub-meshes can be constructed through the Coons surface interpolation. Figure 15e shows the modeling result of every closed loop. The sub-meshes are smooth but split from each other. Finally, the sub-meshes are combined to obtain a complete orebody model, as shown in Figure 15f.

As shown in Figure 16, the method is applied to complex orebodies with different shapes, and good orebody models are obtained, showing that this method has strong adaptability.

4.2. Discussion

The above experimental results show that the proposed method can deal with the contour polylines obtained by geological logging in mine production with good effects. Firstly, the constructed model conforms to the geological trend and is smooth in every mesh. Secondly, the closed-loop division can be carried out automatically, which will greatly improve the efficiency of modeling. Thirdly, since modeling between different regions will not affect each other, it can be carried out at the same time, which will greatly reduce the overall modeling time. However, this method still has some limitations that need to be further studied and solved. Firstly, due to the inherent limitations of the Coons technology, the model constructed by Coons surface interpolation is difficult to update, and it is difficult to take faulting crosscutting into account, which will affect the accuracy of the model. In mine production, many orebodies are cross-cut by faults. For these orebodies, we should choose other modeling methods. Secondly, due to the complexity of the data in actual production, when this method is applied to some complex contour polylines, as shown in Figure 15, it can not complete all closed-loop divisions well, which needs to be specified manually. Finally, before Coons surface interpolation, all regions except simple single-sided regions need to be transformed into four-sided regions, which may affect the final modeling results by introducing numerical instability issues when adding virtual edges and vertexes. Therefore, to further improve the efficiency and accuracy of the modeling of cross-contour polylines, it is necessary to develop a closed-loop division method with higher precision and an n-sided regions expansion method with less effect on interpolation modeling in the future.

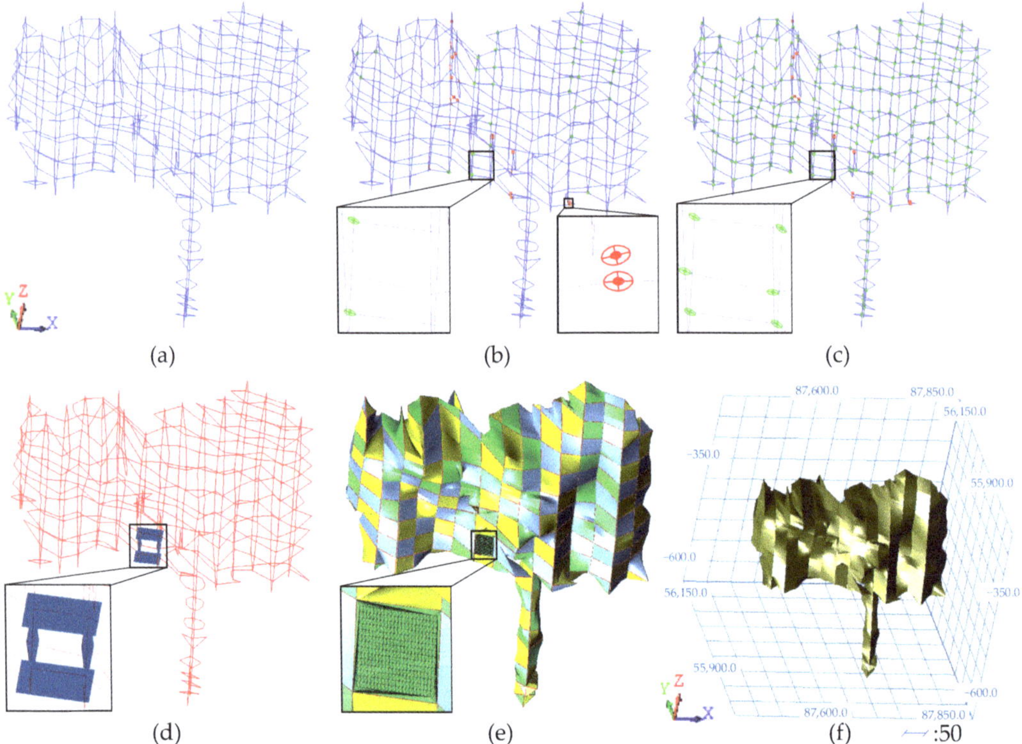

Figure 15. Experimental results of the fourth set of contour polylines: (**a**) the original contour polylines; (**b**) original intersections; (**c**) processed intersections; (**d**) closed-loop division; (**e**) single closed-loop modeling; (**f**) sub-mesh combination.

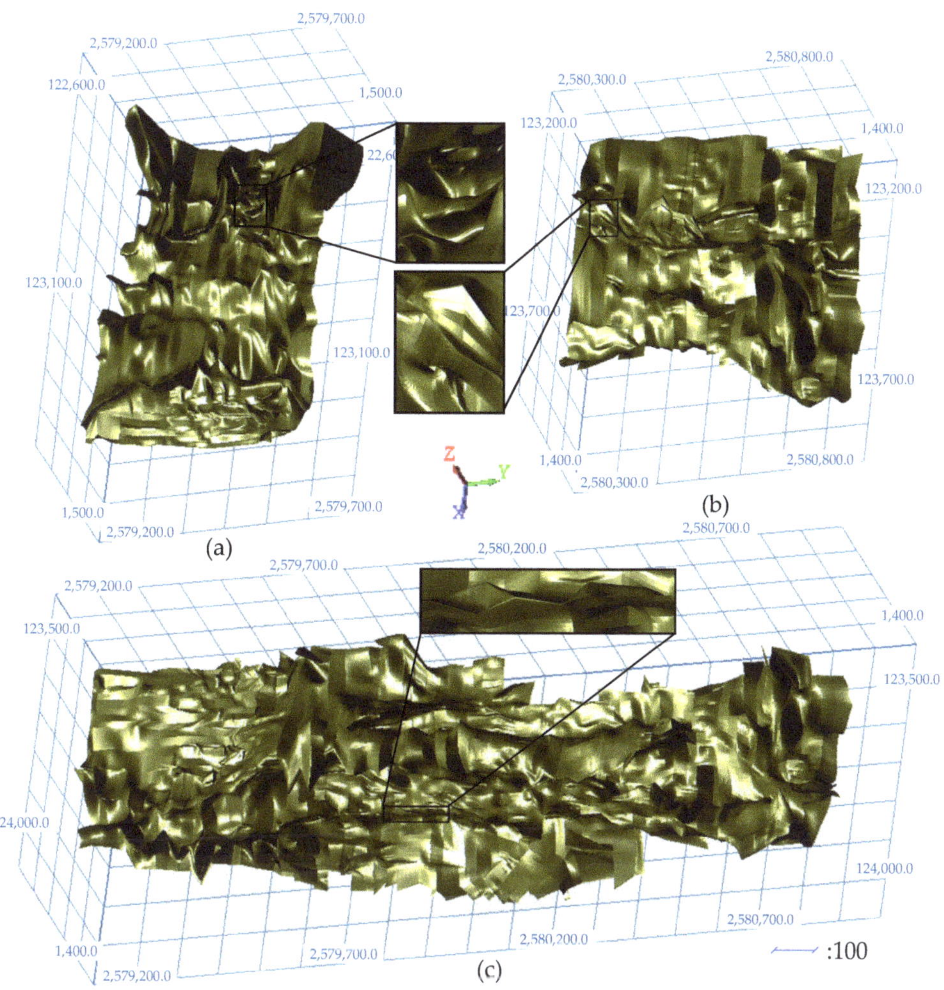

Figure 16. Experimental results of the fifth set of contour polylines: (**a–c**) the orebody models with different complex shapes.

5. Conclusions

We propose a modeling method based on bicubic Coons surface interpolation, which needs the contour polyline to be in an approximate plane within the tolerance range. The innovation of this paper is that we proposed a creative method of automatically dividing the closed loops, which will greatly reduce the modeling time. Additionally, the constructed sub-meshes can be combined without manual intervention. This method solves the problem that it is difficult to divide closed loops based on contour polylines with complex shapes, and it greatly improves the efficiency of modeling based on complex cross-contour polylines. Firstly, the contour polylines are cut by the approximate planes to divide the subspaces. Then, the divided polylines are grouped and connected according to the specified tolerance to form closed loops. Secondly, the Coons surface interpolation is used for modeling based on the closed-loop information. Finally, the constructed sub-meshes are combined to form a complete orebody model. Through experiments with multiple groups of cross-contour polylines, the results show that the proposed method can accurately divide closed loops and model the contour polylines with good effects.

Author Contributions: Conceptualization, Z.W., L.B., D.Z. and Q.T.; methodology, Z.W., L.B., D.Z., J.Z. and Q.T.; software, Z.W., L.B., D.Z., J.Z. and Q.T.; formal analysis, Z.W., L.B., and D.Z.; data curation, Z.W., and D.Z.; writing—original draft preparation, Z.W.; writing—review and editing, Z.W., L.B. and D.Z.; visualization, Z.W., L.B., D.Z., and J.Z.; project administration, L.B., and M.J.; funding acquisition, D.Z. and M.J. All authors have read and agreed to the published version of the manuscript.

Funding: This research was funded by the National Natural Science Foundation of China (52104171), the National Key R&D Program of China (2019YFC0605304), and the China Postdoctoral Science Foundation (2022T150740).

Acknowledgments: We thank the reviewers for their comments and suggestions to improve the quality of the paper.

Conflicts of Interest: The authors declare no conflict of interest.

References

1. Wang, B.; Shi, B.; Song, Z. A simple approach to 3D geological modelling and visualization. *Bull. Eng. Geol. Environ.* **2009**, *68*, 559–565.
2. Jin, B.-X.; Fang, Y.-M.; Song, W.-W. 3D visualization model and key techniques for digital mine. *Trans. Nonferrous Met. Soc. China* **2011**, *21*, s748–s752. [CrossRef]
3. Mallet, J.-L. *Geomodeling*; Oxford University Press: New York, NY, USA, 2002; p. 624.
4. Jessell, M. Three-dimensional geological modelling of potential-field data. *Comput. Geosci.* **2001**, *27*, 455–465. [CrossRef]
5. Feltrin, L.; Mclellan, J.G.; Oliver, N. Modelling the giant, Zn–Pb–Ag Century deposit, Queensland, Australia. *Comput. Geosci.* **2009**, *35*, 108–133. [CrossRef]
6. Vollgger, S.A.; Cruden, A.R.; Ailleres, L.; Cowan, E.J. Regional dome evolution and its control on ore-grade distribution: Insights from 3D implicit modelling of the Navachab gold deposit, Namibia. *Ore Geol. Rev.* **2015**, *69*, 268–284. [CrossRef]
7. Tungyshbayeva, Z.; Royer, J.J.; Zhautikov, T.M. 3D modeling and resources estimation of a gold deposit, Zhungarie province, Kazakhstan. In Proceedings of the 13th SGA Biennial Meeting on Mineral Resources in a Sustainable World, Nancy, France, 24–27 August 2015; pp. 1759–1761.
8. Leapfrog. 2013. Available online: http://www.leapfrog3d.com (accessed on 1 February 2022).
9. Royer, J.J.; Mejia, P.; Caumon, G.; Collon-Drouaillet, P. 3&4D geomodeling applied to mineral resources exploration—A new tool for targeting deposits. In Proceedings of the 12th SGA Biennial Meeting, Uppsala, Sweden, 12–15 August 2013.
10. Coons, S.A. *Surfaces for Computer-Aided Design of Space Forms*; Technical Report MAC-TR-41; Massachusetts Institute of Technology: Cambridge, MA, USA, 1967.
11. Zhong, D.-Y.; Wang, L.-G.; Bi, L.; Jia, M.-T. Implicit modeling of complex orebody with constraints of geological rules. *Trans. Nonferrous Met. Soc. China* **2019**, *29*, 2392–2399. [CrossRef]
12. Hodgkinson, J.H.; Elmouttie, M. Cousins, siblings and twins: A review of the geological model's place in the digital mine. *Resources* **2020**, *9*, 24. [CrossRef]
13. Kong, T.; Zhang, Y.; Fu, X. The model of feature extraction for free-form surface based on topological transformation. *Appl. Math Model* **2018**, *64*, 386–397. [CrossRef]
14. Yamany, S.; Farag, A. Surface signatures: An orientation independent free-form surface representation scheme for the purpose of objects registration and matching. *IEEE Trans. Pattern Anal. Mach. Intell.* **2002**, *24*, 1105–1120. [CrossRef]
15. Saini, D.; Kumar, S. Free-form surface reconstruction from arbitrary perspective images. In Proceedings of the IEEE International Advance Computing Conference, ITM, Gurgaon, India, 21–22 February 2014; pp. 1054–1059. [CrossRef]
16. Turk, G.; O'Brien, J.F. Modelling with implicit surfaces that interpolate. *ACM Trans. Graph.* **2002**, *21*, 855–873. [CrossRef]
17. Yuan, Z.; Yu, Y.; Wang, W. Object-space multiphase implicit functions. *ACM Trans. Graph.* **2012**, *31*, 1–10. [CrossRef]
18. Barthe, L.; Dodgson, N.A.; Sabin, M.A.; Wyvill, B.; Gaildrat, V. Two-dimensional Potential Fields for Advanced Implicit Modeling Operators. *Comput. Graph. Forum* **2003**, *22*, 23–33. [CrossRef]
19. Farin, G. Smooth interpolation to scattered 3D data. In *Surfaces in Computer Aided Geometric Design (Oberwolfach, 1982)*; Barnhill, R.E., Boehm, W., Eds.; North-Holland: Amsterdam, The Netherlands, 1983; pp. 43–63.
20. Cheng, F.H.; Wang, X.F.; Barsky, B.A. Quadratic B-spline curve interpolation. *Comput. Math. Appl.* **2001**, *41*, 39–50. [CrossRef]
21. Hu, S.-M.; Tai, C.-L.; Zhang, S.-H. An extension algorithm for B-splines by curve unclamping. *Comput. Des.* **2002**, *34*, 415–419. [CrossRef]
22. Park, H. B-spline surface fitting based on adaptive knot placement using dominant columns. *Comput. Des.* **2011**, *43*, 258–264. [CrossRef]

23. Krishnamurthy, A.; Khardekar, R.; McMains, S.; Haller, K.; Elber, G. Performing efficient NURBS modeling operations on the GPU. *IEEE Trans. Vis. Comput. Graph.* **2009**, *15*, 530–543. [CrossRef]
24. Selimovic, I. Improved algorithms for the projection of points on NURBS curves and surfaces. *Comput. Aided Geom. Des.* **2006**, *23*, 439–445. [CrossRef]
25. Wang, Q.; Hua, W.; Li, G.Q.; Bao, H.J. Generalized NURBS curves and surfaces. In Proceedings of the International Conference on Geometric Modeling and Processing, Beijing, China, 13–15 April 2004; pp. 365–368.
26. Randrianarivony, M. On global continuity of Coons mappings in patching CAD surfaces. *Comput. Des.* **2009**, *41*, 782–791. [CrossRef]
27. Farin, G.; Hansford, D. Discrete coons patches. *Comput. Aided Geom. Des.* **1999**, *16*, 691–700. [CrossRef]
28. Hugentobler, M.; Schneider, B. Breaklines in Coons surfaces over triangles for the use in terrain modelling. *Comput. Geosci.* **2005**, *31*, 45–54. [CrossRef]
29. Sapidis, N.S.; Besl, P.J. Direct construction of polynomial surfaces from dense range images through region growing. *ACM Trans. Graph.* **1995**, *14*, 171–200. [CrossRef]
30. Hoppe, H.; Derose, T.; Duchamp, T.; McDonald, J.; Stuetzle, W. Surface reconstruction from unorganized points. *ACM SIGGRAPH Comput. Graph.* **1992**, *26*, 71–78. [CrossRef]
31. Edelsbrunner, H.; Mücke, E. Three-dimensional alpha shapes. *ACM Trans. Graph.* **1994**, *13*, 75–82. [CrossRef]
32. Fuchs, H.; Kedem, Z.M.; Uselton, S.P. Optimal surface reconstruction from planar contours. *Commun. ACM* **1977**, *20*, 693–702. [CrossRef]
33. Keppel, E. Approximating complex surfaces by triangulation of contour lines. *IBM J. Res. Dev.* **1975**, *19*, 2–11. [CrossRef]
34. Floater, M.S.; Reimers, M. Meshless parameterization and surface reconstruction. *Comput. Aided Geom. Des.* **2001**, *18*, 77–92. [CrossRef]
35. Macedonio, G.; Pareschi, M. An algorithm for the triangulation of arbitrarily distributed points: Applications to volume estimate and terrain fitting. *Comput. Geosci.* **1991**, *17*, 859–874. [CrossRef]
36. Meyers, D.; Skinner, S.; Sloan, K. Surfaces from contours. *ACM Trans. Graph. (TOG)* **1992**, *11*, 228–258. [CrossRef]
37. Ekoule, A.B.; Peyrin, F.; Odet, C.L. A triangulation algorithm from arbitrary shaped multiple planar contours. *ACM Trans. Graph.* **1991**, *10*, 182–199. [CrossRef]
38. Jones, M.W.; Min, C. A new approach to the construction of surfaces from contour data. *Comput. Graph. Forum* **2010**, *13*, 75–84. [CrossRef]
39. Zhong, D.-Y.; Wang, L.-G.; Jia, M.-T.; Bi, L.; Zhang, J. Orebody modeling from non-parallel cross sections with geometry constraints. *Minerals* **2019**, *9*, 229. [CrossRef]
40. Wu, Z.; Zhong, D.; Li, Z.; Wang, L.; Bi, L. Orebody modeling method based on the normal estimation of cross-contour polylines. *Mathematics* **2022**, *10*, 473. [CrossRef]
41. Song, R.; Nan, J. The design and implementation of a 3D orebody wire-frame modeling prototype system. In Proceedings of the Second International Conference on Information and Computing Science, Manchester, UK, 21–22 May 2009; Volume 2, pp. 357–360. [CrossRef]
42. Jessell, M.; Ailleres, L.; De Kemp, E.; Lindsay, M.; Wellmann, F.; Hillier, M.; Laurent, G.; Carmichael, T.; Martin, R. Next generation three-dimensional geologic modeling and inversion. In *Building Exploration Capability for the 21st Century*; Kelley, K.D., Golden, H.C., Eds.; Soc Economic Geologists, Inc.: Littleton, CO, USA, 2014; pp. 261–272.
43. Knight, R.H. Orebody solid modelling accuracy—A comparison of explicit and implicit modelling techniques using a practical example from the Hope Bay District, Nunavut, Canada. In Proceedings of the 6th International Mining Geology Conference, Darwin, Australia, 21–23 August 2006; pp. 175–183.
44. Lipuš, B.; Guid, N. A new implicit blending technique for volumetric modelling. *Vis. Comput.* **2005**, *21*, 83–91. [CrossRef]

Article

Three-Dimensional Mineral Prospectivity Modeling for Delineation of Deep-Seated Skarn-Type Mineralization in Xuancheng–Magushan Area, China

Fandong Meng [1,2], Xiaohui Li [1,2,*], Yuheng Chen [1,2], Rui Ye [1,2] and Feng Yuan [1,2]

[1] Ore Deposit and Exploration Centre (ODEC), School of Resources and Environmental Engineering, Hefei University of Technology, Hefei 230009, China

[2] Anhui Province Engineering Research Center for Mineral Resources and Mine Environments, Hefei 230009, China

* Correspondence: lxhlixiaohui@163.com

Abstract: The Middle–Lower Yangtze River Metallogenic Belt is an important copper and iron polymetallic metallogenic belt in China. Today's economic development is inseparable from the support of metal mineral resources. With the continuous exploitation of shallow and easily identifiable mines in China, the prospecting work of deep and hidden mines is very important. Mineral prospectivity modeling (MPM) is an important means to improve the efficiency of mineral exploration. With the increase in resource demands and exploration difficulty, the traditional 2DMPM is often difficult to use to reflect the information of deep mineral deposits. More large-scale deposits are needed to carry out 3DMPM research. With the rise of artificial intelligence, the combination of machine learning and geological big data has become a hot issue in the field of 3DMPM. In this paper, a case study of 3DMPM is carried out based on the Xuancheng–Magushan area's actual data. Two machine learning methods, the random forest and the logistic regression, are selected for comparison. The results show that the 3DMPM based on random forest method performs better than the logistic regression method. It can better characterize the corresponding relationship between the geological structure combination and the metallogenic distribution, and the accuracy in the test set reaches 96.63%. This means that the random forest model could provide more effective and accurate support for integrating predictive data during 3DMPM. Finally, five prospecting targets with good metallogenic potential are delineated in the deep area of the Xuancheng–Magushan area for future exploration.

Keywords: 3D mineral prospectivity modeling; random forest; logistic regression; Xuancheng–Magushan area

Citation: Meng, F.; Li, X.; Chen, Y.; Ye, R.; Yuan, F. Three-Dimensional Mineral Prospectivity Modeling for Delineation of Deep-Seated Skarn-Type Mineralization in Xuancheng–Magushan Area, China. *Minerals* **2022**, *12*, 1174. https://doi.org/10.3390/min12091174

Academic Editor: Behnam Sadeghi

Received: 31 July 2022
Accepted: 11 September 2022
Published: 18 September 2022

Publisher's Note: MDPI stays neutral with regard to jurisdictional claims in published maps and institutional affiliations.

1. Introduction

A huge prospecting potential is in concealed and deep mines, especially in the so-called "Second depth space" (500 m–2000 m); there are very likely to be abundant mineral resources there [1,2]. At present, some domestic and foreign examples of deep mineral exploration have proved the views of experts and scholars [3,4]. Although a series of deep mine exploration results show the prospecting potential of concealed and deep mines, there are also many problems, such as difficulty in exploration and imperfect exploration methods [5–7]. Therefore, more reasonable and effective technology is needed at this stage to adapt to the prospecting work in the large-scale Quaternary strata coverage area and the lower-cost method to find hidden and deep mines.

In recent years, with the development of computer technology and the support of geophysical methods, 3D modeling technology can fully integrate multivariate and multidimensional data to accurately depict deep geological structures [8–11]. At present, the wide application of artificial intelligence, especially machine learning technology, can provide a new way to process massive geological big data. Compared with traditional methods, machine learning often has higher prediction accuracy, especially for geological data with massive and high-dimensional characteristics, which can effectively explore the complex nonlinear relationship between ore-control characteristics and ore-forming mechanisms. At present, machine learning methods include the probabilistic neural network, the support vector machine, the random forest, adaptive learning, the restricted Boltzmann machine, etc.; most of them have been applied and developed in the field of 2DMPM. Oh et al. [12] analyzed the potential of hydrothermal gold–silver mineral deposits in the Taebaeksan mineralized district, Korea, and the Artificial neural network (ANN) method and selected factors related to the occurrence of gold and silver minerals as ore-control factors, including magnetic anomaly geophysical data, geological and fault structure geological data, geochemical data, etc. Good results have been achieved [12]. Xiong et al. [13] identified multiple geochemical anomalies related to Fe polymetallic mineralization in the southwestern Fujian district (China) by using the limited Boltzmann machine. The research shows that most of the known skarn-type iron deposits are located in geochemical anomaly areas, which can provide reference for further exploration [13]. In order to effectively delineate favorable exploration targets for Cu-Au mineralization in the Moalleman District, NE Iran, Ghezelbash et al. [14] integrated several effective evidence layers such as geochemistry, geology, structure, and hydrothermal alteration in the study area; used SVM with radial basis function kernel to predict mineralization; and delineated the metallogenic prospect area [14]. However, the above methods are only based on two-dimensional geological data for prediction, which cannot fully characterize the multiple geological characteristics and may be difficult to make fine prediction of deep mines and hidden mines. The combination of 3D technology and artificial intelligence is beneficial to more fully excavate and integrate 3D prediction information and achieve more accurate positioning and quantitative predictions of deeply hidden ore bodies [15–18].

Compared with other mineralized areas in the Middle-Lower Yangtze River Metallogenic Belt, the Quaternary strata in the Xuancheng–Magushan area within the Middle-Lower Yangtze River Metallogenic Belt have a large coverage area and shallow geological exploration. The deep geological structure is not yet clear. It is difficult to describe the deep geological structure in this area in detail, which seriously affects the research of deep ore prospecting and prediction there [19]. Aiming at the Xuancheng–Magushan area, this paper firstly builds a 3D geological model that can accurately describe the deep geological structure with the support of geophysical methods and geological data. Based on this, two machine learning methods, the logistic regression model and the random forest model, were used to predict skarn deposits in the study area in three dimensions. Then, we divide the training set and the test set according to the data, the former trains the model, and the latter evaluates the performance of the model. The optimal results were selected to delineate the prospecting. Finally, the target area is expected to provide a new prospecting direction for further deep prospecting and exploration work in this area.

2. Methods

2.1. 3D Mineral Prospectivity Modeling

In recent years, the MPM has become an important means of prospecting and exploration. It can guide on-site prospecting work, thereby reducing the risk of prospecting. With the development of computer technology, a quantitative-based MPM method system has been put forward at home and abroad, which promotes the development of MPM from qualitative to quantitative and can more accurately delineate the metallogenic target area [20–24]. However, the above-mentioned quantitative MPM methods are mainly oriented towards the traditional two-dimensional prediction, which mostly uses two-dimensional geological data. However, the deep metal mineral resources have experienced multiple periods of geological evolution, resulting in weak surface indication information and complex geological structures. It is difficult to indicate prospecting work with traditional prediction methods based on two-dimensional geological data [25]. As deep ores and hidden ores have become the focus of prospecting in recent years, the research on the quantitative prediction of mineralization has moved from "two-dimensional" to "three-dimensional" [26].

The rise of artificial intelligence also provides a new way to process and mine massive geological data contained in 3D models [27]. Machine learning simulates human learning behavior through computers. It utilizes its nonlinear learning ability to characterize potential complex geological features by continuously training models and fitting parameters. In recent years, many scholars have begun to try to carry out 3DMPM research, including using the evidence weight model, the logistic regression model, the random forest model, and the artificial neural network model [28–31]. The above methods have shown good research potential in the field of 3DMPM. They can effectively process massive multi-dimensional geological data and have become an important development trend in this field.

In this paper, 3D geological modeling, 3D spatial analysis and 3DMPM based on machine learning are integrated. First, a 3D geological model is established based on geological data, and then a variety of 3D spatial analysis methods are used to analyze the 3D geological model and relevant metallogenic indicative characteristics, so as to obtain quantitative ore control and indicative characteristic information. Then, the prediction method based on machine learning is used to predict the mineralization of the deep edge of the mining area, and its effect is evaluated. Finally, the prediction results are used to divide the metallogenic prospective area, to realize the positioning and quantitative prediction of the hidden ore bodies at the deep edge of the known deposits. The forecast flow chart is shown in Figure 1.

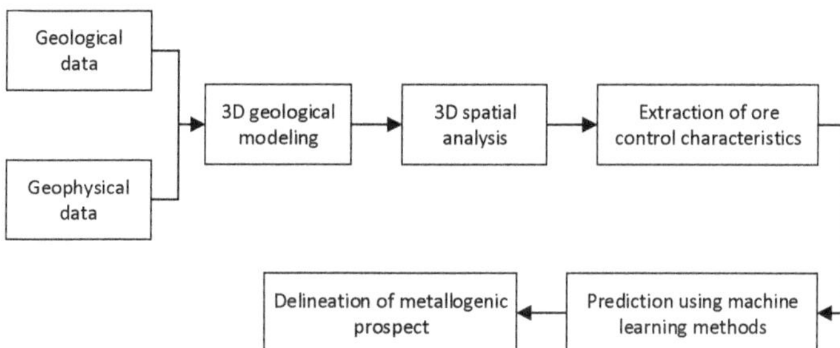

Figure 1. Workflow of three-dimension prospectivity mapping.

2.2. Logistic Regression Algorithm

Logistic regression is a representative algorithm in machine learning. This algorithm has been applied in many fields such as medicine, biology, and geology [32–34]. It can calculate the correlation between the independent input variable and the dependent variable through the regression principle and calculate the specific probability value of the dependent variable belonging to a certain category according to the existing state of the independent variable. As a multivariate nonlinear regression model, it can better fit the nonlinear relationship between various ore-controlling characteristics and metallogenic facts [35,36]. In this paper, metallogenic facts are used as the dependent variable, and various ore -controlling factors related to the metallogenic mechanism are used as the independent variables. The logistic regression method calculates the probability of ore bodies' existence in the corresponding blocks.

$$P(Z) = \frac{1}{(1 + e^{-Z})} \tag{1}$$

$$Z = \alpha + \beta_i x_j \tag{2}$$

In the above formulas: $P(Z)$ is the favorable degree of mineralization, x_i is the i-th ore control or indicator element, $(i = 1, 2, \ldots, n)$, α is a constant, β_i is a regression factor, that is, each control contribution of ore elements to the existence of ore bodies. It can be

forward modeling method to verify the rationality of the 3D geological model and modify and improve the 3D geological model. The modeling results are shown in Figure 5. The relevant modeling results will provide an important data basis for 3DMPM research.

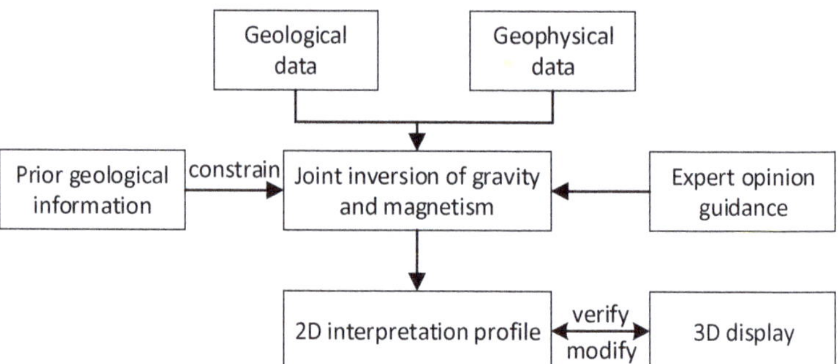

Figure 4. Gravity and magnetic inversion interpretation and profile-verification flow chart.

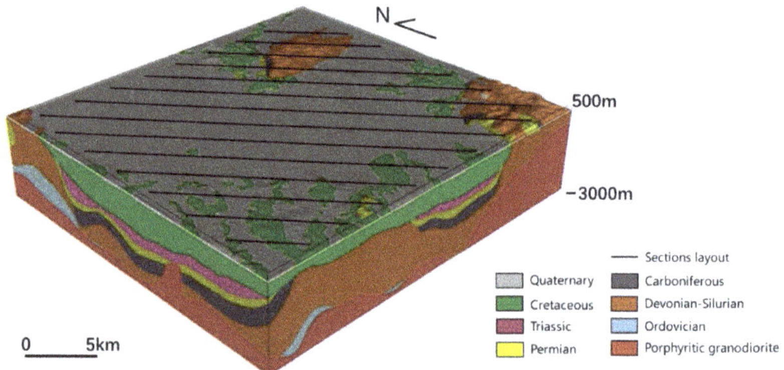

Figure 5. 3D model of the Xuancheng–Magushan Area. (Modified from Ye. [43] and Hu et al. [45]).

3.3. 3D Prospectivity Modeling Model and 3D Prediction Data Set Construction

The 3DMPM method is mainly based on expert experience, the metallogenic model, and the exploration model summarized by the predecessors to obtain the MPM model. Various spatial analysis methods are used to analyze the deep 3D geological model and related metallogenic indicative features and obtain quantitative results. Based on this information, prediction information is constructed. Finally, the metallogenic favorable degree is calculated. The prospecting target area is delineated for the position with the high metallogenic favorable degree. Thus, it provides a new quantitative prediction support for ore prospecting on the deep edge of the mining area.

We took the skarn type deposit represented by Magushan Cu-Mo deposit in the study area as the research object. Firstly, according to the geological and metallogenic character-istics of the Magushan skarn Cu-Mo deposit [43], the metallogenic law and prospecting signs of the skarn copper deposit in the study area were summarized. A 3DMPM model was constructed. It includes prediction elements such as the Carboniferous stratigraphic contact surface, the Permian stratigraphic contact surface, the Triassic stratigraphic contact surface, the rock mass contact zone, and the diorite uplift position.

After that, combined with the 3DMPM model in the study area, the 3D geological model was further analyzed in 3D space. The 3D prediction elements were extracted. The 3D geological body surface extraction method is mainly used for the Triassic stratigraphic contact surface, and rock mass contact zone are extracted, respectively; the 3D geological structure surface analysis function extracts the uplift position of the diorite rock mass. The analysis and extraction methods are shown in Table 1:

Table 1. 3DMPM model and analysis and extraction method of ore control and indicator elements. (Modified from Ye [43]).

Classification	Exploration Criteria	Spatial Analysis Methods
Strata	Carboniferous stratigraphic contact surface distance field	3D geological body surface extraction function 3D Distance Field Analysis
	Permian stratigraphic contact surface distance field	3D geological body surface extraction function 3D Distance Field Analysis
	Triassic stratigraphic contact surface distance field	3D geological body surface extraction function 3D Distance Field Analysis
Intrusions	Rock mass contact zone distance field	3D geological body surface extraction function 3D Distance Field Analysis
Structures	Distance field of diorite uplift location	3D Mathematical Morphological Methods 3D Distance Field Analysis

Based on the constructed 3D block model of the Xuancheng–Magushan area, this paper constructs the sample data period. The parameters of the 3D model are defined as shown in Table 2. The predicted depth is in the shallow space range of −3000 m. A single predicted cubic unit is defined as 100 m × 100 m × 25 m. The predicted space has 7.0735 million cubic units (Figure 6).

Table 2. Definition of spatial parameters for 3DMPM in Xuancheng–Magushan Area.

Parameter	Value (m)
North-south extent (x axis)	23,500
East-west extent (y axis)	21,500
Vertical extent (z axis)	500~−3000
X axis block size	100
Y axis block size	100
Z axis block size	25

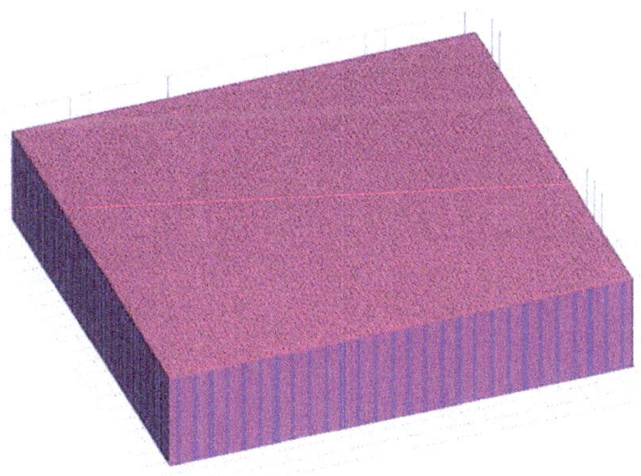

Figure 6. Block model of Xuancheng–Magushan Area.

Based on the geological and metallogenic characteristics of the Magushan skarn Cu-Mo deposit, this study summarizes the metallogenic regularity and prospecting markers of the skarn copper deposit in the study area. Prediction factors include the stratigraphic contact surface, the Triassic stratigraphic contact surface, the rock mass contact zone, and the diorite uplift position. A sample dataset for MPM was constructed by combining the metallogenic facts. In order to verify the generalization ability of the prediction model in the study area, a north–south division was made according to the known ore body locations. The south is used as a training area for the model to learn nonlinear ore-controlling characteristics, and the north is used as a test area to test the model's performance. There are 730 known ore body unit blocks in the study area, all of which are used as positive sample units, of which 614 were placed in the training set, and 116 were placed in the test set. To ensure a balance of positive and negative samples, 1500 non-ore body units around the known ore body were selected as negative samples. Of these, 1200 were put into the training set, and 300 were put into the test set.

4. Prospectivity Modeling Process and Results

4.1. Predictive Model Building

In order to fully explore the nonlinear relationship between the 3D ore-controlling factors and the ore-forming facts, based on the sample data set established above, this paper selects two machine learning methods, logical regression and random forest, to carry out 3D ore-forming predictions in the deep part of the mining area.

In addition to the support of a large number of effective datasets, the machine learning model also needs to set the model's parameters for the current dataset, which is an important factor in determining the model's performance. The random forest algorithm includes the two most important parameters: the number of decision trees M and the number of attributes K in the randomly selected attribute set. In this paper, the sampling dataset will be used to determine the appropriate number of decision trees and attributes of the random forest classification model using cross-validation. Due to the regression model adopted in this paper, after obtaining the error estimates of the results of each cross-validation set, the standard deviation is taken as the evaluation standard to evaluate the consistency of the model on different data sets (Figure 7).

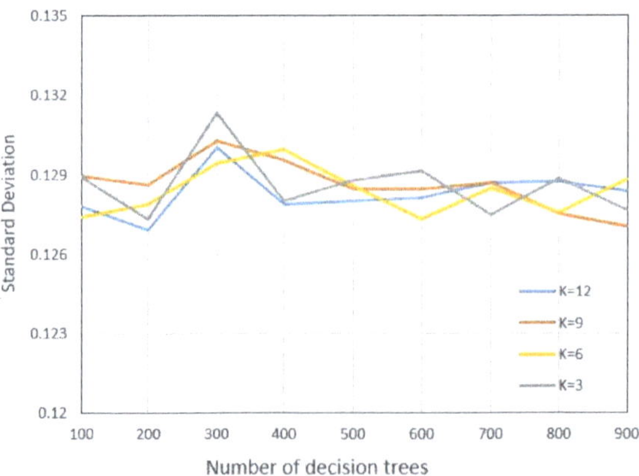

Figure 7. Standard Deviation maps of random forest algorithm under different parameters.

According to the results, this paper uses logistic regression and random forest (M = 200, K = 12) methods to carry out a 3DMPM on the deep edge of the Xuancheng–Magushan area and obtains the distribution map of favorable areas (Figure 8).

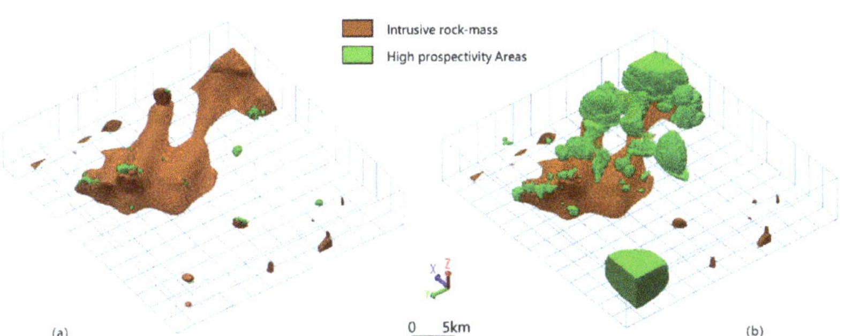

Figure 8. Distribution map of favorable areas, (**a**) Random forest model results; (**b**) Logistic regression model results.

4.2. Model Performance Analysis

The confusion matrix is a standard format for expressing the accuracy evaluation. It is often used in binary classification scenarios. Each column of the matrix represents the prediction of the sample, and each row of the matrix represents the real situation of the sample. To more intuitively express the quality of the model's performance, we extend three metrics from the matrix: precision, recall, and specificity. The trained model is used in the test set divided above to test the performance of the model. According to the results, the blocks with favorable degrees of mineralization greater than 0.5 predicted in the test set are selected as favorable units for mineralization. Finally, we compare the real value and the predicted value of each block in the test set and use these three prediction indicators to compare the model (Table 3).

Table 3. Comparison of performance indicators.

Models	Accuracy	Recall	Speciality
Logistic regression	90.625%	83.62%	93.33%
Random forest	96.63%	93.97%	97.67%

Comparing the three performance indicators, it can be concluded that the random forest model performs better than the logistic regression model, which can effectively distinguish non-ore body units in the case of predicting more known ore bodies in the test set and has a good generalization ability.

The ROC curve is also often used in the performance evaluation of the two-class network [46]. It can indicate the ability to identify the sample at a certain threshold. The vertical and horizontal coordinates of the points on the curve represent the true positive rate (TPR) and the false positive rate (FPR) of the output results under different thresholds, respectively. The ROC curve indicates the percentage of true positive units in the known mineralization units in the different positive prediction ranges of the model. The area under the curve is called the AUC value. The larger the AUC value, the better the model effect. This paper compares the ROC curves of the two models (Figure 9) and finds that the image of the MPM method based on the random forest is more inclined to the upper left corner than the logistic regression model. The AUC values of the two models are 0.989 and 0.969, indicating the random forest model has better performance and more reliable results.

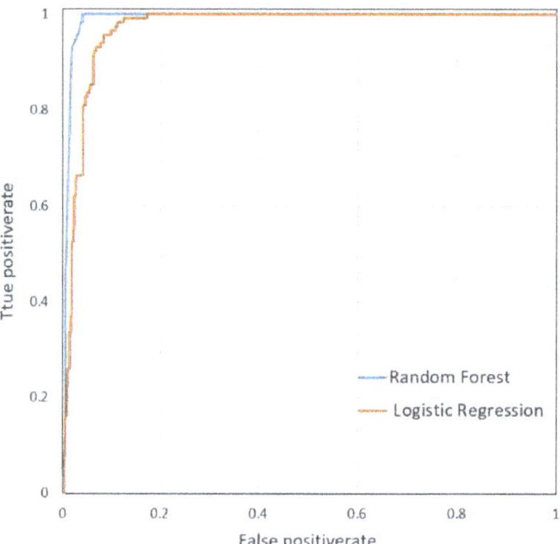

Figure 9. Comparison of ROC curves.

The performance of the two models was further quantitatively evaluated by plotting the capture efficiency curves [47,48] (Figure 10). First, the predicted metallogenic favorableness of all blocks is sorted in descending order. Then various thresholds are set according to the sorting results to reclassify the unit blocks in the study area. Finally, the capture efficiency is calculated by counting the number of known ore body units in different sections. The calculation process of the capture efficiency is to perform the statistical calculation on all blocks in the study area. From the capture efficiency curve, it can be obtained that the blocks in the top 4‰ of the metallogenic favorable degree predicted by the random forest model in the study area can cover all the known ore bodies. In the logistic regression model results, only the blocks in the top 20‰ of the favorable degree of mineralization in the study area can cover all known ore bodies. It can be shown that the random forest can contain more known ore body units in the block unit with high posterior probability and can screen out the metallogenic prospect area more finely.

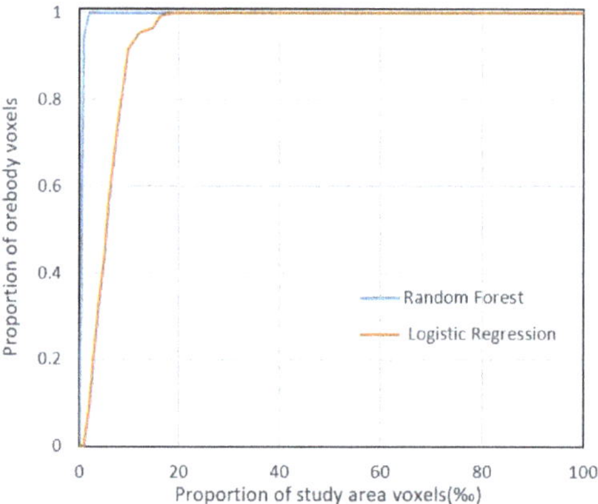

Figure 10. Capture efficiency curves.

5. Discussion

After analyzing the indicators of the logistic regression model and the random forest model, it can be seen that the prediction results of the random forest model are better. The accuracy of the random forest model in the test set is 96.63%, which is higher than that of the logistic regression model by 6.005%, i.e., 10.35% higher in recall and 4.34% higher in specificity, indicating that random forest can better characterize the characteristics of the mineral control in the study area. At the same time, compared with logistic regression, the random forest model can better identify ore body characteristics and can cover more known ore body units in the same number of block units with high metallogenic favorable degrees. By comparing the distributions and shapes of favorable areas predicted by the two methods, it can be seen that the random forest can constrain the specific locations of the prospectivity targets more finely, thereby effectively improving the efficiency of prospecting and exploration.

In this paper, the prediction results of random forest are used to delineate the metallogenic target area, and the unit block with a metallogenic favorable degree greater than 0.5 is selected as the potential metallogenic unit.

According to the prediction results, there are 7652 favorable areas in the study area, accounting for 1.08 % of the whole study area, including 96.71 % of the known ore bodies. Therefore, the random forest model can not only effectively identify the known ore bodies, but also screen out Blocks with greater metallogenic potential. Then five metallogenic potential areas are divided (Figure 11).

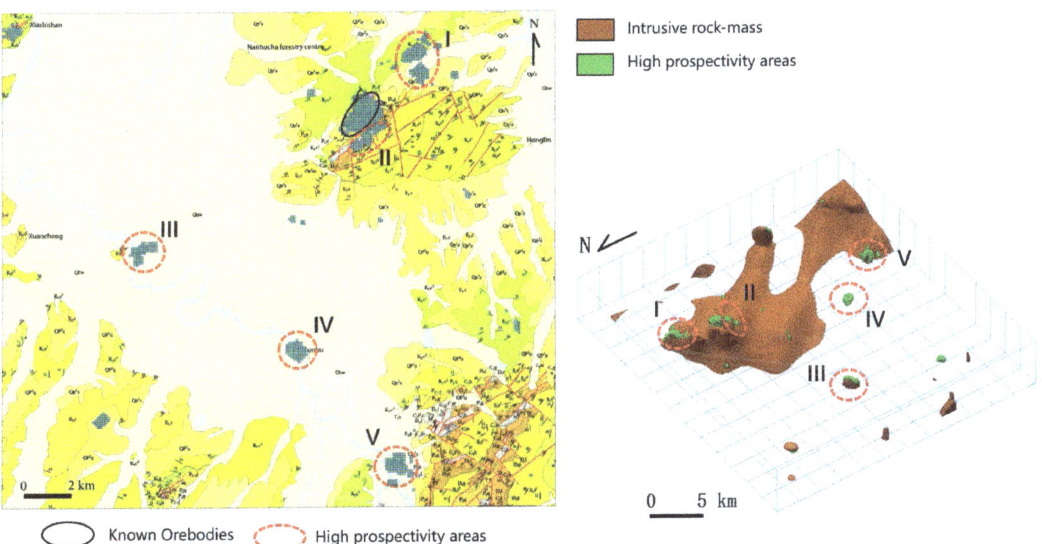

Figure 11. Delineation of prospecting target areas.

The five metallogenic prospective areas classified in this paper all have high metallogenic potential. The No. I and No. II target areas are located in the prospecting area of Magushan. The burial depth of the No. I target area is about −900 m~−1200 m, and the burial depth of the No. II target area is about −1500 m~−2000 m. The target area is located in the middle of the high gravity anomaly in Magushan as a whole, with a trend near east–west, and the isolines on the north and south sides change rapidly; in terms of the aeromagnetization pole anomaly, the Magushan anomaly clearly shows a high magnetic anomaly, with a trend near the pear-shaped distribution in the north and the south: the contour changes smoothly, the gradient changes rapidly on the north side, and extends to the south, showing the subsidence direction of the concealed rock mass. The measurement anomalies of 1:200,000 water system sediments show that Cu, Hg, and W are anomalous in the vicinity of the Magushan deposit. The No. III target area is located on the surface of the high-density body, and the burial depth is about −2100 m~−2800 m. The No. IV target area is generally controlled by structural forms, such as the uplift and depression of the rock mass, and the buried depth is about −1100 m~−1500m. The No. V target area is located at the intersection of the faults, and there are certain magnetic anomalies on the surface of this area, and the burial depth is about −2100 m~−2900 m. Therefore, the five prospectivity targets classified in this paper can be the priority exploration targets for future mineral exploration in this area.

6. Conclusions

(1) The 3DMPM is an important tool for deep targets delineation for future exploration. This paper delineates five prospectivity targets with good mineralization potentials in the deep area of the Xuancheng–Magushan area, which can be used for future exploration.

(2) In the Xuancheng–Magushan area, the favorable areas divided by the random forest model contain 96.71% of known ore bodies and only account for 1.08% of the study area, which can show that the random forest model can perform better than the logistic regression model in the 3DMPM using the dataset of the study area. It means that the random forest model could provide more effective and accurate support for integrating predictive data during the 3DMPM.

Author Contributions: Conceptualization, F.M., X.L. and F.Y.; Methodology, F.M. and X.L.; Software, X.L.; Validation, X.L. and Y.C.; Formal Analysis, F.M., Y.C. and R.Y.; Data Curation, Y.C. and R.Y.; Writing—Original Draft Preparation, F.M.; Writing-Review & Editing, F.M., X.L. and F.Y.; Visualization, Y.C. and R.Y.; Supervision, X.L. and F.Y.; Project Administration, F.Y. All authors have read and agreed to the published version of the manuscript.

Funding: This research was funded by [National Natural Science Foundation of China] grant number [42072321], [National Natural Science Foundation of China] grant number [41820104007] and [National Key R&D Program of China] grant number [2016YFC0600209].

Data Availability Statement: Data not available due to legal restrictions.

Conflicts of Interest: The authors declare no conflict of interest.

References

1. Teng, J.W.; Yang, L.Q.; Liu, H.C.; Yan, Y.F.; Yang, H.; Zhang, H.S.; Zhang, Y.Q.; Tian, Y. Geodynamical responses for formation and concentration of metall minerals in the second deep space of lithosphere. *Chin. J. Geophys.* **2009**, *52*, 1734–1756. (In Chinese with English Abstract)
2. Yan, J.Y.; Teng, J.W.; Lv, Q.T. Geophysical Exploration and Application of Deep Metal Mineral Resources. *Prog. Geophys.* **2008**, *23*, 871–891. (In Chinese with English Abstract)
3. Heinson, G.S.; Direen, N.G.; Gill, R.M. Magnetotelluric evidence for a deep-crustal mineralizing system beneath the Olympic Dam iron oxide copper-gold deposit, southern Australia. *Geology* **2006**, *34*, 573–576. [CrossRef]
4. Hu, X.Y.; Li, X.H.; Yuan, F.; Alison, O.; Simon, M.J.; Li, Y.; Dai, W.Q.; Zhou, T.F. Numerical modeling of ore-forming processes within the Chating Cu-Au porphyry-type deposit, China: Implications for the longevity of hydrothermal systems and potential uses in mineral exploration. *Ore Geol. Rev.* **2020**, *116*, 103230. [CrossRef]
5. Liu, L.M.; Peng, S. Key strategies for predictive exploration in mature environment: Model innovation, exploration technology optimization and information integration. *J. Cent. South Univ. Technol. (Engl. Ed.)* **2005**, *12*, 186–191. [CrossRef]
6. Niu, S.Y.; Wang, B.D.; Sun, A.Q.; Chen, C.; Wang, Z.L.; Ma, B.J.; Wang, W.S.; Jiang, X.P.; Zhao, Y.L.; Gao, Y.C.; et al. Analysis of the ore-controlling structure of the Shihu gold deposit, Hebei Province and deep-seated ore-prospecting prediction. *Chin. J. Geochem.* **2009**, *28*, 386–396. [CrossRef]
7. Zhai, Y.S.; Deng, J.; Wang, J.P.; Peng, R.M.; Liu, J.J.; Yang, L.Q. Researches on deep ore prospecting. *Miner. Depos.* **2004**, *23*, 142–149 (In Chinese with English Abstract)
8. Li, X.Y.; Yuan, Y.; Zhang, M.M.; Jia, C.; Jowitt, S.M.; Ord, A.; Zhou, T.F.; Dai, W.Q. 3D computational simulation-based mineral prospectivity modeling for exploration for concealed Fe–Cu skarn-type mineralization within the Yueshan orefield, Anqing district, Anhui Province, China. *Ore Geol. Rev.* **2018**, *105*, 1–17. [CrossRef]
9. Payne, C.E.; Cunningham, F.; Peters, K.J.; Nielsen, S.; Puccioni, E.; Wildman, C.; Partington, G.A. From 2D to 3D: Prospectivity modelling in the Taupo volcanic zone, New Zealand. *Ore Geol. Rev.* **2015**, *71*, 558–577. [CrossRef]
10. Nielsen, S.H.; Cunningham, F.; Hay, R.; Partington, G.; Stokes, M. 3D prospectivity modelling of orogenic gold in the Marymia Inlier, Western Australia. *Ore Geol. Rev.* **2015**, *71*, 578–591. [CrossRef]
11. Xiao, K.Y.; Li, N.; Sun, L.; Zhou, W.; Li, Y. Large Scale 3D Mineral Prediction Methods and Channels Based on 3D Information Technology. *J. Geol.* **2012**, *36*, 229–236. (In Chinese)

12. Oh, H.J.; Lee, S. Application of artificial neural network for gold-silver deposits potential mapping: A case study of Koreal. *Nat. Resour. Res.* **2010**, *19*, 103–124. [CrossRef]

13. Xiong, Y.H.; Zuo, R.G. Recognition of geochemical anomalies using a deep autoencoder network. *Comput. Geosci.* **2016**, *86*, 75–82. [CrossRef]

14. Ghezelbash, R.; Maghsoudi, A.; Bigdeli, A.; Carranza, E.J.M. Regional-Scale Mineral Prospectivity Mapping: Support Vector Machines and an Improved Data-Driven Multi-criteria Decision-Making Technique. *Nat. Resour. Res.* **2021**, *30*, 1977–2005. [CrossRef]

15. Abedi, M.; Norouzi, G.H. Integration of various geophysical data with geological and geochemical data to determine additional drilling for copper exploration. *J. Appl. Geophys.* **2012**, *83*, 35–45. [CrossRef]

16. Leite, E.P.; de Souza Filho, C.R. Artificial neural networks applied to mineral potential mapping for copper-gold mineralizations in the Carajás Mineral Province, Brazil. *Geophys. Prospect.* **2019**, *57*, 1049–1065. [CrossRef]

17. Li, X.L.; Li, L.H.; Zhang, B.L.; Guo, Q.J. Hybrid self-adaptive learning based particle swarm optimization and support vector regression model for grade estimation. *Neurocomputing* **2013**, *118*, 179–190. [CrossRef]

18. Zuo, R.G.; Carranza, E. Support vector machine: A tool for mapping mineral prospectivity. *Comput. Geosci.* **2011**, *37*, 1967–1975. [CrossRef]

19. Bian, Y.C. On the origin of Magushan Cu–Mo deposit in South Anhui. *J. Geol.* **1995**, *19*, 17–20, (In Chinese with English Abstract).

20. Xiao, K.Y.; Ding, J.H.; Liu, R. The discussion of three-part form of non-fuel mineralresource assessment. *Ceological Rev.* **2006**, *52*, 793–798. (In Chinese with English Abstract)

21. Zhao, P.D. Three Component quantitative resource prediction and assessment: Theory and practice of digital mineral prospecting. *Earth Sci.-J. China Univ. Geosci.* **2002**, *27*, 482–489. (In Chinese with English Abstract)

22. Agterberg, F.P. Multivariate prediction equations in geology. *J. Int. Assoc. Math. Geol.* **1970**, *2*, 319–324. [CrossRef]

23. Agterberg, F.P.; Bonham-Carter, G.F.; Cheng, Q.; Wright, D.F. Weights of evidence modeling and weighted logistic regression for mineral potential mapping. In *Computers in Geology-25 Years of Progress*; Davis, J.C., Herzfeld, U.C., Eds.; Oxford University Press: New York, NY, USA, 1993; pp. 13–32.

24. Wang, S.C. The New Development of Theory and Method of Synthetic Information Mineral Resources Prognosis. *Geol. Bull. China* **2010**, *29*, 1399–1403. (In Chinese with English Abstract)

25. Zhao, P.D. Quantitative Mineral Prediction and Deep Mineral Exploration. *Earth Sci. Front.* **2007**, *14*, 309–318. (In Chinese with English Abstract)

26. Yuan, F.; Li, X.H.; Zhang, M.M.; Zhou, T.F.; Gao, D.M.; Hong, D.L.; Liu, X.M.; Wang, Q.N.; Zhu, J.B. Three Dimension Prospectivity Modelling Based on Integrated Geoinformation for Prediction of Buried Ore Bodies. *Acta Geol. Sin.* **2014**, *88*, 630–643. (In Chinese with English Abstract)

27. Caté, A.; Perozzi, L.; Gloaguen, E.; Blouin, M. Machine learning as a tool for geologists. *Lead. Edge* **2017**, *36*, 215. [CrossRef]

28. Li, X.H.; Yuan, F.; Zhang, M.M.; Jia, C.; Jowitt, S.M.; Ord, A.; Zheng, T.K.; Hu, X.Y.; Li, Y. Three-dimensional mineral prospectivity modeling for targeting of concealed mineralization within the Zhonggu iron orefield, Ningwu Basin, China. *Ore Geol. Rev.* **2015**, *71*, 633–654. [CrossRef]

29. Sun, T.; Chen, F.; Zhong, L.X.; Liu, W.M.; Wang, Y. GIS-basedmineral prospectivity mapping using machine learning methods: Acase study from Tongling ore district, eastern China. *Ore Geol. Rev.* **2019**, *109*, 26–49. [CrossRef]

30. Tao, J.T.; Yuan, F.; Zhang, N.N.; Chang, J.Y. Three-Dimensional Prospectivity Modeling of Honghai Volcanogenic Massive Sulfide Cu–Zn Deposit, Eastern Tianshan, Northwestern China Using Weights of Evidence and Fuzzy Logic. *Math. Geosci.* **2021**, *53*, 131–162. [CrossRef]

31. Xiang, J.; Xiao, K.Y.; Carranza, E.J.M.; Chen, J.P.; Li, S. 3D Mineral Prospectivity Mapping with Random Forests: A Case Study of Tongling, Anhui, China. *Nat. Resour. Res.* **2020**, *29*, 395–414. [CrossRef]

32. Mokhtari, A.R. Hydrothermal alteration mapping through multivariate logistic regression analysis of lithogeochemical data. *J. Geochem. Explor.* **2014**, *145*, 207–212. [CrossRef]

33. Marian, F.; Piotr, C.; Sławomir, B.; Bogusław, R.; Mirosław, K. Logistic Regression Model for Determination of the Age of Brown Hare (Lepus europaeus Pall.) Based on Body Weight. *Animals* **2022**, *12*, 529.

34. Yuko, K.; Koichi, S.; Takeshi, I.; Koichi, T.; Tetsuya, T. Predictors for development of palbociclib-induced neutropenia in breast cancer patients as determined by ordered logistic regression analysis. *Sci. Rep.* **2021**, *11*, 20055.

35. Carranza, E.J.M.; Hale, M. Logistic Regression for Geologically Constrained Mapping of Gold Potential, Baguio District, Philippines. *Explor. Min. Geol.* **2001**, *10*, 165–175. [CrossRef]

36. Xiong, Y.H.; Zuo, R.G. GIS-based rare events logistic regression for mineral prospectivity mapping. *Comput. Geosci.* **2018**, *111*, 18–25. [CrossRef]

37. Breiman, L. Bagging predictors. *Mach. Learn.* **1996**, *24*, 123–140. [CrossRef]

38. Ho, T.K. The random subspace method for constructing decision forests. *IEEE Trans. Pattern Anal. Mach. Intell.* **1998**, *20*, 832–844.

39. Hu, X.Y.; Li, X.Y.; Yuan, F.; Ord, A.; Jowitt, S.M.; Li, Y.; Dai, W.Q.; Ye, R.; Zhou, T.F. Numerical Simulation Based Targeting of the Magushan Skarn Cu–Mo Deposit, Middle-Lower Yangtze Metallogenic Belt, China. *Minerals* **2019**, *9*, 588. [CrossRef]

40. Zhou, T.F.; Fan, Y.; Yuan, F. Advances on petrogenesis and metallogenic study of the mineralization belt of the Middle and Lower Reaches of the Yangtze River area. *Acta Petrol. Sin.* **2008**, *24*, 1665–1678. (In Chinese with English Abstract)

41. Chang, Y.F.; Liu, X.P.; Wu, Y.C. *The Cu-Fe Metallogenic Belt in the Middle-Lower Reaches of Yangtze River*; Geological Publish House: Beijing, China, 1991; p. 379. (In Chinese)
42. Mao, J.W.; Xie, G.Q.; Duan, C.; Pirajno, F.; Ishiyama, D.; Chen, Y.C. A tectono-genetic model for porphyry–skarn–stratabound Cu–Au–Mo–Fe and magnetite–apatite deposits along the Middle–Lower Yangtze River Valley, Eastern China. *Ore Geol. Rev.* **2011**, *43*, 294–314. [CrossRef]
43. Ye, R. 3D Geological Modeling and Mineral Prospectivity Modeling of Magushan Ore Field in Nanling-Xuancheng Ore Concentration Area. Ph.D. Thesis, Hefei University of Technology, Hefei, China, 2020. (In Chinese with English Abstract)
44. Hong, D.J.; Huang, Z.Z.; Chan, S.W.; Wang, X.H. Geological characteristics and exploration directions of the Cu-polymetallic ore deposits in the Magushan-Qiaomaishanareas in Xuancheng, Anhui Province. *East China Geol.* **2017**, *38*, 28–36, (In Chinese with English Abstract).
45. Hu, X.Y.; Li, X.Y.; Yuan, F.; Jowitt, S.M.; Ord, A.; Ye, R.; Li, Y.; Dai, W.Q.; Li, X.L. 3D Numerical Simulation-Based Targeting of Skarn Type Mineralization within the Xuancheng-Magushan Orefield, Middle-Lower Yangtze Metallogenic Belt, China. *Lithosphere* **2020**, *1*, 8351536. [CrossRef]
46. Zhang, Z.J.; Zuo, R.G.; Xiong, Y.H. A comparative study of fuzzy weights of evidence and random forests for mapping mineral prospectivity for skarn-type Fe deposits in the southwestern Fujian metallogenic belt, China. *Sci. China (Earth Sci.)* **2016**, *59*, 556–572. [CrossRef]
47. Fabbri, A.G.; Chung, C. On blind tests and spatial prediction models. *Nat. Resour. Res.* **2008**, *17*, 107–118. [CrossRef]
48. Porwal, A.; Gonzalez-Alvarez, I.; Markwitz, V.; McCuaig, T.; Mamuse, A. Weights-of evidence and logistic regression modelling of magmatic nickel sulfide prospectivity in the Yilgarn Craton, Western Australia. *Ore Geol. Rev.* **2010**, *38*, 184–196. [CrossRef]

Article

3D Mineral Prospectivity Mapping of Zaozigou Gold Deposit, West Qinling, China: Machine Learning-Based Mineral Prediction

Yunhui Kong [1], Guodong Chen [1,*], Bingli Liu [1,2], Miao Xie [1], Zhengbo Yu [1], Cheng Li [1,3], Yixiao Wu [1], Yaxin Gao [1], Shuai Zha [1], Hanyuan Zhang [1], Lu Wang [1] and Rui Tang [1]

[1] Geomathematics Key Laboratory of Sichuan Province, Chengdu University of Technology, Chengdu 610059, China
[2] Key Laboratory of Geochemical Exploration, Institute of Geophysical and Geochemical Exploration, CAGS, Langfang 065000, China
[3] Institute of Mineral Resources, Chinese Academy of Geological Sciences, Beijing 100037, China
* Correspondence: chengd@cdut.edu.cn; Tel./Fax: +86-028-8407-3610

Abstract: This paper focuses on researching the scientific problem of deep extraction and inference of favorable geological and geochemical information about mineralization at depth, based on which a deep mineral resources prediction model is established and machine learning approaches are used to carry out deep quantitative mineral resources prediction. The main contents include: (i) discussing the method of 3D geochemical anomaly extraction under the multi-fractal content-volume (C-V) models, extracting the 12 element anomalies and constructing a 3D geochemical anomaly data volume model for laying the data foundation for researching geochemical element distribution and association; (ii) extracting the element association characteristics of primary geochemical halos and inferring deep metallogenic factors based on compositional data analysis (CoDA), including quantitatively extracting the geochemical element associations corresponding to ore-bearing structures (Sb-Hg) based on a data-driven CoDA framework, quantitatively identifying the front halo element association (As-Sb-Hg), near-ore halo element association (Au-Ag-Cu-Pb-Zn) and tail halo element association (W-Mo-Co-Bi), which provide quantitative indicators for the primary haloes' structural analysis at depth; (iii) establishing a deep geological and geochemical mineral resources prediction model, which is constructed by five quantitative mineralization indicators as input variables: fracture buffer zone, element association (Sb-Hg) of ore-bearing structures, metallogenic element Au anomaly, near-ore halo element association Au-Ag-Cu-Pb-Zn and the ratio of front halo to tail halo (As-Sb-Hg)/(W-Mo-Bi); and (iv) three-dimensional MPM based on the maximum entropy model (MaxEnt) and Gaussian mixture model (GMM), and delineating exploration targets at depth. The results show that the C-V model can identify the geological element distribution and the CoDA method can extract geochemical element associations in 3D space reliably, and the machine learning methods of MaxEnt and GMM have high performance in 3D MPM.

Keywords: 3D mineral prospectivity mapping; quantitative mineral resources prediction model; maximum entropy model; Gaussian mixture model; Zaozigou gold deposit

Citation: Kong, Y.; Chen, G.; Liu, B.; Xie, M.; Yu, Z.; Li, C.; Wu, Y.; Gao, Y.; Zha, S.; Zhang, H.; et al. 3D Mineral Prospectivity Mapping of Zaozigou Gold Deposit, West Qinling, China: Machine Learning-Based Mineral Prediction. *Minerals* 2022, 12, 1361. https://doi.org/10.3390/min12111361

Academic Editor: Martiya Sadeghi

Received: 31 July 2022
Accepted: 16 October 2022
Published: 26 October 2022

Publisher's Note: MDPI stays neutral with regard to jurisdictional claims in published maps and institutional affiliations.

1. Introduction

The main task of quantitative mineral prediction is to conduct comprehensive analysis of geological, geophysical, geochemical and remote sensing data and drilling engineering data in the study area based on the research of the geological background and metallogenic regularity, and then construct mathematical models to effectively extract and identify favorable information on mineralization, carry out data fusion of quantitative mineralization information, construct mineral prediction models, perform potential mineral resources quantitative assessment and exploration targets delineation [1–3].

Quantitative mineral resource prediction is developing toward deep 3D quantitative prediction. The framework of 3D quantitative mineral prediction has been basically completed, among which the representative works include Xiao [4–10], who studied the characteristics of large-scale three-dimensional mineralization prediction and summarized the combined prediction process of mine concession structure, mineral exploration model, metallogenic series and quantitative analysis methods, which laid the foundation for three-dimensional mineral resource prediction; Chen [11–14] proposed the "cubic prediction model" for mineralization prediction by 3D-comprehensive 3D modeling of geology, geophysics, geochemistry and remote sensing [15–17]; Mao [18–22] proposed a three-dimensional prediction process for deep mineral resources prediction, the three-dimensional morphological analysis of geological bodies, geological anomaly extraction and three-dimensional quantitative prediction methods [23]; Yuan [24,25] proposed a "four-step" three-dimensional mineralization prediction method, which summarizes the three-dimensional mineralization prediction work into four key steps: three-dimensional geological database construction, three-dimensional geological modeling and mapping, three-dimensional prediction information deep mining and three-dimensional geosciences data fusion and prediction.

In this study, three-dimensional quantitative mineral resources prediction is executed from a geological and geochemical perspective. It is a common understanding in the last century that the distribution of rock geochemical elements follows normal or lognormal distribution, based on which classical statistical theories and methods have been widely applied in the field of geochemical data analysis [26]. Since the 1990s, fractal and multifractal methods have been developed rapidly, with representative achievement from Cheng's team, who proposed the content-area (C-A) model [26,27], the spectrum-area (S-A) model [28] and Local Singularity Analysis (LSA) [26,29–32], and it is believed that the multifractal distribution has the ability to simultaneously portray the normal distribution, lognormal distribution, Zipf's law, Pareto's law, etc., and can be used as the basic law of geochemical element distribution [26,29]. Based on the theoretical research of fractal and multifractal methods, its application in geochemical spatial distribution pattern research is blossoming [10,33–37]. The multifractal-related methods have achieved many successful cases in regional geochemical anomaly extraction [32,38–45]. With the improvement of 3D software and hardware performance, as well as the development of 3D modeling technology, fractal models used in 3D data have made some progress [34,46,47]. Three-dimensional primary halo geochemical survey is an effective tool for deep mineral resources prediction. In fact, the spatial zonation pattern of primary haloes is the pattern of element clustering regularity in 3D space, and the anomalous structure of elements is an intuitive indicator of deep mineral prediction. Especially, Cheng [27] proposed that the distribution of geochemical elements in either horizontal or vertical directions is consistent with the fractal distribution, based on which Afzal [34] proposed the C-V model, which is a powerful tool to delineate the nonlinear characteristics of element distribution in a 3D space.

Geochemical data are typically compositional data, which gives rise to the problem of "closure effects" due to the lack of scale consistency in the covariance matrix of the compositional data [48–50], and, in practice, classical geostatistical methods that ignore the compositional properties of geochemical data often yield poor results [51–60]. Since the introduction of Aitchison geometry and the application of additive log-ratio transformation (alr) [61], the theory and methods of log-ratio transformation of compositional data have been gradually improved, making the application of classical statistical methods more reasonable [33,35]. The geochemical elemental association extraction method based on Sequential Binary Partition (SBP) [50] can design elemental associations based on the geological background and mineralization knowledge, which is easy to interpret, and provides solutions for geochemical inference of lithology, tectonics, alteration, mineralization, etc. Therefore, the CoDA method provides new ideas for analyzing the relationship of geochemical data and extracts geochemical associations for evaluating mineral resources rapidly [51,55,62–67].

Geosciences is a data-intensive science, and geological survey and mineral exploration have accumulated a large amount of multi-source geosciences data, and the introduction of big data and its application of machine learning-related methods can effectively support mineral resources exploration [26,68–73]. During the past decade or so, a large number of achievements related to machine learning have been published in geosciences [37,57,74–85]. For example, based on metallogenic regularity, an intelligent geology and intelligent mine model is established on a big data platform combined with high-performance computers [70,86]; big data methods are used to realize the automatic collection of geological prospecting thematic data, machine learning- and deep learning-related methods development for deep mining of geological data, intelligent prospecting, etc. [37,87–89]; and three-dimensional prediction methods of hidden orebodies can be based on deep learning [17,72,90–98]. Numerous studies have shown that big data and machine learning have good performance in mineralization prediction, and machine learning-related algorithm research has greatly enriched the approaches of geological data processing and analysis. Meanwhile, data cleaning, data screening and deep mining of geological big data can enrich the information sources for mineralization information inference and quantitative mineral prediction at depth.

This study focuses on the scientific problems of deep extraction and inference of favorable geological and geochemical information about mineralization at depth and quantitative mineralization prediction at large depth. Specifically, the C-V model will be used to extract the spatial distribution pattern of primary halo geochemical element anomalies, and the CoDA will be employed to identify and infer the element associations for tracing the extension of deep ore-controlling information. Moreover, the machine learning methods of the MaxEnt model and GMM will be carried out for mineral resources prediction at the large depth of the Zaozigou gold deposit.

2. Geological Setting and Datasets

2.1. Geological Setting

West Qinling is located at the western part of the Qinling orogenic belt, with the Qilian orogenic belt to the north, the Qaidam block to the west and the songpan block to the south (Figure 1a) [99–102].

The geotectonic position of the Xiahe–Hezuo area is located at the northwestern part of the West Qinling orogenic belt and at the western extension of the Qinling–Qilian–Kunlun central orogenic belt, whose complex geological structure creates a superior mineralization environment [103].

The Zaozigou gold deposit is located within the West Qinling fold belt and is a typical epithermal-type gold deposit in the Xiahe–Hezuo area (Figure 1a). The main controlling factors for mineral resources formation within the region are tectonic movement and magmatism [104], with the regional tectonics spreading in a NW direction with developing folds and fractures. Complex geological structure and magmatism are dominated by Yanshan period intermediate-acid intrusive rocks, which are widely distributed in the form of the batholith, stock, apophysis and veins [104] (Figure 1b).

The Triassic strata are the main stratigraphy for gold deposits. The genesis and spatial-temporal evolution of the intermediate-acidic dike is closely related to gold mineralization in the area, and during the mineralization process, the magmatic rocks not only provide the mineralized material, but also their internal environment is very suitable as an ore-bearing space, which can be regarded as a significant mineralization indicator for gold mineralization [105].

The spatial distribution of mineral deposits is directly controlled by the geotectonic position in the Xiahe–Hezuo area, which plays a major role in the formation of different types of gold deposits and is the boundary of the belt from a spatial perspective. The most important types of mineral-controlling structures are fractures and folds in this area [106,107]. The main orebodies of the Zaozigou deposit are produced in NE, NW and near-SN oriented fracture zones, with two apparent mineralization periods, and post-formation fractures have modified and destroyed the orebodies [108–112].

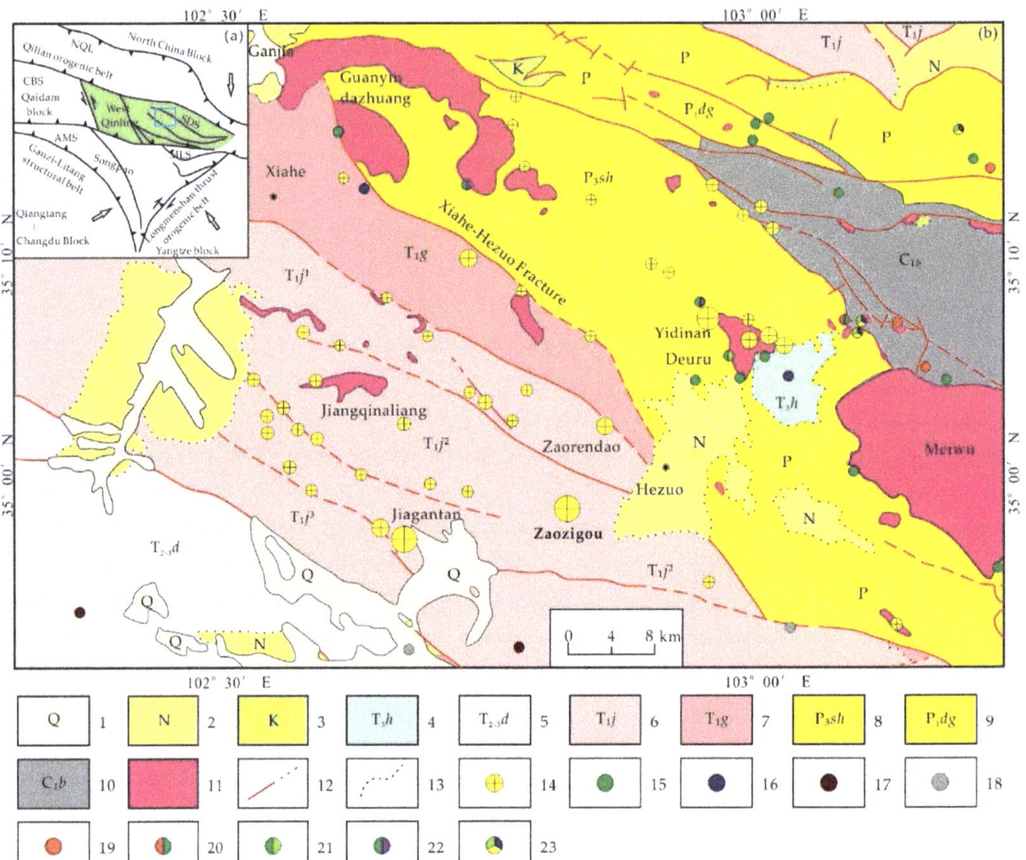

Figure 1. (**a**) Geotectonic location of the study area. NQL: North Qinling tectonic belt; SDS: Shangdan suture zone; CBS: Caibei suture zone; AMS: A'nyemaqen suture zone. (**b**) Geological map of Xiahe–Hezuo area (modified from [104]). 1. Quaternary; 2. Neogene; 3. Cretaceous; 4. Upper Triassic Huari group; 5. Middle-Lower Triassic Daheba group; 6. Lower Triassic Jiangligou group; 7. Lower Triassic Guomugou group; 8. Upper Permian Shiguan group; 9. Lower Permian Daguanshan group; 10. Upper Carboniferous Badu group; 11. Intermediate-acid intrusive rocks; 12. Fracture; 13. Angular unconformity; 14. Gold deposit; 15. Copper deposit; 16. Lead deposit; 17. Antimony deposit; 18. Mercury deposit; 19. Iron deposit; 20. Iron-copper deposit; 21. Copper-molybdenum deposit; 22. Copper-tungsten deposit.

The Zaozigou gold deposit is a typical representative of gold deposits associated with intermediate-acid dike rocks in the south of the Xiahe–Hezuo fracture. It is located approximately 9 km southwest of Hezuo city, Gansu Province, with convenient access to the mine site (Figure 2a). The main ore-bearing position is between Gully 1 and Gully 4, with a total area of approximately 2.6 km^2 (Figure 2b).

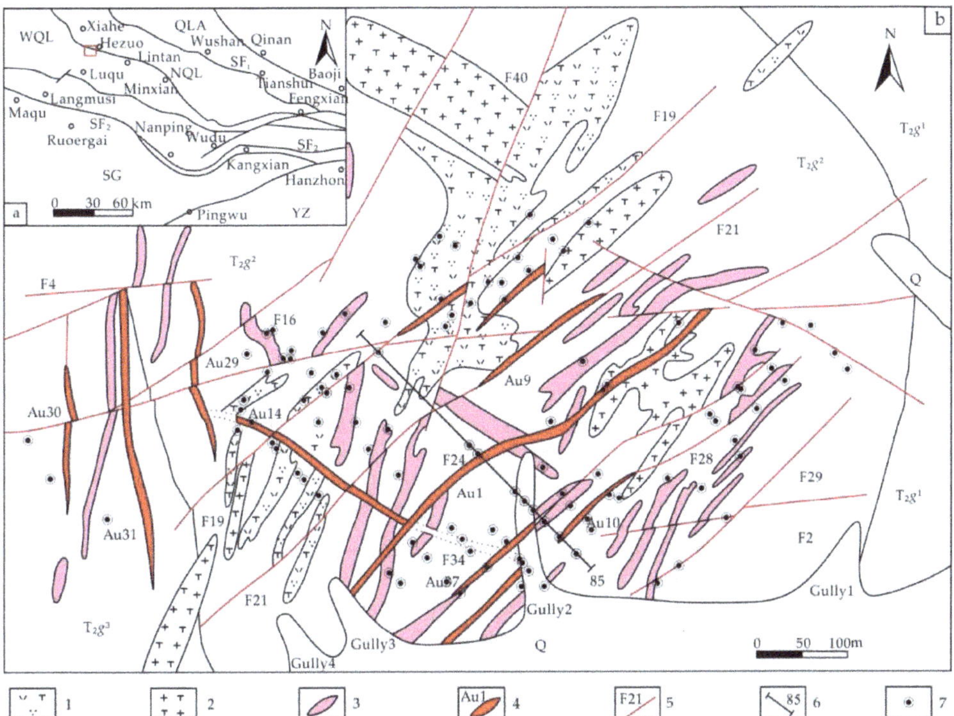

⌄ T	1	+ T	2	⬭	3	Au1 ⬭	4	F21 ╱	5	╲85	6	⊙	7

Figure 2. (**a**) Regional location of study area. QLA = Qilian tectonic belt; WQL: West Qinling tectonic belt; NQL: North Qaidam tectonic belt; SF1: Wushan–Tianshui–Shangdan suture zone; SF2: Maqu–Nanping–Lueyang suture zone; SG: Songpan–Ganzi tectonic belt; YZ: Yangtze block. (**b**) Geological map of Zaozigou gold deposit (modified from [113]). 1. Quartz diorite porphyrite; 2. Granodiorite-porphyry; 3. Plagioclase granite porphyry; 4. Gold orebody; 5. Fracture; 6. Mineral exploration line; 7. Drilling Hole.

There are only Triassic and Quaternary strata exposed in the Zaozigou gold deposit (Figure 2b). The strata of the area are mainly the lower part of the Gulangdi group (T_2g^1) of the Middle Triassic formation, and the Quaternary (Q_h^{alp}) is developed in the intermountain valley. The lower part of the Gulangdi group (T_2g^1) is the main ore-bearing rock, which is a set of fine clastic rocks consisting of siliceous slate, clastic feldspathic fine sandstone with interbedded siltstone slate and argillaceous slate. The formation is composed of a large sedimentary cycle from bottom to top consisting of siltstone → argillaceous slate → calcareous slate [108,114].

The fractures are well developed and have a complex morphology, forming an intersecting trend, indicating that the area has experienced multiple phases of geological activity. Fractures strictly control the distribution of orebodies and vein rocks within the deposit. They can be classified into five groups of orientations, namely NW, N-S, NE, E-W and NNE [115,116]. Fracture structures are the structural surfaces of mineralization and these structural surfaces control the spreading characteristics of the orebodies [117].

Intermediate-acid dike, including fine crystalline diorite, diorite porphyrite, biotite dioritic porphyrite and quartz diorite porphyrite densely produced with a porphyritic structure, spreads in NNE direction, turning to a nearly N-S direction in the western part of the deposit and a few NW directions. Influenced by the regional multi-period deep fracture activity, the multi-period magmatism has overlapped on multi-phases mineralization within the deposit [118].

There are 147 gold orebodies that have been found in the Zaozigou gold deposit, of which 17 orebodies are main orebodies with gold reserves greater than 1 tonne and the total gold reserves amount to more than 100 tonnes [113]. According to the spatial distribution and combination of the mineralization, the Zaozigou deposit can be divided into eastern and western ore groups.

The eastern ore group is mainly located between Gully 1 and Gully 3 with the strike of NE orientation, containing Au1 controlled by F24, Au9 controlled by F21 and Au15 controlled by F25. These orebodies extend over 1000m long and 300m wide, with a NW direction tendency and steep dip near the ground surface, locally nearly upright. In the deep, these orebodies have been staggered by a gently dipping fracture, causing the tendency to change to a SE orientation (Figure 2b). In addition, orebodies M4 and M6 are laying underground, with the strike of NWW orientation, the tendency of SW orientation and a dip of 8°~26°. These orebodies cross the NE-striking orebodies obliquely, staggering them, and their own mineralization behavior occurs simultaneously [119].

The western ore group is mainly distributed between Gully 3 and Gully 4, spreading in a nearly N-S direction, with Au29~Au31 as the main orebodies. The orebodies extend over 1000 m long and are wider than 500 m, with a strike of 350°~10°, varying tendency and dips greater than 75°, locally subvertical. These orebodies extend, and the mineralization is weaker in the steeper parts and stronger in the shallower parts.

2.2. Datesets Description

Historical geological and geochemical data were completed, including geological reports, geological exploration maps, drills geochemical data, etc., from the Development Research Center of China Geological Survey and No. 3 Geological and Mineral Exploration team, Gansu Provincial Bureau of Geology and Mineral Exploration and Development; the coordinate system used in the mine-scale is Gaussian Kruger projection coordinates.

This study collects data from 72 drillings in the Zaozigou gold deposit and establishes a drilling location database, an assay database, an inclinometry database and a lithology database. The 3D model of drillings is constructed based on the drill hole data database (Figure 3).

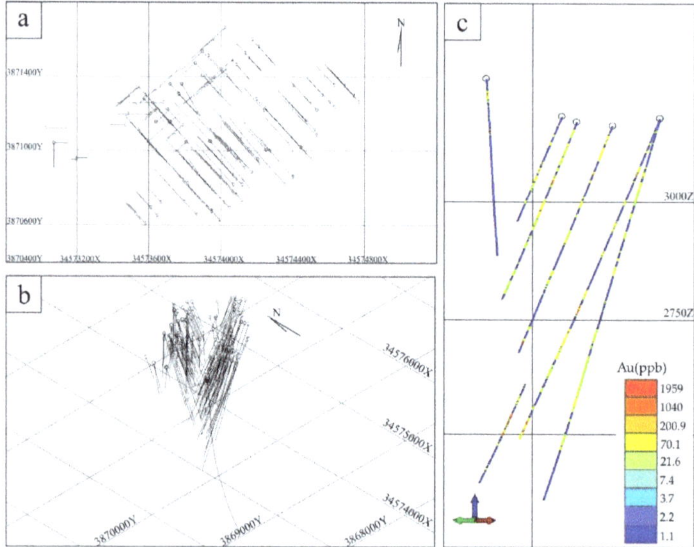

Figure 3. 3D model of Drillings in Zaozigou gold deposit. (**a**) Plan view of drillings distribution. (**b**) Side view of drillings distribution. (**c**) Sample grade distribution of drillings.

The primary geochemical halo data from the drillings are collected from the "Zaozigou gold successive resources exploration project in Hezuo city, Gansu Province", with a total of 72 drillings and 5028 samples with 12 elements of Ag, As, Au, Cu, Hg, Pb, Zn, Sb, W, Bi, Co and Mo (the element detection methods can be seen in the literature [113]). The sampling method was the continuous picking method with sample intervals generally within 10m. Some orebodies or strongly altered areas were sampled at decreased intervals.

3. Methodology

3.1. Concentration-Volume (C-V) Model

Geochemical anomalies are a concept relative to geochemical background, and the multifractal approach provides an effective tool to separate anomalies from the background. For hydrothermal mineralization processes, multi-phase mineralization is common, which will result in multi-phase superposed element spatial distribution [38,120,121]. The C-V model can process the nonlinear primary geochemical halo data with the following equation:

$$V(\rho < v) \propto kv^{-D_1} \quad V(\rho \geq v) \propto kv^{-D_2} \tag{1}$$

where $V(\rho < v)$ and $V(\rho \geq v)$ are the volumes of the element content less than v and greater than v; D_1 and D_2 are the fractal dimension values, also called the singularity index; k is a constant coefficient, which can be calculated by the least square method; v is the threshold value of element contents, and several element content intervals are divided by v. The curves $V(\rho < v)$ and $V(\rho \geq v)$ of the volumes corresponding to different v follow a power-law relationship. Taking the natural logarithm of both sides of the equation, the *lg-lg* plots have a linear relationship in different v intervals. The fractal dimensions in different v intervals can be calculated by the least square method. The geochemical anomalies and backgrounds can be extracted by different fractal dimensions [39].

In practice, the C-V model is used on the primary geochemical halo data volume model for anomaly identification, where the threshold of fractal dimensions may indicate the boundary between different mineralization.

3.2. Compositional Data Analysis

Geochemical data, as typical of compositional data, should be properly transformed before data analysis [122–125] to eliminate the effects of "closure effects". The "opened (transformed)" data can often be analyzed by classical statistical methods to obtain performance improvements [36,53,61,122].

3.2.1. Central Log-Ratio Transformation (clr)

The calculation of this method is: (i) calculate the geometric mean of all compositional subvectors; (ii) divide each subvector by the geometric mean separately; (iii) take the natural logarithm. Its calculation formula is shown as follows.

$$\text{clr}(x) = \left[ln\frac{x_1}{\sqrt[D]{\Pi_{i=1}^{D} x_i}}, ln\frac{x_2}{\sqrt[D]{\Pi_{i=1}^{D} x_i}}, \cdots, ln\frac{x_D}{\sqrt[D]{\Pi_{i=1}^{D} x_i}} \right] \tag{2}$$

3.2.2. Sequential Binary Partition (SBP)

The common isometric log-ratio transformation (ilr) is difficult to interpret. Egozcue [50] proposed the sequential binary partition (SBP) technique based on the ilr transformation, which can provide geochemical interpretation reasonably [66,123,124].

The sequential binary partition technique performs non-overlapping dichotomous classification continuously by the relative information between variables. In practice, positive (+) and negative (−) signs are used to represent two different classifications of compositional variables, and '0' is used to represent the unconcerned variables in one time. By performing continuous non-overlapping dichotomy in a simplex space, a basis vector is formed and then transformed. The results, called compositional balance, can be geologically interpreted according to dichotomy clusters. Especially, the technique provides an important tool to identify element associations [66,123,124,126].

The relevant formula is [50]:

$$B_i = \sqrt{\frac{r_i s_i}{r_i + s_i}} \, ln \frac{\left(\prod_+ x_j \right)^{\frac{1}{r}}}{\left(\prod_- x_k \right)^{\frac{1}{s}}} \quad i = 1, 2, \ldots, D-1; \ j = 1, 2, \ldots, r; \ k = 1, 2, \ldots, s \quad (3)$$

where B_i denotes the new compositional vector defined by the standard orthogonal basis, $\prod_+ x_j$ is the product of the variables labeled (+) involved in the ith binary partition and $\prod_- x_k$ is the product of the variables labeled (−) involved in the ith binary partition.

3.2.3. Geochemical Compositional Data Analysis Framework Based on CoDA

The log-ratio transformation of geochemical data can solve the problem of the "closure effect" caused by the lack of scale consistency in the covariance matrix of the compositional data. The clr-biplot and compositional balance methods developed based on the clr and SBP have their advantages in the geochemical associations' extraction, especially in inferring lithology, faults, alteration and mineralization, and the corresponding frameworks have been well applied [65,66,124].

This study uses a data-driven and knowledge-driven framework of compositional data analysis to identify the geochemical associations of the primary halo. The data-driven framework infers the data characteristics by measuring elemental statistical correlations to gain the element associations, while the knowledge-driven framework is based on geological and geochemical understanding to design the element associations (Figure 4).

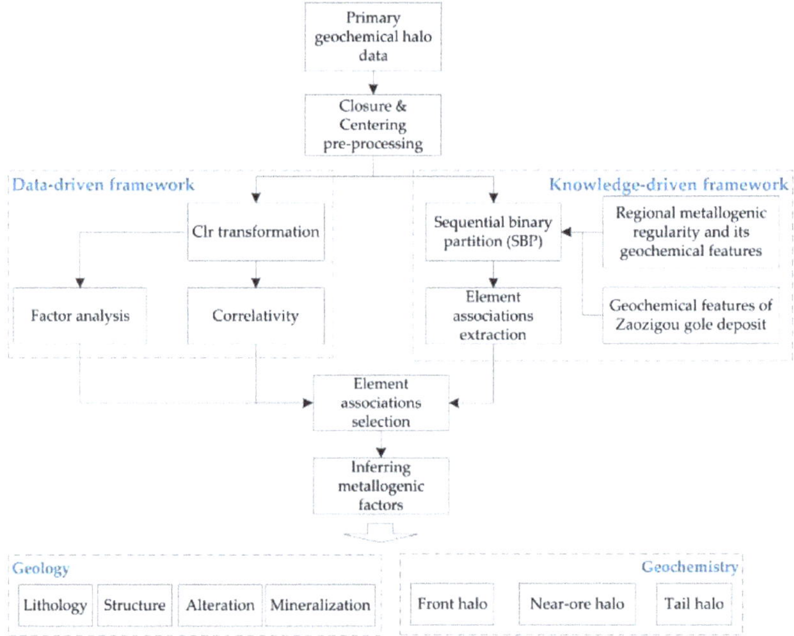

Figure 4. Geochemical CoDA framework.

In Figure 4, the closure and centering are necessary preprocess steps, which can can be seen in the literature [51]. The data-driven framework in this study used clr transformation to "open" geochemical data, and the factor analysis and correlativity methods are used to explore the relationship among elements and extract element associations. Meanwhile, the knowledge-driven framework designs the element associations through deep research of the geological features, such as the element concentrations in different rocks, the primary geochemical halo associations, and so on. Then, the SBP is performed to quantitatively extract the value of corresponding geochemical associations. Eventually, the results of the CoDA can be employed to infer metallogenic information in an unknown area.

3.3. Machine Learning-Based Quantitative Mineral Prediction Methods

Machine learning algorithms could highlight hidden details in datasets without explicit search and have the ability to identify complex spatial patterns [127]. Given this, scholars have attempted to use machine learning algorithms to extract mineralization information by integrating multi-source geosciences data for identifying mineralization-related anomalies and their variation that cannot be detected by traditional methods [128–135]. Machine learning algorithms can be roughly categorized as supervised learning algorithms and unsupervised algorithms. For better validating the suitability of 3D MPM, this study discusses the application of the supervised algorithm MaxEnt model and unsupervised algorithm GMM.

3.3.1. MaxEnt Model

The principle of MaxEnt is a criterion of probabilistic model learning for making predictions based on incomplete information. The main idea is that, when predicting the probability distribution of a random event, all known constraints need to be satisfied without subjective assumptions so that the probability distribution is most uniform, the prediction risk is minimal and the entropy value of the probability distribution is maximum [136].

Let $T = \{(x_1, y_1), (x_2, y_2), \dots (x_n, y_n)\}$ be the training dataset and $f_i(x, y), i = 1, 2, \dots, n$ be the eigenfunction, and the learning process of the MaxEnt model is equivalent to the constrained optimization problem:

$$\max_{P \in C} H(P) = - \sum_{x,y} \widetilde{P}(x) P(y|x) ln(P(y|x))$$
$$s.t.\, E_P(f_i) = E_{\widetilde{P}}(f_i)\, i = 1, 2, \dots, n \tag{4}$$
$$\sum_y P(y|x) = 1$$

where $E_{\widetilde{P}}(f_i) = \sum_{x,y} \widetilde{P}(x, y) f_i(x, y)$ is the expected value of n eigenfunction $f_i(x, y)$ related to the empirical distribution $\widetilde{P}(x, y)$; $E_{\widetilde{P}}(f_i) = \sum_{x,y} \widetilde{P}(x) P(y|x) f_i(x, y)$ is the expected value of n eigenfunctions $f_i(x, y)$ related to the $P(Y|X)$ with the empirical distribution $\widetilde{P}(X)$.

Following the custom of optimization problems, the problem of finding the maximum value is rewritten as the equivalent problem of finding the minimum value:

$$\min_{P \in C} -H(P) = \sum_{x,y} \widetilde{P}(x) P(y|x) ln(P(y|x))$$
$$s.t.\, E_P(f_i) - E_{\widetilde{P}}(f_i) = 0\, i = 1, 2, \dots, n \tag{5}$$
$$\sum_y P(y|x) = 1$$

The solution resulting from solving the constrained optimization problem is the solution learned by the MaxEnt model. However, the empirical distribution expectation is usually not equal to the true expectation but will be approximate to the true expectation. If the solution is solved strictly according to the above constraints, it is easy to cause overfitting of the training data during the learning process. Therefore, the constraints can be appropriately relaxed in practice, such as replacing the above equation with $E_P(f_i) - E_{\widetilde{P}}(f_i) \leq \beta_i$, β_i is the modulation multiplier, which is a constant.

3.3.2. Gaussian Mixture Model (GMM)

GMM is a quantified model which generated through Gaussian probability density function fitting. This model decomposes objective distribution into several Gaussian probability density functions. By using enough Gaussian functions and tuning the parameters, the model can generate a very complex probability density to approximate to almost any continuous probability. Theoretically, any objective distribution can be fitted by a combination of multiple Gaussian density functions, and the higher the number of probability density sub-functions, the more closely it approximates to the actual data distribution. GMM has the advantages of good flexibility, is not limited by the sample size and can accurately describe the data structure.

When the dataset $X = (X_1, X_2, \ldots, X_T)$ of the training data can be divided into k classes and each class obeys Gaussian distribution, the k order probability distribution of the GMM is

$$P(X|\theta) = \sum_{i=1}^{k} w_i P(X|\theta_i) \tag{6}$$

$$P(X|\theta_i) = \frac{1}{(2\pi)^{\frac{D}{2}} |\Sigma_i|^{\frac{1}{2}}} exp\left(-\frac{1}{2}(X - \mu_i)^T \sum_i^{-1}(X - \mu_i)\right) \tag{7}$$

$$\sum_{i=1}^{k} w_i = 1, \ w_i > 0 \tag{8}$$

where $P(X|\theta_i)$ is the probability density of the ith Gaussian model, w_i is the weight of the ith model in the whole GMM, μ_i is the mean vector and $\sum_i$ is the covariance matrix. After the parameter initialization, the Expectation-Maximization (EM) algorithm based on the maximum likelihood estimation is often chosen for the parameter estimation of the GMM [137].

In the geological point of view, the GMM is an unsupervised machine learning algorithm; it divides the data into two categories (mineralization and non-mineralization) and then makes category judgments on samples by learning information about unlabeled samples.

As for the dataset $D = \{x_1, x_2, \ldots, x_m\}$, which contains k Gaussian mixture components, the algorithm steps are as follows.

1. Initializing the k multivariate Gaussian distributions and their weights.
2. Estimating the posterior probability of each sample generated by each component according to Bayes' theorem.
3. Updating the mean vector, covariance matrix and the weights according to the step 2.
4. Repeating steps 2–3 until the increase in the likelihood function has been less than the convergence threshold, or the maximum number of iterations is reached.
5. For each sample point, calculating its posterior probability of belonging to each cluster according to Bayes' theorem and classifying the sample into the cluster with the largest posterior probability.

4. Results

4.1. Three-Dimensional Primary Geochemical Halo Anomaly Data Volume Modeling Based on the C-V Model

This section adopts the multifractal C-V model to analyze the spatial anomalous structure of elements and provides single-element anomaly signatures for the subsequent deep prediction.

Taking the ore-forming element Au as an example:

Using the ordinary kriging interpolation method [138–140], with the experimental variogram fitted to build a 3D data volume model of Au (Figure 5), the Au content value observably does not obey the normal distribution (Figure 6a). Therefore, the multifractal C-V model was carried out to identify Au anomalies.

Through counting the variation of volume with Au content, the lgAu-lgV scatter diagram was generated and lg-lg lines were fitted by the least square method (Figure 3b). Meanwhile, the fractal dimension can be obtained as follows:

The slopes of the lines correspond to fractal dimensions, and the inflection points are threshold values of geochemical abnormal intensity as shown in Table 1 and Figure 6b.

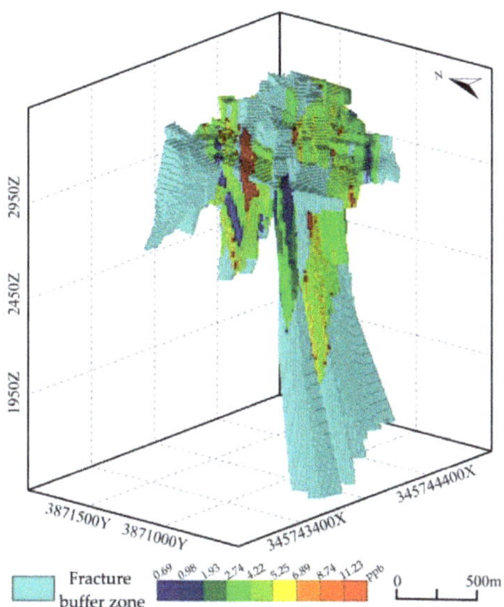

Figure 5. Three-dimensional data volume model of Au.

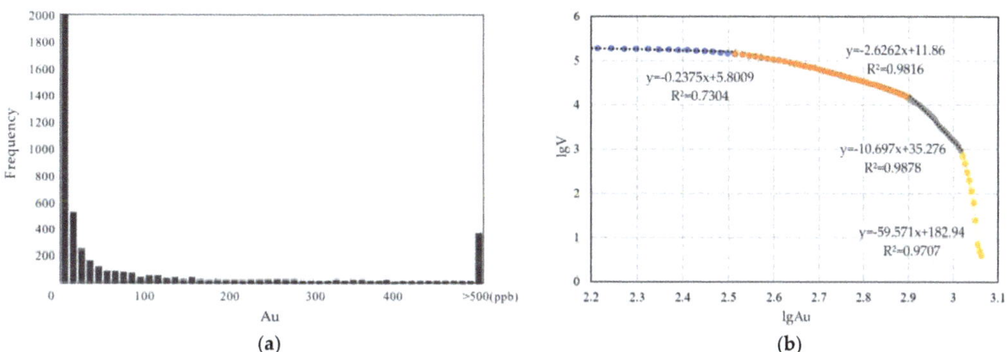

(a)   (b)

Figure 6. (**a**) Histogram of the frequency distribution of Au. (**b**) lgAu-lgV curve of the 3D data volume model.

Table 1. Fractal characteristics of 3D Au data volume model.

Anomaly Classes	Fractal Dimension	R Square (R^2)	Inflection Point	Au (ppb)
Background area	0.2375	0.7304	1.0876	12.2349
Outer anomalies	2.6262	0.9816	2.5137	326.3623
Middle anomaly	10.697	0.9878	2.9049	803.3411
Internal anomaly	59.571	0.9707	3.0202	1047.6108

The orebody is compared with the outside anomalies, central anomaly and internal anomaly area by superposition, as shown in Figure 7b,c, and the three show a good spatial correlation with the orebody.

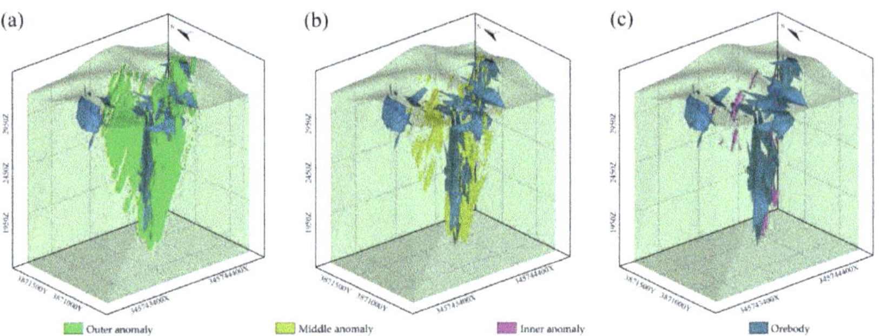

Figure 7. Three-dimensional model of Au anomaly data volume in Zaozigou gold deposit. (**a**) outer anomalies of Au. (**b**) Middle anomalies of Au. (**c**) Inner anomalies of Au.

The middle and inner anomalies are mainly distributed inside and around the orebody, which accurately reflects the spatial spreading of the orebody and its trend.

On this basis, this study analyzed and visualized the 3D anomalous structures of the remaining 11 elements using the above model (Figure 8; Table 2).

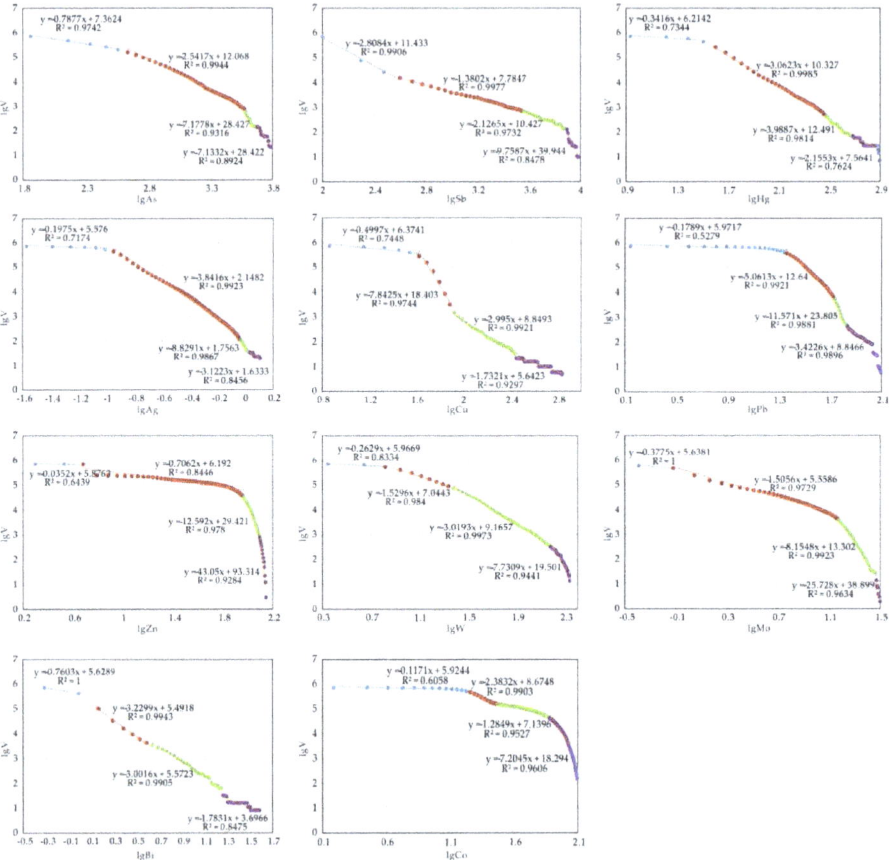

Figure 8. lg-lg graph of the 3D data volume model.

Table 2. Fractal characteristics of remaining elements in 3D data volume model.

Element	Anomaly Classes	Fractal Dimension	R Square (R^2)	Inflection Point	Cut-Off Value
As	Background area	0.7877	0.9742	1.8621	72.7949
	Outer anomalies	2.5417	0.9944	2.5595	362.653
	Middle anomaly	7.1778	0.9316	3.5762	3768.486
	Internal anomaly	7.1332	0.8924	3.6731	4710.524
Sb	Background area	2.8084	0.9906	2.0003	100.091
	Outer anomalies	1.3802	0.9977	2.6022	400.0882
	Middle anomaly	2.1265	0.9732	−0.4559	3500.06
	Internal anomaly	9.7587	0.8478	−0.1079	7800
Hg	Background area	0.3416	0.7344	0.9318	8.5462
	Outer anomalies	3.0623	0.9985	1.5123	32.5296
	Middle anomaly	3.9887	0.9814	2.4599	288.3531
	Internal anomaly	2.1553	0.7624	2.6814	480.2207
Ag	Background area	0.1975	0.7174	−1.5708	0.0289
	Outer anomalies	3.8416	0.9923	−1.0079	0.0982
	Middle anomaly	8.8291	0.9867	−0.0403	0.9114
	Internal anomaly	−3.1223	0.8456	0.0287	1.0683
Cu	Background area	0.4997	0.7448	0.864	7.312
	Outer anomalies	7.8425	0.9744	1.6271	42.372
	Middle anomaly	2.995	0.9921	1.9266	84.444
	Internal anomaly	1.7321	0.9297	2.4374	273.768
Pb	Background area	0.1789	0.5279	0.1436	1.392
	Outer anomalies	5.0613	0.9921	1.3684	23.356
	Middle anomaly	11.571	0.9881	1.7353	54.364
	Internal anomaly	3.4226	0.9896	1.8362	68.576
Zn	Background area	0.0352	0.6439	0.2996	1.993
	Outer anomalies	0.7062	0.8446	0.7904	6.172
	Middle anomaly	12.592	0.978	1.9663	92.538
	Internal anomaly	43.05	0.9284	2.0954	124.577
W	Background area	0.2629	0.8334	0.3484	2.2303
	Outer anomalies	1.5296	0.984	0.8215	6.6297
	Middle anomaly	3.0193	0.9973	1.343	22.0275
	Internal anomaly	7.7309	0.9441	2.175	149.6098
Mo	Background area	0.3775	1	−0.3907	0.4067
	Outer anomalies	1.5056	0.9729	−0.1184	0.7614
	Middle anomaly	8.1548	0.9923	1.1745	14.9466
	Internal anomaly	25.728	0.9634	1.4696	29.4864
Bi	Background area	0.7603	1	−0.3087	0.4913
	Outer anomalies	3.2299	0.9943	−0.0116	0.9737
	Middle anomaly	3.0016	0.9905	1.2518	3.8681
	Internal anomaly	1.7831	0.8475	1.5919	17.8579
Co	Background area	0.1171	0.6058	0.1909	1.5524
	Outer anomalies	2.3832	0.9903	1.2625	18.3011
	Middle anomaly	1.2849	0.9527	1.4756	29.8963
	Internal anomaly	7.2045	0.9606	1.8749	74.9889

Note: The cut-off value unit of Hg and Ag is ppb, others are ppm.

Figure 9 shows that the middle anomalies of As and Sb are mainly distributed near the elevation of 1700~1900 m. The middle anomalies of Ag, Cu, Pb and Zn have close relationship to the orebody. W, Mo, Co and Bi have two concentrations, the first one is located near the surface and the second one is distributed near the elevation of 2500 m. The Zaozigou gold deposit has multi-phase mineralization, forming a complicated spatial distribution of elements, while the C-V model can better identify the anomaly for recognizing the pattern of the primary geochemical halo in the Zaozigou gold deposit.

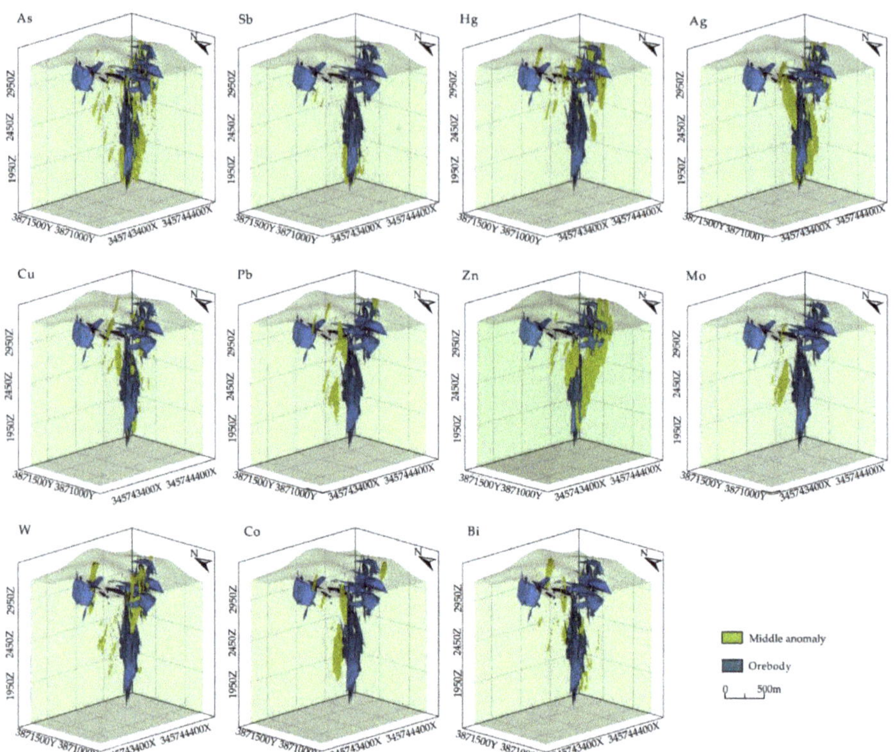

Figure 9. The middle anomalies distribution in Zaozigou gold deposit.

4.2. Data-Driven CoDA and Its Based Element Association Extraction

4.2.1. Correlation Analysis

The primary halos were processed by cluster analysis (Figure 10). The elements can be roughly divided into two groups: Au, As, Sb, Ag, W, Hg and Cu, Zn, Bi, Pb, Co, Mo. Among them, the elements most closely associated with Au are As, Sb and Ag; the Au, As, Sb and Cl are moderate volatile elements [141] often associated in gold mineralization like Hg (indicator of volcanism). Ag is often associated to gold as electrum. Hg should be the front halo indicating element of the gold orebody; the correlation coefficient between Au and As reaches 0.8, and most of the As exists within arsenopyrite, which is an important gold-bearing mineral. Therefore, this Au-As-Sb should be the element association of mineralization reflecting a middle- and low-temperature metallogenic environment.

4.2.2. Element Associations Identification Based on clr-Biplot

Geochemical data are typically compositional data, and if traditional multivariate statistical methods (e.g., principal component analysis, factor analysis, etc.) are applied directly to the raw geochemical data, it may lead to erroneous results. Therefore, the raw data should be properly transformed before data analysis is performed.

Data from 12 geochemical elements were clr-transformed, and the skewness values of the clr-transformed data were statistically obtained (Figure 11). Compared with the raw data, clr-transformed data has the lower skewness value around zero, indicating that the data distribution after clr transformation tends to be more normal in character (Figures 11 and 12).

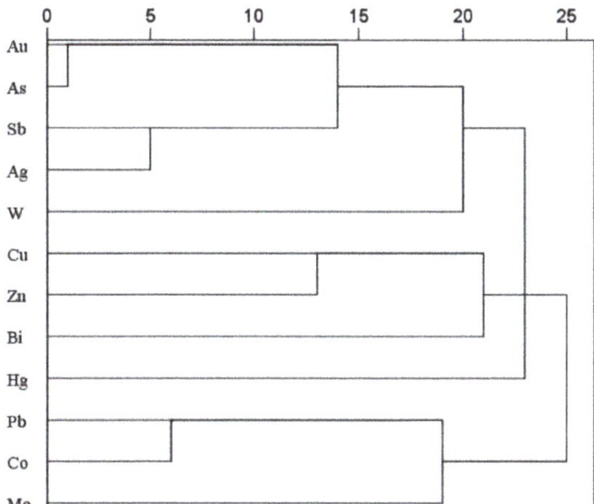

Figure 10. Hierarchical dendrogram of cluster analysis in Zaozigou gold deposit.

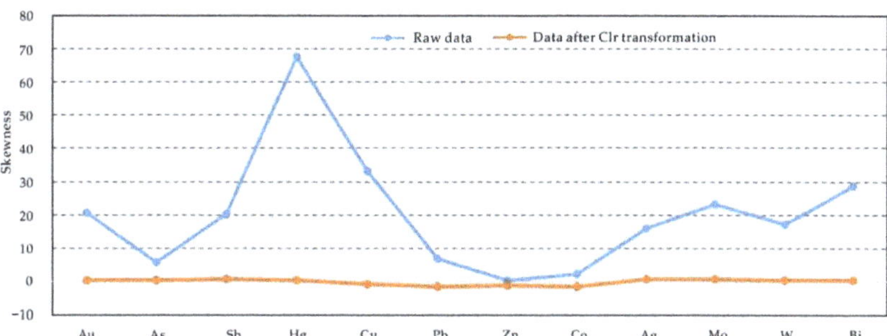

Figure 11. Comparison of the skewness values of elemental data before and after clr transformation.

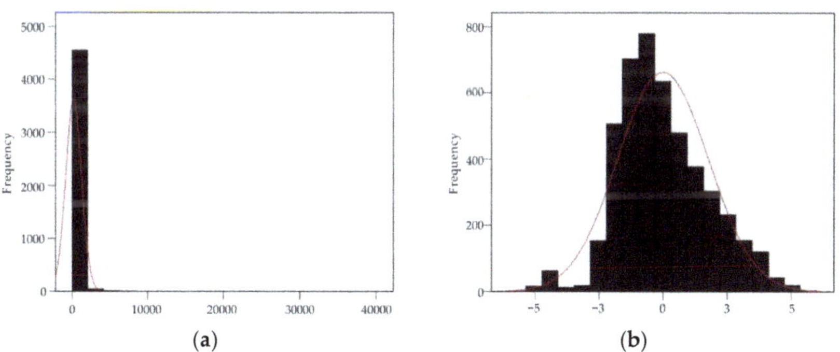

Figure 12. Histogram of Au. (**a**) Raw data. (**b**) clr-transformed data.

Factor analysis (FA) is used to extract element associations. Four factors are extracted as element associations to indicate different geological and geochemical meanings. From the view of the loadings of FA, factor F1 (34.95%, five variables) represents Cu, Pb, Zn, Ag and Bi, which are a group of medium-temperature elements; factor F2 (12.83%, three variables) is the element association of Au, As and Sb, representing the Au-polymetallic sulfide phase, which is the most dominant phase of gold mineralization; factor F3 (11.13%, one variable) indicates Mo, which is a high-temperature element and may be related to magmatism; factor F4 (9.26%, two variables) is Sb and Hg association, where Sb mainly exists within the form of stibnite within the quartz-stibnite veins and Hg is closely related to fractures (Figure 13). The distribution of each factor in the three-dimensional space is shown in Figure 14a.

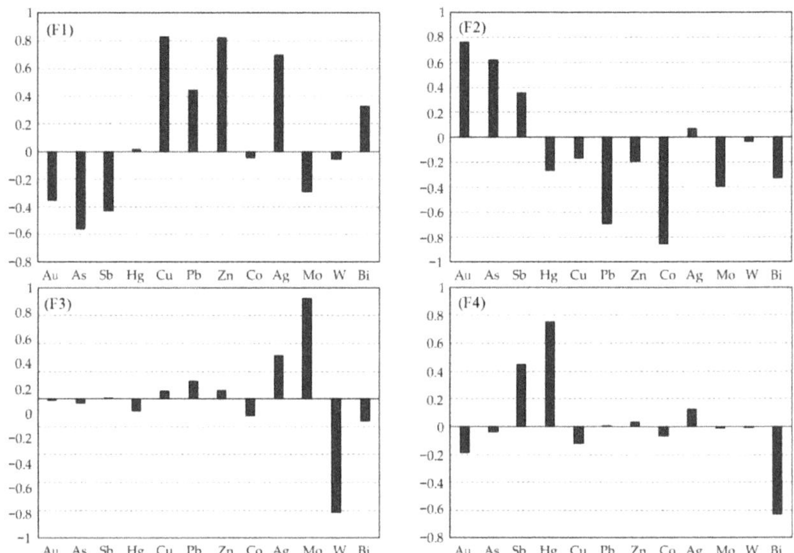

Figure 13. Distribution of element loadings in each factor.

To analyze and show the geological meaning of each factor more clearly, the 85# profile was selected for analysis by profile cutting (Figure 14). It can be intuitively understood that the F2 factor is closely related to mineralization, and its spatial distribution is well matched with the known orebodies in the exploration profile. The F4 factor is closely related to ore-bearing fractures, so that the Sb-Hg element association could be used as evidence of deep fracture extension.

4.3. Knowledge-Driven CoDA and Its Based Element Association Extraction

From the anomaly data volume model based on the C-V method in this paper, it is clear to recognize the distribution of each element in a three-dimensional space.

1. In the mid-shallow part and the deeper part of elevation at 2500 m, As, Sb and Hg are concentrated as the front halo of the orebodies. The anomalies of Au, Ag, Cu, Pb and Zn linked to sulphurs (pyrite, arsenopyrite, galena and covelline) and are superposed with the orebodies and can be regarded as the near-ore halo elements. At the shallow position of 2500 m, the anomaly locations of W, Mo and Bi linked to magmatism are at the tail of the orebodies, which can be regarded as the tail halo element association.
2. In terms of Au, its anomalies are controlled by fractures observably, and observable fractures certainly cross-cut orebodies. Especially, the abnormal intensity is larger along the depth indicating the deep mineralization.

Based on the analysis above, the knowledge of the element associations of the front halo, near-ore halo and tail halo can be summarized. Moreover, corresponding element associations are quantitatively extracted as compositional balances by the knowledge-driven CoDA framework (Figure 4), that is, the front halo association is B1 (As-Sb-Hg vs Au-Ag-Cu-Pb-Zn-W-Bi-Co-Mo), the near-ore halo association is B2 (Au-Ag-Cu-Pb-Zn vs As-Sb-Hg-W-Bi-Co-Mo), and the tail halo association is B3 (W-Bi-Co-Mo vs As-Sb-Hg-Au-Ag-Cu-Pb-Zn) (Figure 15).

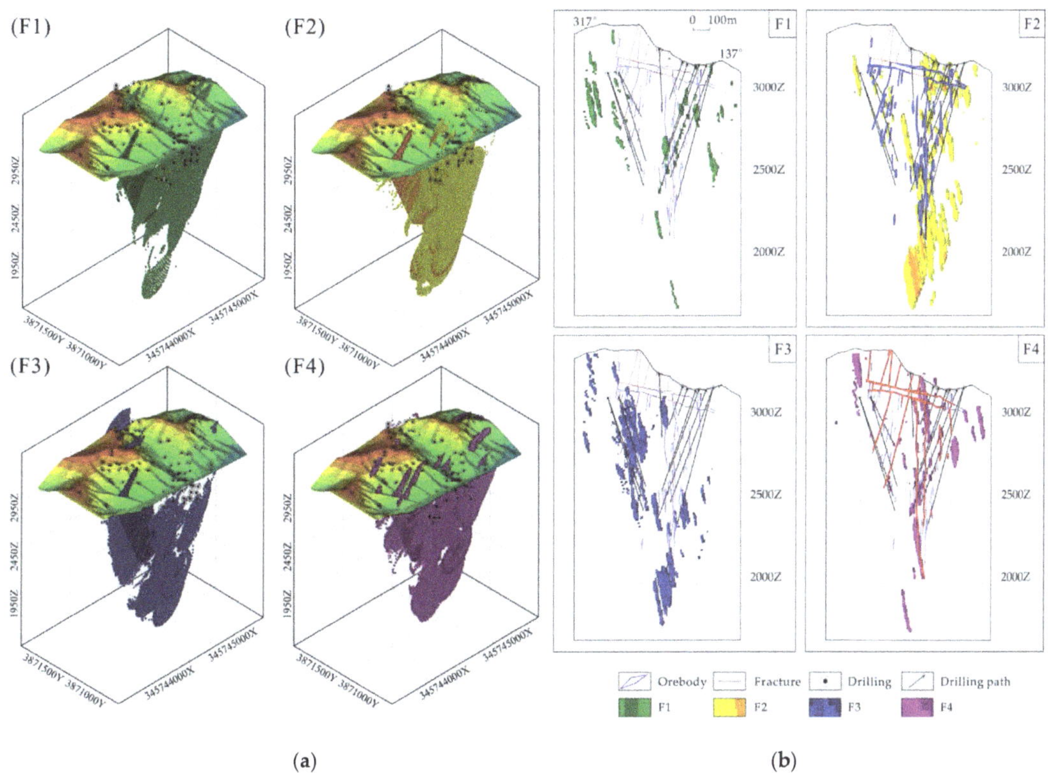

Figure 14. (**a**) Factor scores distribution of each factor. (**b**) Distribution of factor scores in the 85# exploration profile.

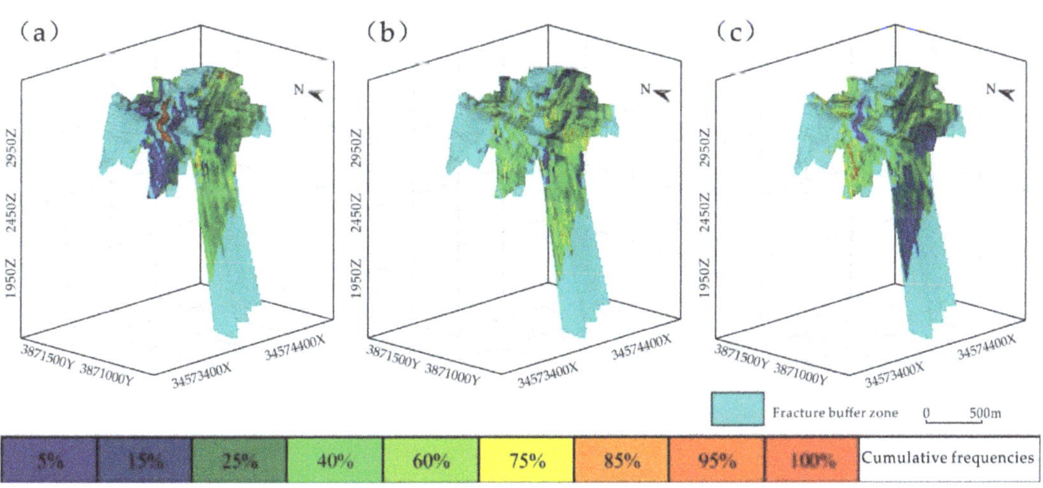

Figure 15. Primary Geochemical halo element associations. (**a**) Front halo association B1; (**b**) Near-ore halo association B2; (**c**) Tail halo association B3.

4.4. Geological and Geochemical Quantitative Prediction Model at Depth of Zaozigou Gold Deposit

The mineral resources prediction model is usually summarized as text, diagrams, and tables by integrating comprehensive metallogenic information, such as orebodies, ore deposits, ore fields and even metallogenic zones. Establishing a mineral resources prediction model is an effective way to discover potential deposits and has significantly important meaning for guiding mineral exploration [142].

Orebodies are strictly controlled by fractures in the Zaozigou gold deposit. The 30 m buffer zone of the fractures can effectively reflect the influence range of the fracture, which can be used as a mineral prediction indicator. Factor F4 is an element association of Sb-Hg, which has close relationship with fractures, and can be used as a favorable indicator for inferring deep fractures [143–146] (Figure 14a).

Geochemical element distribution, association and zonation are favorable indicators for mineral resources prediction. The geochemical anomalies are extracted by the multiple fractal C-V model in Section 4.1, among which the middle anomaly of Au can well reflect the spatial distribution of orebodies (Figure 7b) and should be used as an important quantitative indicator. Near-ore halo element association B2 is extracted by the knowledge-driven CoDA and also can express the location of orebodies well; it should be another mineral prediction indicator (Figure 15). The ratio of front halo to tail halo is an important geochemical parameter for predicting orebodies, and B1/B3 is regarded as a prediction indicator accordingly [143] (Figure 16).

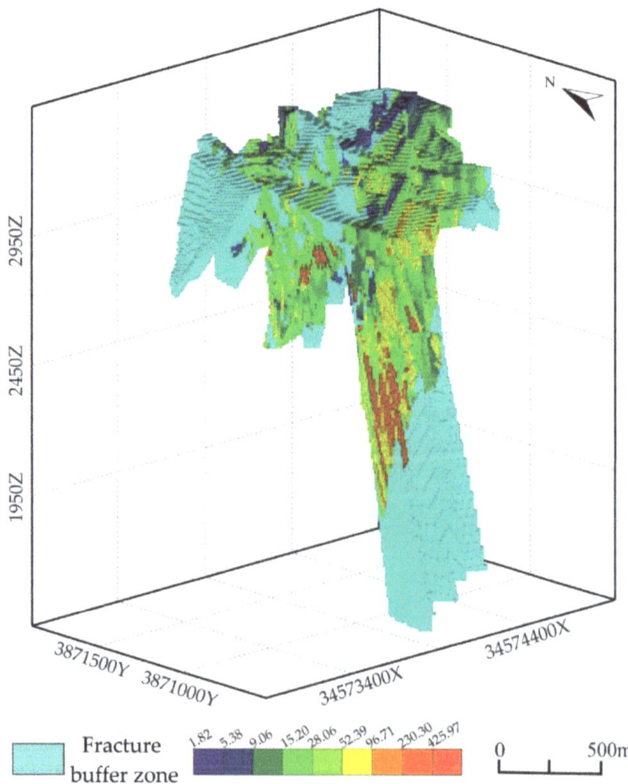

Figure 16. Spatial distribution of As-Sb-Hg(B1)/W-Bi-Co-Mo(B3).

In summary, the geological and geochemical quantitative prediction model at the depth of the Zaozigou gold deposit is constructed as in Table 3.

Table 3. Geological and geochemical quantitative mineral resource prediction model at depth of Zaozigou gold deposit.

Ore-Forming Factor	Description	Prediction Indicator	Variables
Geology	Fracture	Influence range of fracture	30 m buffer zone
		Element association of fracture	Hg-Sb (F4)
Geochemistry	Ore-forming element	Geochemical anomaly	Au
	Primary geochemical halo	Element association of near-ore halo	Au-Ag-Cu-Pb-Zn (B2)
		Geochemical parameter (Front halo/tail halo)	As-Sb-Hg(B1)/W-Bi-Co-Mo (B3)

4.5. Three-Dimensional MPM Based on Machine Learning

To address the scientific problem of quantitative mineralization prediction at large depths, the previous section quantitatively extracted the deep geochemical mineralization signatures and constructed a geological and geochemical quantitative mineral resource prediction model at depth. In this section, the MaxEnt model and GMM are applied to carry out the 3D MPM to quantitatively predict deep mineral resources, and the uncertainty evaluation of the two models is performed for improving the accuracy of mineralization prediction.

4.5.1. Training Sample Construction

The MaxEnt model is a supervised machine learning algorithm, which requires learning an optimal model from a given training dataset and using this model to output the corresponding result for classification.

For mineralization prediction, the target variable of supervised learning (i.e., the label of training samples) is either mineralized or nonmineralized (denoted by 1 and 0, respectively). A total of 39,306 positive samples were extracted from known orebodies and 49,686 negative samples were extracted from the nonmineralized position confirmed by drillings, and these were used as the training dataset. In contrast to mineralization, which generates in a concentrated way in a limited space, the non-mineralization is a widespread phenomenon, and negative samples are selected to be distributed as randomly and uniformly as possible in the wall rock without mineralization and alteration throughout the study area [33] (Figure 17).

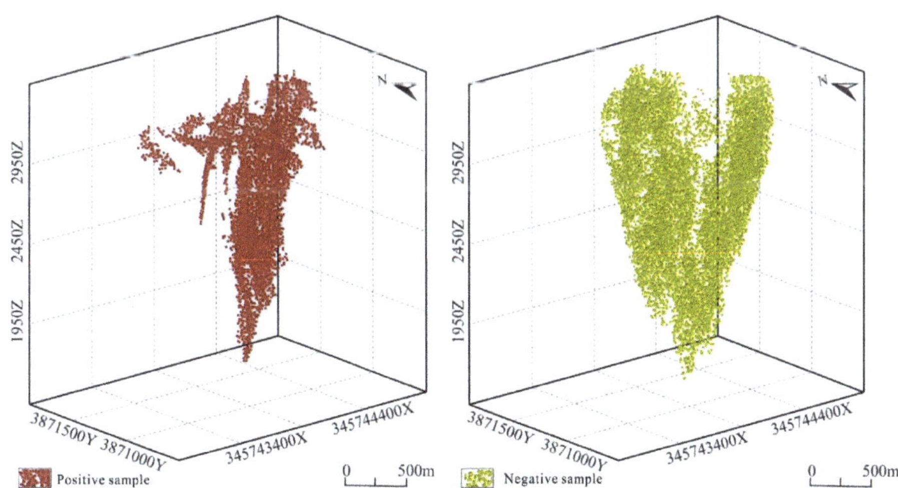

Figure 17. Distribution of positive samples and negative samples.

4.5.2. Three-Dimensional MPM and Uncertainty Evaluation of MaxEnt Model

The MaxEnt method originated from statistical mechanics and was developed by Phillips et al. using JAVA. This study uses version 3.4.1 of MaxEnt software (https://biodiversityinformatics.amnh.org/open_source/maxent/, (accessed on 15 June 2022)) to carry out the 3D MPM.

The MaxEnt method estimates the probability of the target variable with the maximum entropy value and is controlled by a set of constraints representing incomplete information about the target distribution. In mineralization prediction, the best interpretation of unknown occurrences by the model is to maximize the entropy value of the probability distribution for estimating the location of orebodies, and many scholars have achieved better results in this regard [65,76,147]

When modeling with MaxEnt software, if the model parameters are not set properly, it may lead to overfitting or redundancy [148]. The overfitting can be controlled by the modulation multiplier β [149], and the best performance of the model is obtained by setting the β value 2~4 [148,150]. Therefore, the study tested different values to find the best β value of 2 for the model to reduce the influence of model overfitting.

Five prediction indicators are integrated into the MaxEnt model as input parameters, and data were randomly selected from the dataset during simulation, with 75% of the dataset as the training data and 25% as the test data. To reduce the randomness of the simulation results, the model is repeated 50 times with a maximum convergence threshold of 0.00001. A maximum background points value of 10,000 is selected, and a logical value format output is chosen for a more favorable interpretation of the results.

The final prediction result of the MaxEnt model is evaluated using the average value of 50 iterations of the simulation, with the contribution rate of each mineral indicator shown in Table 4.

Table 4. Contribution rate of prediction indicators.

Prediction Indicator	Rate of Contribution (%)
Au	77.6
30m buffer zone	13
Au-Ag-Cu-Pb-Zn (B2)	7.7
Hg-Sb (F4)	0.9
As-Sb-Hg(B1)/W-Bi-Co-Mo(B3)	0.8

The AUC value of the test dataset is 0.844 and the AUC value of the training dataset is 0.848, so the MaxEnt model has high accuracy in mineral resources prediction at depth (Figure 18).

The output logical probability of the MaxEnt model is in the range of 0.000804~0.927941, which is mapped to 0 and 1, and then the 3D MPM is formed (Figure 19).

Although the mineral prospectivity map shows a good relationship between high probabilities and known gold orebodies (Figure 19), it is difficult to determine a certain logistic probabilities value as the prediction threshold value.

We take the ratio of prediction volume to orebodies occupied volume as a parameter. Observably, it must be that reverse variation of this parameter follows the greater logistic probabilities (Figure 20). The high potential area (logical probability > 0.525) is defined by the logistic probability of 0.525, which covers 80% of the known orebodies; the medium potential area (0.3 < logical probability < 0.525) is defined by the logistic probability of 0.3, which is another inflection point and covers all the known orebodies (Figure 20). The spatial distribution of mineralization potential areas by the MaxEnt model are shown in Figure 21a, based on which two mineral exploration targets are circled (Figure 21b).

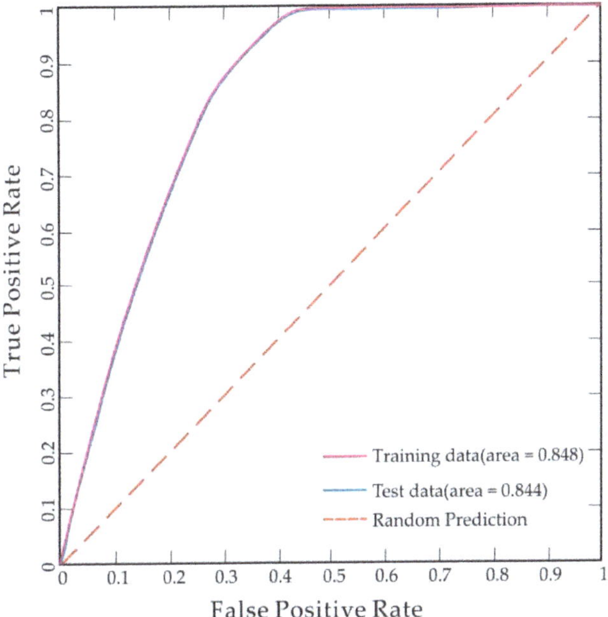

Figure 18. ROC curve of MaxEnt model.

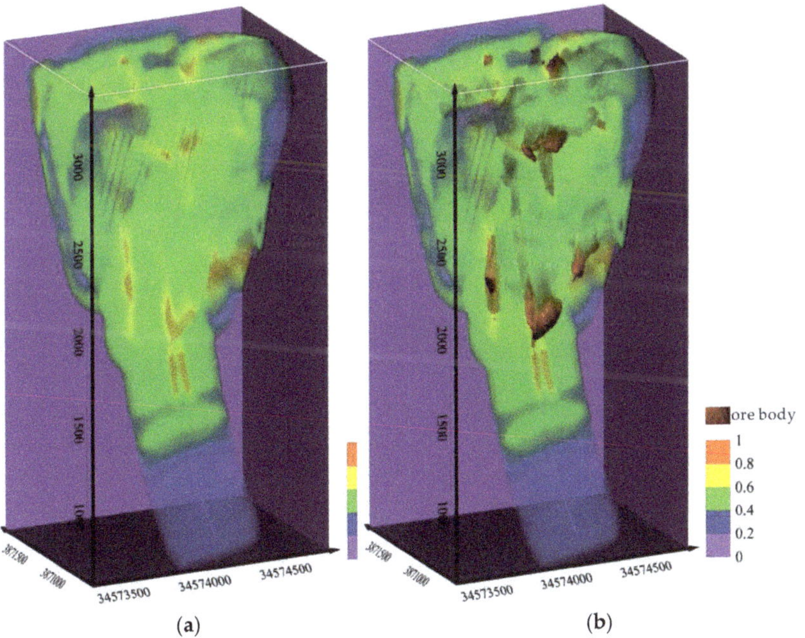

Figure 19. Three-dimensional MPM by MaxEnt model. (**a**) Three-dimensional MPM; (**b**) three-dimensional MPM with orebodies.

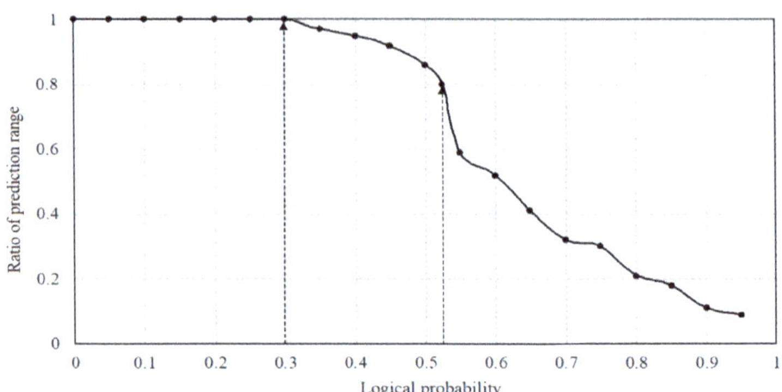

Figure 20. Logical probability versus the ratio of prediction volume to orebody occupied volume.

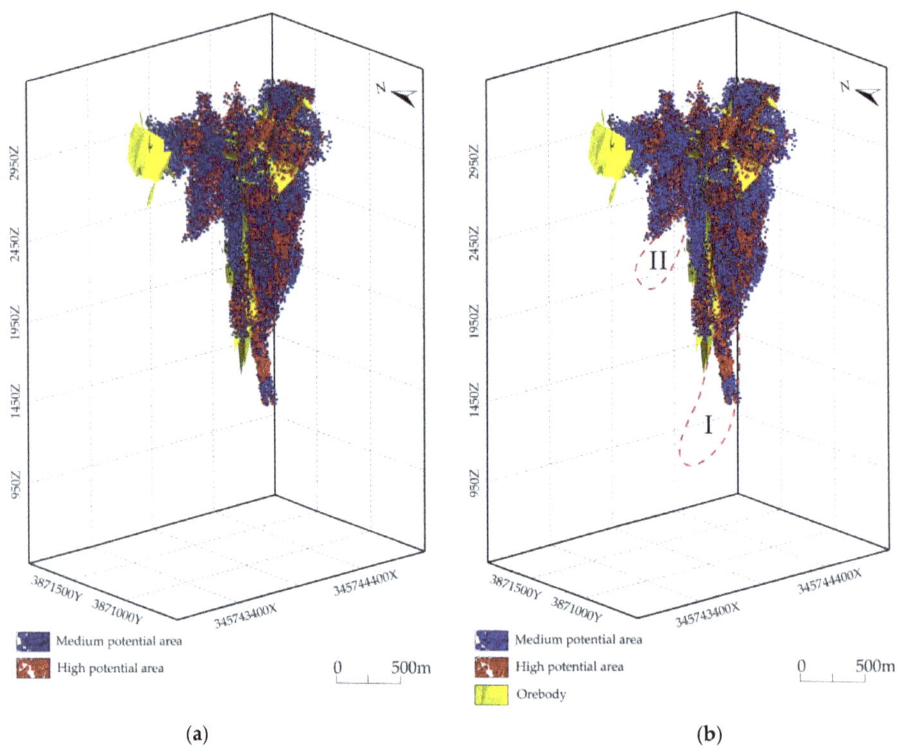

Figure 21. (**a**) The distribution of mineralization potential areas. (**b**) MaxEnt model-based exploration targets.

4.5.3. Three-Dimensional MPM and Uncertainty Evaluation of GMM

When training and testing the model, the labeled data (which is used only in the evaluation) is divided into 75% for the training dataset and 25% for the test dataset, and the GMM is used to learn the information of the training dataset, and then the ROC curve is used for performance evaluation of the training dataset and test dataset, respectively. The AUC value of the test dataset is 0.75 and the AUC value of the training dataset is 0.75 (Figure 22).

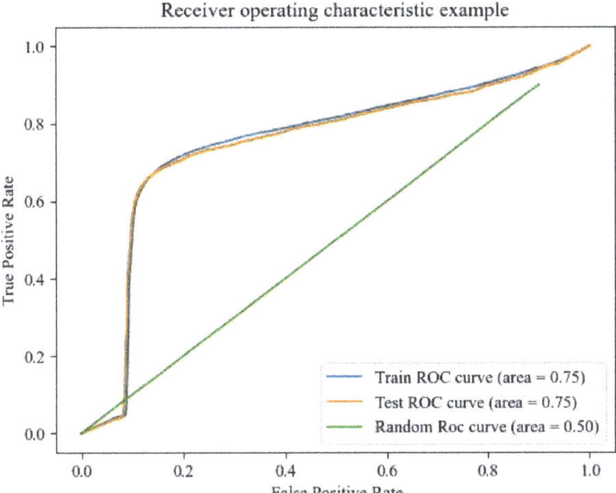

Figure 22. ROC curve of GMM.

From the point of the AUC value, 0.75 is not a high value, which may indicate that unsupervised training methods have a shortage in prediction with big data.

However, from the mineral prospectivity map point of view, the prediction area has covered orebodies well, and it still has a certain indication function in the mineralization prediction. Finally, two mineral exploration targets are delineated at depth (Figure 23).

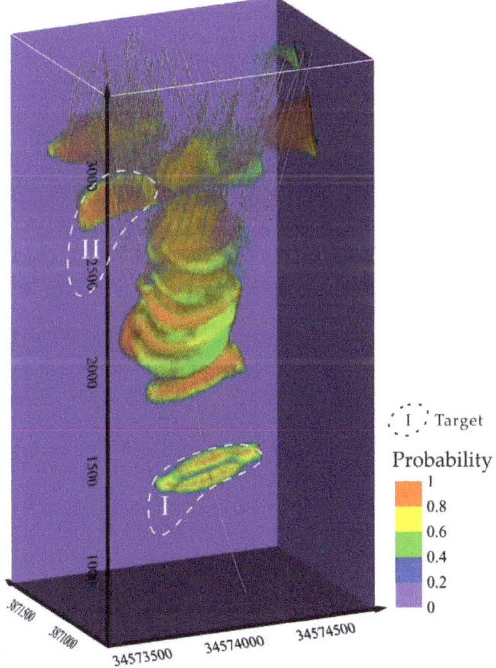

Figure 23. Mineral resources prediction at large depth based on GMM.

5. Discussion

This study employed the geostatistical interpolation method to build a 3D geochemical model and geochemical anomaly model. In addition to deterministic modeling of 3D geology and geochemistry [151,152], geostatistical techniques also include uncertainty modeling of spatial distribution of subsurface heterogeneous structures and dynamic processes of fluid migration [153]. In view of the 3D heterogeneous structure, the multi-point geostatistical method can be used to overcome the shortage of traditional geostatistical simulations in delineating the geometric continuity of geological structures [154–159]. Meanwhile, traditional geostatistical simulation has the limitations of large computation, complicated parameterization and difficult to characterize multi-scale data. The application of machine learning and deep learning methods to reconstruct geological and geochemical structures can improve the simulation efficiency and can accurately express complex heterogeneous spatial structures [160,161], which deserves further research work.

The machine learning methods of the MaxEnt model and GMM are carried out for 3D MPM in the Zaozigou gold deposit in this study. Compared with the GMM, the MaxEnt model has a higher precision in detection of ore-induced anomalies, which demonstrates a higher reliability of 3D MPM (Figures 18 and 22). The prediction results of the two methods express a high correlation with the known orebodies, based on which two mineral exploration targets are circled (Figures 21 and 23).

Target I of the MaxEnt model is located at an elevation of 1600~2000 m, belonging to the NE-orientation orebody group, which should be the extension of the Au1 orebody. The Au and Sb concentration in this position (Figure 9) and the ratio of front halo to tail halo has been increasing (Figure 16). Additionally, it appears that the high logical probability calculated by the MaxEnt model and GMM at this position indicates the Au1 orebody may extend deeper or a concealed orebody exists therein. Meanwhile, the Target I of GMM is located at the elevation of about 1300 m, reflecting the weak anomaly in the deep drill.

Target II of the two methods is similarly located at the NW-orientation orebody group at the elevation of about 2500 m, where the fractures distribute complexly and the anomalies of tail halo elements and front halo elements overlapped (Figure 9).

6. Conclusions

In this paper, the three-dimensional primary halo anomaly data volume model is built based on the multifractal C-V model, which fully considers the nonlinear characteristics of the primary geochemical data. The C-V model is a three-dimensional extension of the two-dimensional multifractal method, according to which the geochemical concentrations are clearly illustrated at depth. The 3D geochemical anomaly data volume model provides an important element distribution indicator to the 3D MPM.

The data-driven CoDA method was performed in this paper by using clr transformation and factor analysis, among which factor F4 is selected as a prediction indicator. The knowledge-driven CoDA method used the SBP approach to extract the element associations of front halo, near-ore halo and tail halo, and the association of near-ore halo and the ratio of front halo to tail halo are selected as the other two prediction indicators. These selected geochemical association indicators are reliable for their good reflection in metallogenic regularity.

From the results of this paper, the MaxEnt model and the GMM are efficient machine learning methods in 3D MPM. By comparing the spatial distribution of the orebodies and the indication of the metallogenic regularity, the delineated mineral exploration targets can be considered as the mineral potential areas for further investigation. However, it must be mentioned that machine learning algorithms have fast and accurate calculation in the case of small data but lack generalization ability compared with deep learning algorithms in big data. As the amount of data gradually increases, the prediction ability of the MaxEnt model and the GMM usually reach the bottleneck, while deep learning can use more parameters to continuously optimize and improve the detection ability of the models. Deep learning-based 3D Mineral Prospectivity Mapping of the Zaozigou gold deposit should be paid more attention in the further research.

Author Contributions: Ideas, Y.K., G.C. and B.L.; Methodology, Y.K., G.C., C.L. and Z.Y.; software, M.X., S.Z. and H.Z.; writing—original draft preparation, Y.K., G.C., Y.W. and Y.G.; writing—review and editing, G.C. and B.L.; visualization, Y.K., M.X., S.Z., H.Z., L.W. and R.T.; supervision, B.L.; funding acquisition, B.L. All authors have read and agreed to the published version of the manuscript.

Funding: This research was funded by the National Key Research and Development Program of China (Grants 2017YFC0601505); the National Natural Science Foundation of China (Grants 41602334; 42072322); the Key Laboratory of Geochemical Exploration, Ministry of Natural Resources (Grant AS2019P02-01); Sichuan Science and Technology Program (Grant 2022NSFSC0510); and the Opening Fund of the Geomathematics Key Laboratory of Sichuan Province (Grant scsxdz2020yb06, scsxdz2021zd04).

Data Availability Statement: Restrictions apply to the availability of these data. Data was obtained from the Development Research Center of China Geological Survey and No. 3 Geological and Mineral Exploration team, Gansu Provincial Bureau of Geology and Mineral Exploration and Development and are available from Bingli Liu with the permission of the Development Research Center of China Geological Survey and No. 3 Geological and Mineral Exploration team, Gansu Provincial Bureau of Geology and Mineral Exploration and Development.

Acknowledgments: The authors thank the anonymous reviewers and the editors for their hard work on this paper. We are grateful to the Development Research Center of China Geological Survey and No. 3 Geological and Mineral Exploration team, Gansu Provincial Bureau of Geology and Mineral Exploration and Development for their data support.

Conflicts of Interest: The authors declare no conflict of interest.

References

1. Wang, S.C.; Yang, Y.H.; Yan, G.S.; Li, J.C. Study of large and Giant Gold Deposits: Quantitative Prognosis Method in China. *Geol. Rev.* **2000**, *46*, 17–24, (In Chinese with English Abstract).
2. Zhao, P.D. "Three-Component" Quantitative Resource Prediction and Assessments: Theory and Practice of Digital Mineral Prospecting. *Earth Sci.* **2002**, *27*, 482–489, (In Chinese with English Abstract).
3. Zhao, P.D. Quantitative mineral prediction and deep mineral exploration. *Earth Sci. Front.* **2007**, *14*, 1–10, (In Chinese with English Abstract).
4. Xiao, K.Y.; Zhao, P.D. A Preliminary Discussion on Basic Problems and Researching Programming of the Large Scale Metallogenetic Prognosis. *Miner. Explor.* **1994**, *3*, 49–56, (In Chinese with English Abstract).
5. Xiao, K.Y.; Zhang, X.H.; Chen, Z.H.; Song, G.Y.; Ge, Y.; Liu, D.L.; Wang, S.L.; Ning, S.N.; Cao, Y. Comparison of Method of Weights of Evidence and Information. *Comput. Tech. Geophys. Geochem. Explor.* **1999**, *21*, 223–226, (In Chinese with English Abstract).
6. Xiao, K.Y.; Zhu, Y.S.; Song, G.Y. Quantitative Evaluation of Mineral Resources GIS. *Geol. China.* **2000**, *278*, 29–32. (In Chinese)
7. Xiao, K.Y.; Chen, X.G.; Li, N.; Zou, W.; Sun, L. 3D visualization technology for geological and mineral exploration evaluation and software development for prospectors. *Miner. Depos.* **2010**, *29*, 758–760. (In Chinese)
8. Xiao, K.Y.; Li, N.; Sun, L.; Zou, W.; Li, Y. Large scale 3D mineral prediction methods and channels based on 3D information technology. *J. Geol.* **2012**, *36*, 229–236, (In Chinese with English Abstract).
9. Xiao, K.Y.; Li, N.; Sun, L.; Zou, W.; Li, Y. 3D Digital Mineral Deposit Model Establishment Method and Its Application. *Miner. Deposits* **2012**, *31*, 929–930. (In Chinese)
10. Xiao, K.Y.; Sun, L.; Li, N.; Wang, K.; Fan, J.F.; Ding, J.H. Mineral resources assessment under the thought of big data. *Geol. Bull. China* **2015**, *34*, 1266–1272, (In Chinese with English Abstract).
11. Chen, J.P.; Chen, Y.; Zeng, M. 3D positioning and quantitative pre-diction of the Koktokay No. 3 pegmatite dike, Xinjiang, China, based on the digital mineral deposit model. *Geol. Bull. China* **2008**, *27*, 552–559, (In Chinese with English Abstract).
12. Chen, J.P.; Lv, P.; Wu, W.; Zhao, J.; Hu, Q. A 3D method for predicting blind orebodies, based on a 3D visualization model and its application. *Earth Sci. Front.* **2007**, *14*, 56–65, (In Chinese with English Abstract). [CrossRef]
13. Chen, J.P.; Wang, C.N.; Shang, B.C.; Shi, R. Three-Dimensional Metallogenic Prediction in Yongmei Region Based on Digital Ore Deposit Model. *Sci. Technol. Manag. Land Resour.* **2012**, *29*, 14–20, (In Chinese with English Abstract).
14. Chen, J.P.; Yu, M.; Yu, P.P.; Shang, B.C.; Zheng, X.; Wang, L.M. Method and Practice of 3D Geological Modeling at Key Metallogenic Belt with Large and Medium Scale. *Acta Geol. Sin.* **2014**, *88*, 1187–1195, (In Chinese with English Abstract).
15. Lv, P. Cube Predicting Model Based 3D Predicting Methods of Blind Orebody and Software Development. Ph.D. Thesis, China University of Geosciences, Beijing, China, 2007. (In Chinese with English Abstract).

16. Zhang, S.H. Deep Learning foe Mineral Prospectivity Mapping of Lala-type Copper Deposit in the Huili Region, Sichuan. Ph.D. Thesis, China University of Geosciences, Beijing, China, 2020. (In Chinese with English Abstract).
17. Zhang, S.H.; Xiao, K.Y. Random Forest—Based Mineralization Prediction of the Lala—Type Cu Deposit in the Huili Area, Sichuan Province. *Geol. Explor.* **2020**, *56*, 239–252, (In Chinese with English Abstract).
18. Mao, X.C.; Tang, Y.H.; Deng, H. Three-dimensional morphological analysis method for geologic bodies and its application. *J. Cent. South Univ. Sci. Technol.* **2012**, *43*, 588–595, (In Chinese with English Abstract).
19. Mao, X.C.; Tang, Y.H.; Lai, J.Q.; Zou, Y.H.; Chen, J.; Peng, S.L.; Shao, Y.J. Three Dimensional Structure of Metallogenic Geologic Bodies in the Fenghuangshan Ore Field and Ore-controlling Geological Factors. *J. Geol.* **2011**, *85*, 1507–1518, (In Chinese with English Abstract).
20. Mao, X.C.; Wang, Q.; Chen, J.; Deng, H.; Liu, Z.K.; Wang, J.L.; Chen, J.P.; Xiao, K.Y. Three-dimensional Modeling of Deep Metallogenic Structure in Northwestern Jiaodong Peninsula and Its Gold Prospecting Significance. *Acta Geol. Sin.* **2020**, *41*, 166–178, (In Chinese with English Abstract).
21. Mao, X.C.; Zhang, M.M.; Deng, H.; Zou, Y.H.; Chen, J. 3D Visualization Prediction Method for Concealed Ore Bodies in the Deep. *Mining Area. J. Geol.* **2016**, *40*, 363–371, (In Chinese with English Abstract).
22. Mao, X.C.; Zou, Y.H.; Chen, J.; Lai, J.Q.; Peng, S.L.; Shao, Y.J.; Shu, Z.M.; Lv, J.W.; Lv, C.Y. Three-dimensional visual prediction of concealed ore bodies in the deep and marginal parts of crisis mines: A case study of the Fenghuangshan ore field in Tongling, Anhui, China. *Geol. Bull. China* **2010**, *29*, 401–413, (In Chinese with English Abstract).
23. Xiang, J.; Chen, J.P.; Xiao, K.Y.; Li, S.; Zhang, Z.P.; Zhang, Y. 3D metallogenic prediction based on machine learning: A case study of the Lala copper deposit in Sichuan Province. *Geol. Bull. China* **2019**, *38*, 2010–2021, (In Chinese with English Abstract).
24. Yuan, F.; Li, X.H.; Zhang, M.M.; Jia, C.; Hu, X.Y. Research Progress of 3D Prospectivity Modeling. *Gansu Geol.* **2018**, *27*, 32–36, (In Chinese with English Abstract).
25. Yuan, F.; Zhang, M.M.; Li, X.H.; Ge, C.; Lu, S.M.; Li, J.S.; Zhou, Y.Z.; Lan, X.Y. Prospectivity modeling: From two dimension to three-dimension. *Acta Petrol. Sin.* **2019**, *35*, 3863–3874, (In Chinese with English Abstract).
26. Cheng, Q.M. What are Mathematical Geosciences and its frontiers? *Earth Sci. Front.* **2021**, *28*, 6–25, (In Chinese with English Abstract).
27. Cheng, Q.; Agterberg, F.P.; Ballantyne, S.B. The separation of geochemical anomalies from background by fractal methods. *J. Geochem. Explor.* **1994**, *51*, 109–130. [CrossRef]
28. Cheng, Q.M. Multifractality and spatial statistics. *Comput. Geosci.* **1999**, *25*, 949–961. [CrossRef]
29. Chen, Z.J.; Cheng, Q.M.; Cheng, J.G.; Xie, S.Y. A novel iterative approach for mapping local singularities from geochemical data. *Nonlinear Process. Geophys.* **2007**, *14*, 317–324. [CrossRef]
30. Cheng, Q.M. Modeling local scaling properties for multiscale mapping. *Vadose Zone J.* **2008**, *7*, 525–532. [CrossRef]
31. Cheng, Q.M.; Agterberg, F. Multifractals and singularity analysis in mineral exploration and environmental assessment. *J. Geochem. Explor.* **2018**, *189*, 1. [CrossRef]
32. Cheng, Q.M. Multifractal and Geostatistic Methods for Characterizing Local Structure and Singularity Properties of Exploration Geochemical Anomalies. *Earth Sci.* **2001**, *26*, 161–166, (In Chinese with English Abstract).
33. Carranza, E.J.M. Geochemical Anomaly and Mineral Prospectivity Mapping in GIS. In *Handbook of Exploration and Environmental Geochemistry*; Elsevier: Amsterdam, The Netherlands, 2009; p. 11.
34. Afzal, P.; Alghalandis, Y.F.; Khakzad, A.; Moarefvand, P.; Omran, N.R. Delineation of mineralization zones in porphyry Cu deposits by fractal concentration–volume modeling. *J. Geochem. Explor.* **2011**, *108*, 220–232. [CrossRef]
35. Zuo, R.G.; Wang, J.; Chen, G.X.; Yang, M.G. Identification of weak anomalies: A multifractal perspective. *J. Geochem. Explor.* **2014**, *148*, 12–24. [CrossRef]
36. Chen, Y.L. Mineral potential mapping with a restricted Boltzmann machine. *Ore Geol. Rev.* **2015**, *71*, 749–760. [CrossRef]
37. Chen, Y.L.; Wu, W. Mapping mineral prospectivity using an extreme learning machine regression. *Ore Geol. Rev.* **2017**, *80*, 200–213. [CrossRef]
38. Cheng, Q.M.; Zhang, S.Y.; Zuo, R.G.; Chen, Z.J.; Xie, S.Y.; Xia, Q.L.; Xu, D.Y.; Yao, L.Q. Progress of multifractal filtering techniques and their applications in geochemical information extraction. *Earth Sci. Front.* **2009**, *16*, 185–198, (In Chinese with English Abstract).
39. Xiang, Z.L. Study on 3D Geological Modeling method and Prospecting Prediction of Deep Comprehensive Information in Mining area. Ph.D. Thesis, Henan Polytechnic University, JiaoZuo, China, 2019. (In Chinese with English Abstract).
40. Chen, Y.Q.; Zhang, S.Y.; Xia, Q.L.; Li, W.C.; Lu, Y.X.; Huang, J.N. Application of Multi Fractal Filtering to Extraction of Geochemical Anomalies from Multi Geochemical Backgrounds: A Case Study of the Southern Section of "Sanjiang Ore-Forming Zone", Southwestern China. *Earth Sci.* **2006**, *31*, 861–866, (In Chinese with English Abstract).
41. Sun, Z.J. Multifractal Method of Geochem Ical Threshold in Mineral Exploration. *Comput. Tech. Geophys. Geochem. Explor.* **2007**, *29*, 54–57+94, (In Chinese with English Abstract).
42. Li, X.H. 3D Prospectivity Modeling for Concealed Orebody and System Development. Ph.D. Thesis, Hefei University Of Technology, Hefei, China, 2015. (In Chinese with English Abstract).
43. Chen, G.X. Identifying Weak But Complex Geophysical and Geochemical Anomalies Caused by Buried Ore Bodies Using Fractal and Wavelet Methods. Ph.D. Thesis, China University of Geosciences, Wuhan, China, 2016. (In Chinese with English Abstract).
44. Liu, S.F. Fractal Analysis on Geochemical Distribution and Anomaly Separation in the Guangxi Zhuang Autonomous Region. Ph.D. Thesis, China University of Geosciences, Beijing, China, 2017. (In Chinese with English Abstract).

45. Carranza, E.J.M.; Zuo, R. Introduction to the thematic issue: Analysis of exploration geochemical data for mapping of anomalies. *Geochem. Explor. Env. Anal.* **2017**, *17*, 183–185. [CrossRef]
46. Zuo, R.G.; Cheng, Q.M.; Xia, Q.L. Application of fractal models to characterization of vertical distribution of geochemical element concentration. *J. Geochem. Explor.* **2009**, *102*, 37–43. [CrossRef]
47. Delavar, S.T.; Afzal, P.; Borg, G.; Rasa, I.; Lotfi, M.; Omran, N.R. Delineation of mineralization zones using concentration-volume fractal method in Pb-Zn carbonate hosted deposits. *J. Geochem. Explor.* **2012**, *118*, 98–110. [CrossRef]
48. Pearson, K. On a Form of Spurious Correlation Which May Arise When Indices Are Used in the Measurement of Organs. *Proc. R. Soc. Lond.* **1897**, *60*, 489–502.
49. Chayes, F. On correlation between variables of constant sum. *J. Geophys. Res.* **1960**, *65*, 4185–4193. [CrossRef]
50. Egozcue, J.J.; Pawlowsky-Glahn, V. Groups of Parts and Their Balances in Compositional Data Analysis. *Math. Geol.* **2005**, *37*, 795–828. [CrossRef]
51. Pawlowsky-Glahn, V.; Egozcue, J.J. Exploring Compositional Data with the CoDa-Dendrogram. *Austrian J. Stat.* **2011**, *40*, 103–113.
52. Zhou, D. Geological Compositional Data Analysis: Difficulties and Solutions. *Earth Sci.* **1998**, *23*, 41–46, (In Chinese with English Abstract).
53. Carranza, E.J.M. Analysis and mapping of geochemical anomalies using logratio-transformed stream sediment data with censored values. *J. Geochem. Explor.* **2011**, *110*, 167–185. [CrossRef]
54. Grunsky, E.C.; Mueller, U.A.; Corrigan, D. A study of the lake sediment geochemistry of the Melville Peninsula using multivariate methods: Applications for predictive geological mapping. *J. Geochem. Explor.* **2014**, *141*, 15–41. [CrossRef]
55. Parent, L.E.; Parent, S.T.; Ziadi, N. Biogeochemistry of soil inorganic and organic phosphorus: A compositional analysis with balances. *J. Geochem. Explor.* **2014**, *141*, 52–60. [CrossRef]
56. Wang, W.L.; Zhao, J.; Cheng, Q.M. Mapping of Fe mineralization-associated geochemical signatures using logratio transformed stream sediment geochemical data in eastern Tianshan, China. *J. Geochem. Explor.* **2014**, *141*, 6–14. [CrossRef]
57. Carranza, E.J.M. Geochemical Mineral Exploration: Should We Use Enrichment Factors or Log-Ratios. *Nat. Resour. Res.* **2016**, *26*, 411–428. [CrossRef]
58. Mckinley, J.M.; Grunsky, E.; Mueller, U. Environmental Monitoring and Peat Assessment Using Multivariate Analysis of Regional-Scale Geochemical Data. *Math. Geosci.* **2017**, *50*, 235–246. [CrossRef]
59. Reimann, C.; Filzmoser, P.; Hron, K.; Kynlová, P.; Garrett, R.G. A new method for correlation analysis of compositional (environmental) data—A worked example. *Sci. Total Environ.* **2017**, *607*, 965–971. [CrossRef] [PubMed]
60. Thiombane, M.; Martín-Fernández, J.; Albanese, S.; Lima, A.; Doherty, A.; Vivo, B.D. Exploratory analysis of multi-element geochemical patterns in soil from the Sarno River Basin (Campania region, southern Italy) through compositional data analysis (CODA). *J. Geochem. Explor.* **2018**, *195*, 110–120. [CrossRef]
61. Aitchison, J. The statistical analysis of compositional data. *J. R. Stat. Soc. B* **1982**, *44*, 139–177. [CrossRef]
62. Buccianti, A.; Lima, A.; Albanese, S.; Cannatelli, C.; Esposito, R.; Vivo, B.D. Exploring topsoil geochemistry from the CoDA (Compositional Data Analysis) perspective: The multi-element data archive of the Campania Region (Southern Italy). *J. Geochem. Explor.* **2015**, *159*, 302–316. [CrossRef]
63. Mckinley, J.M.; Hron, K.; Grunsky, E.C.; Reimann, C.; Tolosana-Delgado, R. The single component geochemical map: Fact or fiction. *J. Geochem. Explor.* **2016**, *162*, 16–28. [CrossRef]
64. Buccianti, A.; Lima, A.; Albanese, S.; DeVivo, B. Measuring the change under compositional data analysis (CoDA): Insight on the dynamics of geochemical systems. *J. Geochem. Explor.* **2018**, *189*, 100–108. [CrossRef]
65. Liu, Y.; Cheng, Q.M.; Zhou, K.F. New Insights into Element Distribution Patterns in Geochemistry: A Perspective from Fractal Density. *Nat. Resour. Res.* **2018**, *28*, 5–29. [CrossRef]
66. Zheng, W.B.; Liu, B.L.; McKinley, J.M.; Cooper, M.R.; Wang, L. Geology and geochemistry-based metallogenic exploration model for the eastern Tethys Himalayan metallogenic belt, Tibet. *J. Geochem. Explor.* **2021**, *224*, 106743. [CrossRef]
67. Liu, Y.; Carranza, E.J.M.; Xia, Q. Developments in Quantitative Assessment and Modeling of Mineral Resource Potential: An Overview. *Nat. Resour. Res.* **2022**, *31*, 1825–1840. [CrossRef]
68. Agterberg, F. Principles of Probabilistic Regional Mineral Resource Estimation. *Earth Sci.* **2011**, *36*, 189–200.
69. Zhou, Y.Z.; Chen, S.; Zhang, Q.; Xiao, F.; Wang, S.G.; Liu, Y.P.; Jiao, S.T. Advances and Prospects of Big Data and Mathematical Geoscience. *Acta Petrol. Sin.* **2018**, *34*, 255–263, (In Chinese with English Abstract).
70. Zhou, Y.Z.; Wang, J.; Zuo, R.G.; Xiao, F.; Shen, W.J.; Wang, S.G. Machine learning, deep learning and Python language in field of geology. *Acta Petrol. Sin.* **2018**, *34*, 3173–3178, (In Chinese with English Abstract).
71. Zhou, Y.Z.; Zuo, R.G.; Liu, G.; Yuan, F.; Mao, X.C.; Guo, Y.J.; Liao, J.; Liu, Y.P. The Great-leap-forward Development of Mathematical Geoscience During 2010–2019: Big Data and Artificial Intelligence Algorithm Are Changing Mathematical Geoscience. *Bull. Miner. Petrol. Geochem.* **2021**, *40*, 556–573+777, (In Chinese with English Abstract).
72. Zuo, R.G. Deep Learning-Based Mining and Integration of Deep-Level Mineralization Information. *Bull. Miner. Petrol. Geochem.* **2019**, *38*, 53–60, (In Chinese with English Abstract).
73. Zuo, R.G.; Peng, Y.; Li, T.; Xiong, Y.H. Challenges of Geological Prospecting Big Data Mining and Integration Using Deep Learning Algorithms. *Earth Sci.* **2021**, *46*, 350–358, (In Chinese with English Abstract).
74. Zuo, R.G.; Carranza, E.J.M. Support vector machine: A tool for mapping mineral prospectivity. *Comput. Geosci.* **2011**, *37*, 1967–1975. [CrossRef]

75. Shabankareh, M.; Hezarkhani, A. Application of support vector machines for copper potential mapping in Kerman region, Iran. *J. Afr. Earth Sci.* **2016**, *128*, 116–126. [CrossRef]

76. Zhang, S.; Xiao, K.Y.; Carranza, E.J.M.; Yang, F. Maximum Entropy and Random Forest Modeling of Mineral Potential: Analysis of Gold Prospectivity in the Hezuo–Meiwu District, West Qinling Orogen, China. *Nat. Resour. Res.* **2018**, *28*, 645–664. [CrossRef]

77. Sun, T.; Chen, F.; Zhong, L.X.; Liu, W.M.; Wang, Y. GIS-based mineral prospectivity mapping using machine learning methods: A case study from Tongling ore district, eastern China. *Ore Geol. Rev.* **2019**, *109*, 26–49. [CrossRef]

78. Wang, J.; Zuo, R.G.; Xiong, Y.H. Mapping Mineral Prospectivity via Semi-supervised Random Forest. *Nat. Resour. Res.* **2019**, *29*, 189–202. [CrossRef]

79. Sun, T.; Li, H.; Wu, K.X.; Chen, F.; Hu, Z.J. Data-Driven Predictive Modelling of Mineral Prospectivity Using Machine Learning and Deep Learning Methods: A Case Study from Southern Jiangxi Province, China. *Minerals* **2020**, *10*, 102. [CrossRef]

80. Wang, Z.Y.; Zuo, R.G.; Dong, Y.N. Mapping Himalayan leucogranites using a hybrid method of metric learning and support vector machine. *Comput. Geosci.* **2020**, *138*, 104455. [CrossRef]

81. Chen, J.; Mao, X.C.; Liu, Z.K.; Deng, H. Three-dimensional Metallogenic Prediction Based on Random Forest Classification Algorithm for the Dayingezhuang Gold Deposit. *Geotecton. Metallog.* **2020**, *44*, 231–241, (In Chinese with English Abstract).

82. Li, C.B.; Xiao, K.Y.; Li, N.; Song, X.L.; Zhang, S.; Wang, K.; Chu, W.K.; Cao, R. A Comparative Study of Support Vector Machine, Random Forest and Artificial Neural Network Machine Learning Algorithms in Geochemical Anomaly Information Extraction. *Acta Geosci. Sin.* **2020**, *41*, 309–319, (In Chinese with English Abstract).

83. Wang, Y.; Zhou, Y.Z.; Xiao, F.; Wang, J.; Wang, K.Q.; Yu, X.T. Numerical Metallogenic Modelling and Support Vector Machine Methods Applied to Predict Deep Mineralization: A Case Study from the Fankou Pb-Zn Ore Deposit in Northern Guangdong. *Geotecton. Metallog.* **2020**, *44*, 222–230, (In Chinese with English Abstract).

84. Li, S.; Chen, J.P.; Liu, C.; Wang, Y. Mineral Prospectivity Prediction via Convolutional Neural Networks Based on Geological Big Data. *J. Earth Sci.* **2021**, *32*, 327–347. [CrossRef]

85. Zhang, S.; Carranza, E.J.M.; Wei, H.T.; Xiao, K.Y.; Yang, F.; Xiang, J.; Zhang, S.H.; Xu, Y. Data-driven Mineral Prospectivity Mapping by Joint Application of Unsupervised Convolutional Auto-encoder Network and Supervised Convolutional Neural Network. *Nat. Resour. Res.* **2021**, *30*, 1011–1031. [CrossRef]

86. Zhou, Y.Z.; Li, P.X.; Wang, S.G.; Xiao, F.; Li, J.Z.; Gao, L. Research Progress on Big Data and Intelligent Modelling of Mineral Deposits. *Bull. Miner. Petrol. Geochem.* **2017**, *36*, 327–331+344, (In Chinese with English Abstract).

87. Singer, D.A.; Kouda, R. Application of a feed forward neural network in the search for Kuroko deposits in the Hokuroku District. *Math. Geol.* **1996**, *28*, 1017–1023. [CrossRef]

88. Rodriguez-Galiano, V.; Sanchez-Castillo, M.; Chica-Olmo, M.; Chica-Rivas, M. Machine learning predictive models for mineral prospectivity: An evaluation of neural networks, random forest, regression trees and support vector machines. *Ore Geol. Rev.* **2015**, *71*, 804–818. [CrossRef]

89. Zuo, R.G.; Xiong, Y.H. Geodata science and geochemical mapping. *J. Geochem. Explor.* **2019**, *209*, 106431. [CrossRef]

90. Xiong, Y.H.; Zuo, R.G. Robust Feature Extraction for Geochemical Anomaly Recognition Using a Stacked Convolutional Denoising Autoencoder. *Math. Geosci.* **2021**, *54*, 623–644. [CrossRef]

91. Gao, Y. Mineral Prospecting Information Mining and Mapping Mineral Prospectivity for Copper Polymetallic Mineralization in Southwest Fujian Province, China. Ph.D. Thesis, China University of Geosciences, Wuhan, China, 2019. (In Chinese with English Abstract).

92. Zuo, R.G. Geodata Science-Based Mineral Prospectivity Mapping: A Review. *Nat. Resour. Res.* **2020**, *29*, 3415–3424. [CrossRef]

93. Li, T.; Zuo, R.G.; Xiong, Y.H.; Peng, Y. Random-Drop Data Augmentation of Deep Convolutional Neural Network for Mineral Prospectivity Mapping. *Nat. Resour. Res.* **2020**, *30*, 27–38. [CrossRef]

94. Cai, H.H.; Xu, Y.Y.; Li, Z.X.; Cao, H.H.; Feng, Y.X.; Chen, S.Q.; Li, Y.S. The division of metallogenic prospective areas based onconvolutional neural network model: A case study of the Daqiao gold polymetallic deposit. *Geol. Bull. China* **2019**, *38*, 1999–2009, (In Chinese with English Abstract).

95. Wang, Z.Y. Mapping of Himalaya Leucogranites Based on Metric Learning. Ph.D. Thesis, China University of Geosciences, Wuhan, China, 2020. (In Chinese with English Abstract).

96. Deng, H.; Wei, Y.F.; Chen, J.; Liu, Z.K.; Yu, S.Y.; Mao, X.C. Three-dimensional prospectivity mapping and quantitative analysis of structural ore-controlling factors in Jiaojia Auore-belt with attention convolutional neural networks. *J. Cent. South Univ.* **2021**, *52*, 3003–3014, (In Chinese with English Abstract).

97. Deng, H.; Zheng, Y.; Chen, J.; Wei, Y.F.; Mao, X.C. Deep Learning-based 3D Prediction Model for the Dayingezhuang Gold Deposit, Shandong Province. *Acta Geosci. Sin.* **2020**, *41*, 157–165, (In Chinese with English Abstract).

98. Zhang, Z.J.; Cheng, Q.M.; Yang, J.; Wu, G.P.; Ge, Y.Z. Machine Learning for Mineral Prospectivity: A Case Study of Iron-polymetallic Mineral Prospectivity in Southwestern Fujian. *Earth Sci. Front.* **2021**, *28*, 221–235, (In Chinese with English Abstract).

99. Feng, Y.M.; Cao, X.Z.; Zhang, E.P.; Hu, Y.X.; Pan, X.P.; Yang, J.L.; Jia, Q.Z.; Li, W.M. Tectonic Evolution Framework and Nature of The West Qinling Orogenic Belt. *Northwest. Geol.* **2003**, *36*, 1–10, (In Chinese with English Abstract).

100. Wei, L.X. Tectonic Evolution and Mineralization of Zaozigou Gold Deposit, Gansu Province. Master's Thesis, China University of Geosciences, Beijing, China, 2015. (In Chinese with English Abstract).

101. Zeng, J.J.; Li, K.N.; Yan, K.; Wei, L.L.; Huo, X.D.; Zhang, J.P. Tectonic Setting and Provenance characteristics of the Lower Triassic Jiangligou Formation in West Qinling—Constraints from Geochemistry of Clastic Rock and zircon U-Pb Geochronology of Detrital Zircon. *Geol. Rev.* **2021**, *67*, 1–15, (In Chinese with English Abstract).
102. Li, Z.B.; Liu, Z.Y.; Li, R. Geochemical Characteristics and metallogenic Potential Analysis of Daheba Formation in Ta-Ga Area of Gansu Province. *Contrib. Geol. Miner. Resour. Res.* **2021**, *36*, 187–194, (In Chinese with English Abstract).
103. Chen, Y.; Wang, K.J. Geological Features and Ore Prospecting Indicators of Sishangou Silver Deposit. *Gansu. Metal.* **2015**, *37*, 108–111, (In Chinese with English Abstract).
104. Di, P.F. Geochemistry and Ore-Forming Mechanism on Zaozigou gold deposit in Xiahe-Hezuo, West Qinling, China. Ph.D. Thesis, Lanzhou University, Lanzhou, China, 2018. (In Chinese with English Abstract).
105. Li, K.N.; Li, H.R.; Liu, B.C.; Yan, K.; Jia, R.Y.; Wei, L.L. Geochemical characteristics of TTG Dick rock and the Relation with Gold Mineralization in West Qinling Mountain. *Sci. Technol. Eng.* **2019**, *19*, 52–65, (In Chinese with English Abstract).
106. Kang, S.S. Geological Characteristics and Prospecting Criteria of Nanmougou Copper Deposit, Gansu Province. *Gansu. Metal.* **2018**, *40*, 79–85, (In Chinese with English Abstract).
107. Kang, S.S.; Dou, X.G.; Zhi, C.; Wu, X.L. Geochemical Characteristics and Genetic Analysis of the Namugou Copper Deposit in Sunan County, Gansu. *Gansu. Metal.* **2019**, *41*, 65–72, (In Chinese with English Abstract).
108. Liu, Y. Relationship between Intermediate-acid Dike Rock and Gold Mineralization of the Zaozigou Deposit, Gansu Province. Master's Thesis, Chang'an University, Xi'an, China, 2013. (In Chinese with English Abstract).
109. Hu, J.Q. Mineral Control Factors, Metallogenic Law and Prospecting Direction of Integrated Gold Mine Exploration Area in Shilijba-Yangshan Area of Gansu Province. *Gansu. Sci. Technol.* **2018**, *34*, 27–33. (In Chinese)
110. Zhao, J.Z.; Chen, G.Z.; Liang, Z.L.; Zhao, J.C. Ore-body Geochemical Features of Zaozigou Gold Deposit. *Gansu. Geol.* **2013**, *22*, 38–43, (In Chinese with English Abstract).
111. Lu, J. Study on Characteristics and Ore-host Regularity of Gold Mineral in the Western Qinling Region, Gansu Province. Master's Thesis, China University of Geosciences, Beijing, China, 2016. (In Chinese with English Abstract).
112. Zhang, Y.N.; Liang, Z.L.; Qiu, K.F.; Ma, S.H.; Wang, J.L. Overview on the Metallogenesis of Zaozigou gold deposit in the West Qinling Orogen. *Miner. Explor.* **2020**, *11*, 28–39, (In Chinese with English Abstract).
113. Tang, L.; Lin, C.G.; Cheng, Z.Z.; Jia, R.Y.; Li, H.R.; Li, K.N. 3D Characteristics of Primary Halo and Deep Prospecting Prediction in The Zaozigou Gold Deposit, Hezuo City, Gansu Province. *Geol. Bull. China* **2020**, *39*, 1173–1181, (In Chinese with English Abstract).
114. Chen, G.Z.; Wang, J.L.; Liang, Z.L.; Li, P.B.; Ma, H.S.; Zhang, Y.N. Analysis of Geological Structures in Zaozigou Gold Deposit of Gansu Province. *Gansu. Geol.* **2013**, *22*, 50–57, (In Chinese with English Abstract).
115. Chen, G.Z.; Liang, M.L.; Wang, J.L.; Zhang, Y.N.; Li, P.B. Characteristics and Deep Prediction of Primary Superimposed Halos in The Zaozigou Gold Deposit of Hezuo, Gansu Province. Geophys. *Geochem. Explor.* **2014**, *38*, 268–277, (In Chinese with English Abstract).
116. Jin, D.G.; Liu, B.C.; Chen, Y.Y.; Liang, Z.L. Spatial Distribution of Gold Bodies in Zaozigou Mine of Gansu Province. *Gansu. Geol.* **2015**, *24*, 25–30+41, (In Chinese with English Abstract).
117. Zhu, F.; Wang, G.W. Study on Grade Model of Gansu Zaozigou Gold Mine Based on Geological Statistics. *Acta Mineral. Sin.* **2015**, *35*, 1065–1066. (In Chinese)
118. Chen, G.Z.; Li, L.N.; Zhang, Y.N.; Ma, H.S.; Liang, Z.L.; Wu, X.M. Characteristics of fluid inclusions and deposit formation in Zaozigou gold mine. In Proceedings of the The 15th Annual Academic Conference of Chinese Society for Mineralogy, Petrology and Geochemistry, Changchun, China, 24 June 2015.
119. Wu, X.M. Study on Geological Characteristics and Metallogenic Regularity of the Gelouang Gold Deposit. Master's Thesis, Lanzhou University, Lanzhou, China, 2018. (In Chinese with English Abstract).
120. Liu, H.T.; Jiang, Y.M.; Wu, T.J.; Chang, D.L. Application of singularity analysis to geochemical anomaly recognition in Chifeng area. *Contrib. Geol. Miner. Resour. Res.* **2011**, *26*, 341–344, (In Chinese with English Abstract).
121. Wang, Y. The Application of Fractal Theory in Geophysical Research. Master's Thesis, Chang'an University, Xi'an, China, 2013. (In Chinese with English Abstract).
122. Filzmoser, P.; Hron, K.; Reimann, C. Univariate statistical analysis of environmental (compositional) data: Problems and possibilities. *Sci. Total Environ.* **2009**, *407*, 6100–6108. [CrossRef]
123. Wang, L.; Liu, B.L.; McKinley, J.M.; Cooper, M.R.; Li, C.; Kong, Y.H.; Shan, M.X. Compositional data analysis of regional geochemical data in the Lhasa area of Tibet, China. *Appl. Geochem.* **2021**, *135*, 105108. [CrossRef]
124. Zheng, W.B.; Liu, B.L.; Tang, J.X.; McKinley, J.M.; Cooper, M.R.; Tang, P.; Lin, B.; Li, C.; Wang, L.; Zhang, D. Exploration indicators of the Jiama porphyry–skarn deposit, southern Tibet, China. *J. Geochem. Explor.* **2022**, *236*, 106982. [CrossRef]
125. Zuo, R.G. Identification of weak geochemical anomalies using robust neighborhood statistics coupled with GIS in covered areas. *J. Geochem. Explor.* **2014**, *136*, 93–101. [CrossRef]
126. Liu, Y. The Study of Regional Geochemistry Data Analysis and Metallogenic Information Fusion Models. Ph.D. Thesis, China University of Geosciences, Wuhan, China, 2015. (In Chinese with English Abstract).
127. Prado, E.; Filho, C.; Carranza, E.; Motta, J.G. Modeling of Cu-Au Prospectivity in the Carajás mineral province (Brazil) through Machine Learning: Dealing with Imbalanced Training Data. *Ore Geol. Rev.* **2020**, *124*, 103611. [CrossRef]
128. Brown, W.M.; Gedeon, T.D.; Groves, D.I.; Barnes, R.G. Artificial neural networks: A new method for mineral prospectivity mapping. *Aust. J. Earth Sci.* **2000**, *47*, 757–770. [CrossRef]

129. Skabar, A. Mineral potential mapping using feed-forward neural networks. In Proceedings of the International Joint Conference on Neural Networks IEEE, Portland, OR, USA, 20–24 July 2003.

130. Leite, E.P.; Filho, C. Probabilistic neural networks applied to mineral potential mapping for platinum group elements in the Serra Leste region, Carajás Mineral Province, Brazil. *Comput. Geosci.* **2009**, *35*, 675–687. [CrossRef]

131. Leite, E.P.; de Souza, C.R. Artificial neural networks applied to mineral potential mapping for copper-gold mineralizations in the Carajás Mineral Province, Brazil. *Geophys. Prospect.* **2009**, *57*, 1049–1065. [CrossRef]

132. Abedi, M.; Norouzi, G.H.; Bahroudi, A. Support vector machine for multi-classification of mineral prospectivity areas. *Comput. Geosci.* **2012**, *9*, 272–283. [CrossRef]

133. Rodriguez-Galiano, V.F.; Chica-Olmo, M.; Chica-Rivas, M. Predictive modelling of gold potential with the integration of multisource information based on random forest: A case study on the Rodalquilar area, Southern Spain. *Int. J. Geogr. Inf. Sci.* **2014**, *28*, 1336–1354. [CrossRef]

134. Gao, Y.; Zhang, Z.J.; Xiong, Y.H.; Zuo, R.G. Mapping mineral prospectivity for Cu polymetallic mineralization in southwest Fujian Province, China. *Ore Geol. Rev.* **2016**, *75*, 16–28. [CrossRef]

135. Geranian, H.; Tabatabaei, S.H.; Asadi, H.H.; Carranza, E.J.M. Application of Discriminant Analysis and Support Vector Machine in Mapping Gold Potential Areas for Further Drilling in the Sari-Gunay Gold Deposit, NW Iran. *Nat. Resour. Res.* **2016**, *25*, 145–159. [CrossRef]

136. Xie, S. Research on the prediction of potential suitable distribution of Biomphalaria straminea in Guangdong Province. Master's Thesis, South China Agricultural University, Guangzhou, China, 2017. (In Chinese with English Abstract).

137. Xu, C. Distribution of Vehicle Free Flow Speeds Based on Gaussian Mixture Model. *J. Highw. Transp. Res. Dev. Chin. Ed.* **2012**, *19*, 132–158, (In Chinese with English Abstract).

138. Matheron, G. Kriging or polynomial interpolation procedures? *Trans. Can. Inst. Min. Metal.* **1967**, *70*, 240–244.

139. Journel, A.G.; Huijbregts, C.J. *Mining Geostatistics*; Academic Press: London, UK, 1978.

140. Cressie, N. The origins of kriging. *Math. Geosci.* **1990**, *22*, 239–252. [CrossRef]

141. Wang, H.; Lineweaver, C.H. Chemical Complementarity between the Gas Phase of the Interstellar Medium and the Rocky Material of Our Planetary System. In Proceedings of the 15th Australian Space Research Conference, Canberra, Australia, 29 September–1 October 2016; pp. 173–182.

142. Ye, T.Z. Theoretical Framework of Synthetic Geological Information Prediction Techniques and Methods for Mineral Deposit Models. *J. Jilin Univ. Earth Sci. Ed.* **2013**, *43*, 1053–1072, (In Chinese with English Abstract).

143. Beus, A.A.; Grigorian, S.V. *Geochemical Exploration Methods for Mineral Deposits*; Applied Publishing Ltd.: Wilmette, IL, USA, 1977; pp. 31–270.

144. Levinson, A.A. *Introduction to Exploration Geochemistry*; Applied Publishing Ltd.: Wilmette, IL, USA, 1974; p. 612.

145. Rabeaut, O.; Legault, M.; Cheilletz, A.; Jébrak, M.; Royer, J.J.; Cheng, L.Z. Gold Potential of a Hidden Archean Fault Zone: The Case of the Cadillac–Larder Lake Fault. *Explor. Min. Geol.* **2010**, *19*, 99–116. [CrossRef]

146. Mejia-Herrera, P.; Royer, J.J.; Caumon, G.; Cheilletz, A. Curvature Attribute from Surface-Restoration as Predictor Variable in Kupferschiefer Copper Potentials: An Example from the Fore-Sudetic Region. *Nat. Resour. Res.* **2015**, *24*, 275–290. [CrossRef]

147. Li, B.B.; Liu, B.L.; Guo, K.; Li, C.; Wang, B. Application of a Maximum Entropy Model for Mineral Prospectivity Maps. *Minerals* **2019**, *9*, 556. [CrossRef]

148. Kong, W.X.; Li, X.H.; Zou, H.F. Optimization of Maximum Entropy Model in Species Distribution Prediction. *J. Appl. Ecol.* **2019**, *30*, 2116–2128, (In Chinese with English Abstract).

149. Elith, J.; Phillips, S.J.; Hastie, T.; Dudik, M.; Chee, Y.E.; Yates, C.J. A statistical explanation of MaxEnt for ecologists. *Divers. Distrib.* **2013**, *17*, 43–57. [CrossRef]

150. Radosavljevic, A.; Anderson, R.P. Making better MAXENT models of species distributions: Complexity, overfitting and evaluation. *J. Biogeogr.* **2013**, *41*, 629–643. [CrossRef]

151. Caumon, G.; Collon-Drouaillet, P.; De Veslud, C.L.C.; Viseur, S.; Sausse, J. Surface-based 3D modeling of geological structures. *Math. Geosci.* **2009**, *41*, 927–945. [CrossRef]

152. Chen, Q.; Liu, G.; Li, X.; Zhang, Z.; Li, Y. A corner-point-grid-based voxelization method for the complex geological structure model with folds. *J. Vis.* **2017**, *20*, 875–888. [CrossRef]

153. Yarus, J.M.; Chambers, R.L. *Stochastic Modeling and Geostatistics: Principles, Methods, and Case Studies*; American Association of Petroleum Geologists: Tulsa, OK, USA, 1994; p. 379.

154. Chen, Q.; Liu, G.; Ma, X.; Li, X.; He, Z. 3D stochastic modeling framework for Quaternary sediments using multiple-point statistics: A case study in Minjiang Estuary area, southeast China. *Comput. Geosci.* **2019**, *136*, 104404. [CrossRef]

155. Chen, Q.; Mariethoz, G.; Liu, G.; Comunian, A.; Ma, X. Locality-based 3-D multiple-point statistics reconstruction using 2-D geological cross sections. *Hydrol. Earth Syst. Sci.* **2018**, *22*, 6547–6566. [CrossRef]

156. Gueting, N.; Caers, J.; Comunian, A.; Vanderborght, J.; Englert, A. Reconstruction of three-dimensional aquifer heterogeneity from two-dimensional geophysical data. *Math. Geosci.* **2018**, *50*, 53–75. [CrossRef]

157. Mahmud, K.; Mariethoz, G.; Baker, A.; Treble, P.C. Hydrological characterization of cave drip waters in a porous limestone: Golgotha Cave, Western Australia. *Hydrol. Earth Syst. Sci.* **2018**, *22*, 977–988. [CrossRef]

158. Feng, W.; Yin, Y.; Zhang, C.; Duan, T.; Zhang, W.; Hou, G.; Zhao, L. A training image optimal selecting method based on composite correlation coefficient ranking for multiple-point geostatistics. *J. Petrol. Sci. Eng.* **2019**, *179*, 292–311. [CrossRef]

159. Cui, Z.; Chen, Q.; Liu, G.; Mariethoz, G.; Ma, X. Hybrid parallel framework for multiple-point geostatistics on tianhe-2: A robust solution for large-scale simulation. *Comput. Geosci.* **2021**, *157*, 104923. [CrossRef]
160. Chen, Q.; Cui, Z.; Liu, G.; Yang, Z.; Ma, X. Deep convolutional generative adversarial networks for modeling complex hydrological structures in Monte-Carlo simulation. *J. Hydrol.* **2022**, *610*, 127970. [CrossRef]
161. Yang, Z.; Chen, Q.; Cui, Z.; Liu, G.; Dong, S.; Tian, Y. Automatic reconstruction method of 3D geological models based on deep convolutional generative adversarial networks. *Comput. Geosci.* **2022**, *26*, 1135–1150. [CrossRef]

Article

3D Mineral Prospectivity Mapping of Zaozigou Gold Deposit, West Qinling, China: Deep Learning-Based Mineral Prediction

Zhengbo Yu [1], Bingli Liu [1,2,*], Miao Xie [1], Yixiao Wu [1], Yunhui Kong [1], Cheng Li [1,3], Guodong Chen [1], Yaxin Gao [1], Shuai Zha [1], Hanyuan Zhang [1], Lu Wang [1] and Rui Tang [1]

[1] Geomathematics Key Laboratory of Sichuan Province, Chengdu University of Technology, Chengdu 610059, China
[2] Key Laboratory of Geochemical Exploration, Institute of Geophysical and Geochemical Exploration, CAGS, Langfang 065000, China
[3] Institute of Mineral Resources, Chinese Academy of Geological Sciences, Beijing 100037, China
* Correspondence: liubingli@cdut.edu.cn; Tel./Fax: +86-028-8407-3610

Abstract: This paper focuses on the scientific problem of quantitative mineralization prediction at large depth in the Zaozigou gold deposit, west Qinling, China. Five geological and geochemical indicators are used to establish geological and geochemical quantitative prediction model. Machine learning and Deep learning algorithms are employed for 3D Mineral Prospectivity Mapping (MPM). Especially, the Student Teacher Ore-induced Anomaly Detection (STOAD) model is proposed based on the knowledge distillation (KD) idea combined with Deep Auto-encoder (DAE) network model. Compared to DAE, STOAD uses three outputs for anomaly detection and can make full use of information from multiple levels of data for greater overall robustness. The results show that the quantitative mineral resources prediction by applying the STOAD model has a good performance, where the value of Area Under Curve (AUC) is 0.97. Finally, three main mineral exploration targets are delineated for further investigation.

Keywords: 3D mineral prospectivity mapping; geological and geochemical quantitative prediction model at depth; Deep auto-encoder network; Student Teacher Ore-induced Anomaly Detection; Zaozigou gold deposit

Citation: Yu, Z.; Liu, B.; Xie, M.; Wu, Y.; Kong, Y.; Li, C.; Chen, G.; Gao, Y.; Zha, S.; Zhang, H.; et al. 3D Mineral Prospectivity Mapping of Zaozigou Gold Deposit, West Qinling, China: Deep Learning-Based Mineral Prediction. *Minerals* **2022**, *12*, 1382. https://doi.org/10.3390/min12111382

Academic Editor: Martiya Sadeghi

Received: 31 July 2022
Accepted: 27 October 2022
Published: 30 October 2022

Publisher's Note: MDPI stays neutral with regard to jurisdictional claims in published maps and institutional affiliations.

1. Introduction

Geosciences is a data-intensive science, and geological survey and mineral exploration has accumulated a large amount of multi-source geosciences data, and the introduction of big data and its application of machine learning and deep learning can effectively support mineral resources exploration [1–6].

3D ore-induced information has multiphase, multisource, and multivariable characteristics, which derives the difficulty of information extraction, identification, and deduction, posing challenges for its classification and prediction. Big data intelligent algorithms in geosciences provide efficient tools for mineral exploration with machine learning and deep learning methods [7–9]. The deep learning algorithms, including Convolutional Neural Networks (CNN), Recurrent Neural Networks (RNN), Multi-Layer perceptron (MLP), Deep Auto-encoder (DAE), and Deep Belief Networks (DBN) [10–16], have been successfully used in geological, geochemical, geophysical, and remote sensing data analysis and quantitative mineral resources prediction [17–31].

Convolutional Neural Network, as one of the most successful neural network models of deep learning methods, has a number of achievements in geosciences, such as lithology identification [32], geological mapping [33], and 3D geological structure inversion [34–36]. In recent years, many scholars have also achieved excellent results by applying CNN to the study of mineral prediction [37–44].

Auto-encoder network is a typical unsupervised learning algorithm for ore-induced anomaly detection. The auto-encoder network contains two parts, that is Encoder and Decoder, which can be used for data dimensional reduction to reconstruct sample population with unknown complex multivariate probability distributions, especially for big data [45,46]. With the rise in deep learning, auto-encoder networks and related variant networks are widely used for image classification [47–49], anomaly detection [34,50], and data generation [51], etc.

The idea of Deep Auto-encoder network for anomaly detection in geosciences is to learn the reconstruction ability of anomaly-free samples, and to achieve anomaly detection by reconstructing anomaly samples with a large error. Related achievements include the integration of auto-encoder network with density-based spatial clustering for geochemical anomaly detection [52], physically constrained variational auto-encoder for geochemical pattern recognition [53], and detection of geochemical anomalies related to mineralization using the GANomaly network [54]. DAE is an unsupervised learning method, which achieves more efficient detection of ore-induced comprehensive anomalies, and does not need to manually label positive and negative samples, saving valuable labor costs.

Knowledge distillation (KD) is an important method to quickly deal with the problems caused by labeled data (positive and negative sample imbalance, number of labeled samples, etc.), which can efficiently achieve information transfer between networks, and it is also called Student–Teacher learning framework. KD is widely used in computer vision, speech recognition, natural language processing, etc., to achieve model compression [55] and knowledge transfer [56]. KD is composed of the Teacher and Student network, the Teacher network usually has certain learning ability through training, and the Student network is used to imitate the ability of the Teacher network and thus achieve knowledge transfer. If the Student network has a smaller network structure compared to the Teacher network, it also achieves model compression. So far, KD methods have been less applied in geological and geochemical data analysis or mineral resources prediction.

The main task of quantitative mineral prediction is to conduct comprehensive analysis of geological, geophysical, geochemical, and remote sensing data and drilling engineering data in the study area based on the research of geological background and metallogenic regularity, and then construct mathematical models to effectively extract and identify favorable information on mineralization, carry out data fusion of quantitative mineralization information, construct mineral prediction models, perform mineral prospectivity mapping, and exploration targets delineation.

Compared with using the DAE to reconstruct the anomaly-free samples directly and then detecting the test samples, the Teacher network of KD model can learn the reconstruction ability of all samples and then use the Student network to learn the reconstruction ability of the Teacher network for the normal samples. The test process achieves the detection of anomaly data by calculating the Mean Square Error (MSE) in the results of Student and Teacher networks. Therefore, we propose the STOAD model based on the module of CNN, the framework of KD and the reconstruction ability of DAE, to achieve 3D Mineral Prospectivity Mapping in Zaozigou gold deposit. Results show the STOAD model performs more efficiently and more robustly compared to the DAE model, and three main mineral exploration targets are delineated.

2. Geological Setting and Datasets

2.1. Geological Setting

The West Qinling is located at the western part of the Qinling orogenic belt, with the Qilian orogenic belt to the north, the Qaidam block to the west, and the songpan block to the south (Figure 1a) [57–60].

The geotectonic position of the Xiahe-Hezuo area is located at the northwestern part of the West Qinling orogenic belt and at the western extension of the Qinling–Qilian–Kunlun central orogenic belt, whose complex geological structure creates a superior mineralization environment [61].

The Zaozigou Gold deposit is located within the West Qinling fold belt and is a typical epithermal-type gold deposit in the Xiahe-Hezuo area (Figure 1a). The main controlling factors for mineral resources formation within the region are tectonic movement and magmatism [62], with the regional tectonics spreading in a NW direction with developing folds and fractures. Complex geological structure and magmatism are dominated by Yanshan period intermediate-acid intrusive rocks, which are widely distributed in the form of the batholith, stock, apophysis, and veins [62] (Figure 1b).

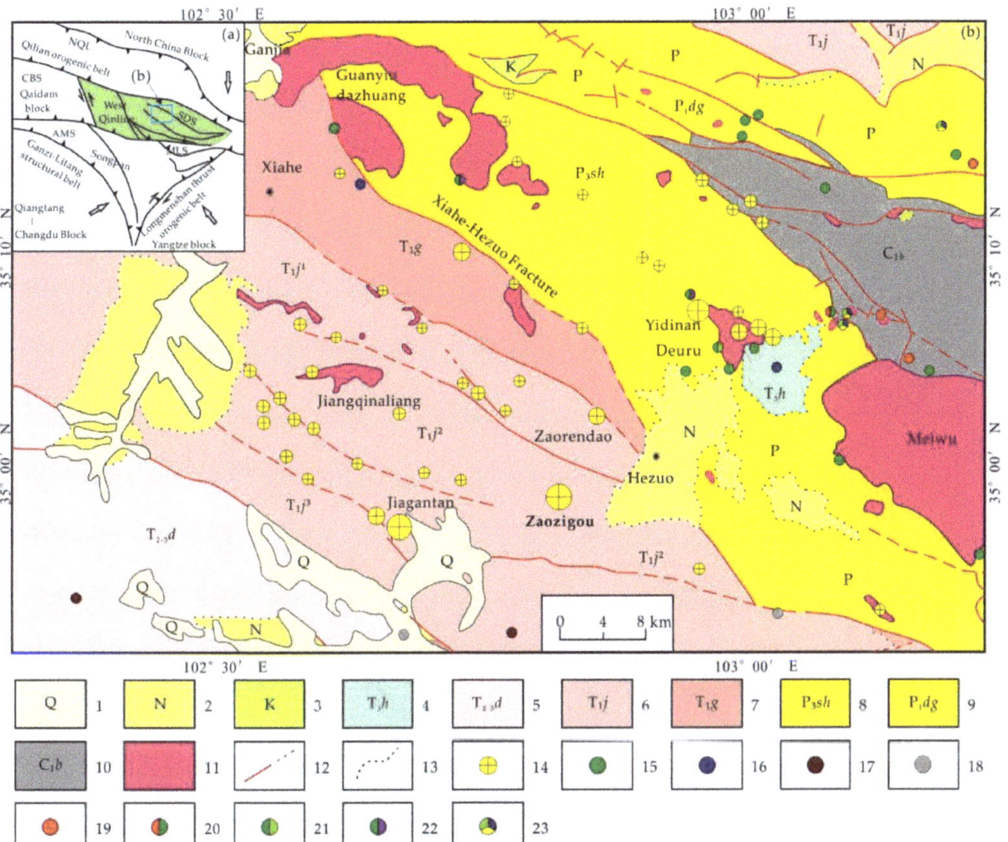

Figure 1. (**a**) Geotectonic location of the study area. NQL: North Qinling tectonic belt; SDS: Shangdan suture zone; CBS: Caibei suture zone; AMS: A'nyemaqen suture zone. (**b**) Geological map of Xiahe-Hezuo area (modified from [62]). 1. Quaternary; 2. Neogene; 3. Cretaceous;4. Upper Triassic Huari group; 5. Middle-Lower Triassic Daheba group; 6. Lower Triassic Jiangligou group; 7. Lower Triassic Guomugou group; 8. Upper Permian Shiguan group; 9. Lower Permian Daguanshan group; 10. Upper Carboniferous Badu group; 11. Intermediate-acid intrusive rocks; 12. Fracture; 13. Angular unconformity; 14. Gold depsit; 15. Copper deposit; 16. Lead depsit; 17. Antimony deposit; 18. Mercury deposit; 19. Iron deposit; 20. Iron–copper deposit; 21. Copper–molybdenum deposit; 22. Copper-tungsten deposit; 23. Polymetallic deposit.

The Triassic strata are the main stratigraphy for gold deposit. The genesis and spatial-temporal evolution of the intermediate-acidic dike is closely related to gold mineralization in the area, and during the mineralization process, the magmatic rocks not only provide the mineralized material, but also their internal environment is very suitable as an ore-bearing space, which can be regarded as a significant mineralization indicator for gold mineralization [63].

The spatial distribution of mineral deposits is directly controlled by the geotectonic position in the Xiahe-Hezuo area, which plays a major role in the formation of different types of gold deposits and is the boundary of the belt from a spatial perspective. The most important types of mineral-controlling structures are fractures and folds in this area [64,65]. The main orebodies of the Zaozigou deposit are produced in NE, NW, and near-SN oriented fracture zones, with two apparent mineralization periods, and post-formation fractures have modified and destroyed the orebodies [66–70].

The Zaozigou gold deposit is a typical representative of gold deposits associated with intermediate-acid dike rocks in the south of the Xiahe-Hezuo fracture. It is located at approximately 9 km southwest of the Hezuo city, Gansu Province, with convenient access to the mine site (Figure 2a). The main ore-bearing position is between Gully 1 and Gully 4, with a total area of approximately 2.6 km^2 (Figure 2b).

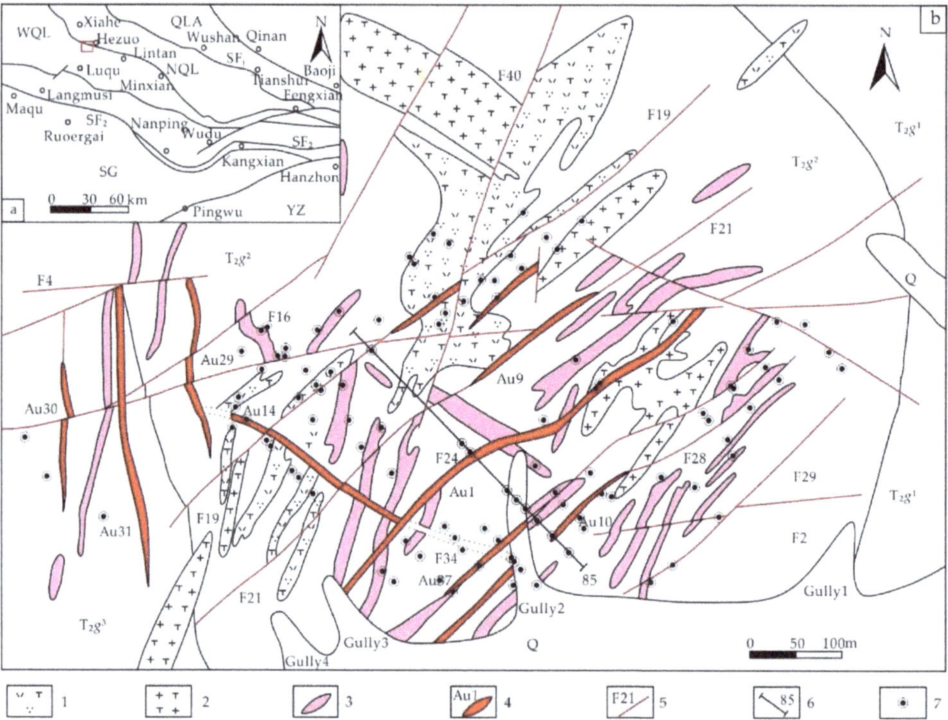

Figure 2. (**a**) Regional location of study area. QLA = Qilian tectonic belt; WQL: West Qinling tectonic belt; NQL: North Qaidam tectonic belt; SF1: Wushan–Tianshui–Shangdan suture zone; SF2: Maqu-Nanping–Lueyang suture zone; SG: Songpan–Ganzi tectonic belt; YZ: Yangtze block. (**b**) Geological map of Zaozigou gold deposit (modified from [71]). 1. Quartz diorite porphyrite; 2. Granodiorite-porphyry; 3. Plagioclase granite porphyry; 4. Gold orebody; 5. Fracture; 6. Mineral exploration line; 7. Drilling Hole.

There are only Triassic and Quaternary strata exposed in the Zaozigou gold deposit (Figure 2b). The strata of the area are mainly the lower part of the Gulangdi group (T_2g^1) of the Middle Triassic formation, and the Quaternary ($Q_h{}^{alp}$) is developed in the intermountain valley. The lower part of the Gulangdi group (T_2g^1) is the main ore-bearing rock, which is a set of fine clastic rocks consisting of siliceous slate, clastic feldspathic fine sandstone with interbedded siltstone slate and argillaceous slate. The formation is composed of a large sedimentary cycle from bottom to top consisting of siltstone → argillaceous slate → calcareous slate [66,72].

The fractures are well developed and have a complex morphology, forming an intersecting trend, indicating that the area has experienced multiple phases of geological activity. Fractures strictly control the distribution of orebodies and vein rocks within the deposit. They can be classified into five groups of orientations, namely NW, N-S, NE, E-W, and NNE [73,74]. Fracture structures are the structural surfaces of mineralization, and these structural surfaces control the spreading characteristics of the orebodies [75].

Intermediate-acid dike, including fine crystalline diorite, diorite porphyrite, biotite dioritic porphyrite, and quartz diorite porphyrite densely produced with a porphyritic structure, spreading in NNE direction, turning to a nearly N-S direction in the western part of the deposit and a few NW directions. Influenced by the regional multi-period deep fracture activity, the multi-period magmatism has overlapped on multi-phases mineralization within the deposit [76].

There are 147 gold orebodies have been found in the Zaozigou gold deposit, of which 17 orebodies are main ore bodies with gold reserves greater than 1 tonne, and the total gold reserves of more than 100 tonnes [71]. According to the spatial distribution and combination of the mineralization, Zaozigou deposit can be divided into eastern and western ore groups.

The Eastern ore group is mainly located between Gully 1 and Gully 3 with the strike of NE orientation, containing Au1 controlled by F24, Au9 controlled by F21, and Au15 controlled by F25. These orebodies extend over 1000 m long and 300 m wide, with a NW direction tendency and steep dip near the ground surface, locally nearly upright. In the deep, these orebodies have been staggered by gently dipping fracture, causing the tendency to change to SE orientation (Figure 2b). In addition, orebodies M4 and M6 are laying underground, with the strike of NWW orientation, the tendency of SW orientation, and the dip of 8°~26°. These orebodies cross the NE-striking orebodies obliquely, staggering them, and their own mineralization behavior occurs simultaneously [77].

The Western ore group is mainly distributed between Gully 3 and Gully 4, spreading in a nearly N-S direction, with August 29~August 31 as the main orebodies. The orebodies extend over 1000 m long and wider than 500 m, with a strike of 350°~10°, varying tendency and dips greater than 75°, locally subvertical. These orebodies extend, and the mineralization is weaker in the steeper parts and stronger in the shallower parts.

2.2. Datesets Description

Historical geological and geochemical data were completed, including geological reports, geological exploration maps, and drills geochemical data, from the Development Research Center of China Geological Survey and No.3 Geological and Mineral Exploration team, Gansu Provincial Bureau of Geology and Mineral Exploration and Development, the coordinate system used in the mine-scale is Gaussian Kruger projection coordinates.

This study collected 72 drillings data in the Zaozigou gold deposit, established a drilling location database, an assay database, an inclinometry database, and a lithology database. The 3D model of drillings was constructed based on the drill hole data database (Figure 3).

The primary geochemical halo data from the drillings were collected from the "Zaozigou gold successive resources exploration project in Hezuo city, Gansu Province", with a total of 72 drillings and 5028 samples with 12 elements of Ag, As, Au, Cu, Hg, Pb, Zn, Sb, W, Bi, Co, and Mo (the element detection methods can refer literature [71]). The sampling method was the continuous picking method with sample intervals generally within 10 m. Some orebodies or strongly altered areas being sampled at decreased intervals.

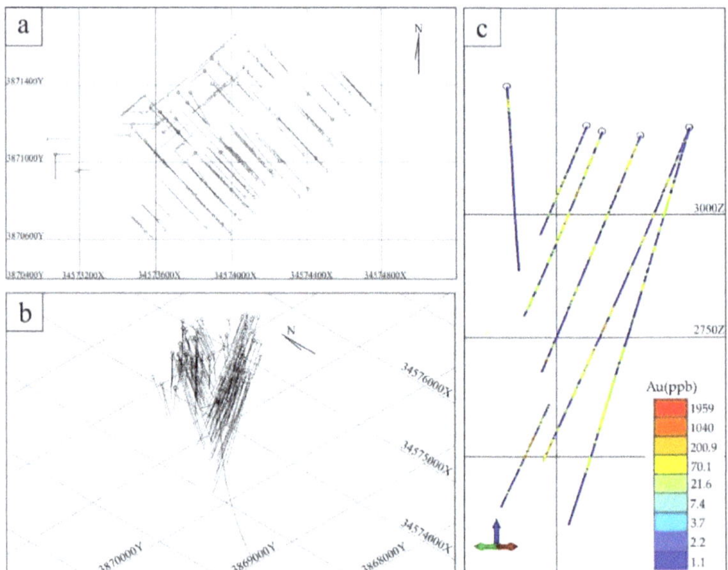

Figure 3. 3D model of Drillings in Zaozigou gold deposit. (**a**) Plan view of drillings distribution. (**b**) Side view of drillings distribution. (**c**) Sample grade distribution of drillings.

3. Methodology

3.1. Deep Auto-Encoder Network

3.1.1. Network Structure Designing

The basic end-to-end structure of DAE and STOAD consists of the input layer (Input), convolutional layer (CONV), activation function layer, and output layer (Output).

DAE is a method that uses neural networks to compress and reconstruct data, and its core is to use an Encoder to compress the original data X to obtain the intermediate variable Z, and then use a Decoder to reconstruct Z to obtain X' (Figure 4). It can be widely applied to tasks such as data dimensional reduction, feature extraction, and anomaly detection.

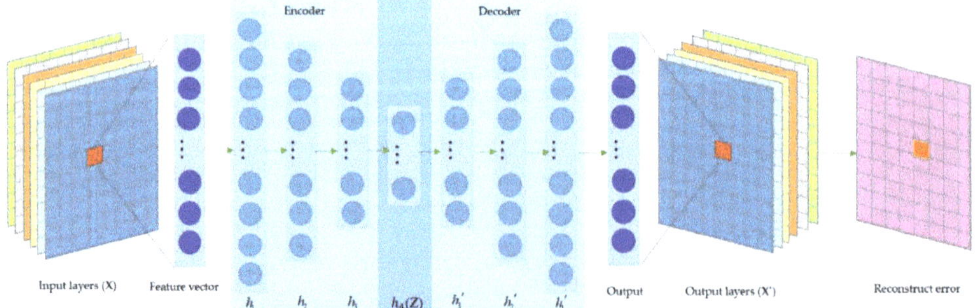

Figure 4. Network structure of DAE.

3.1.2. Model Training

In the DAE model, the objective function is to measure the error between the reconstructed sample and the real sample and back propagate to adjust the network parameters. Mean Square Error (MSE) is usually used as the objective function in the reconstruction problem, and the optimization process is mainly divided into three steps: (i) forward propagation for calculating the output, (ii) calculating the loss of reconstruction, and

(iii) back propagation to calculating the gradient according to the loss for optimizing the model parameters.

$$x^l = \sigma\left(z^l\right) = \sigma\left(W^l x^{l-1} + b^l\right) \tag{1}$$

Equation (1) express the calculation process from the $l - 1$th layer to the lth layer of CNN, where x^l represents the features of the lth layer (e.g., x^1 represents the input data, x^2 represents the output obtained after the first layer of convolution and activation calculation), b^l represents the bias term of the lth layer, W^l represents the weight parameter of the convolution of the lth layer, Z^l represents the intermediate variables of the lth layer without the activation function calculation, and σ is the activation function. Rectified Linear Unit (ReLU) function is used as the activation function in this paper.

Assuming that the DAE model has a total of L layers, the errors of the input data and the reconstructed data are calculated by MSE to obtain the loss value.

$$Loss = J(W, b, x, y) = \frac{1}{2} \parallel \sigma\left(z^L\right) - x^1 \parallel_2^2 = \frac{1}{2} \parallel \sigma\left(W^L x^{L-1} + b^L\right) - x^1 \parallel_2^2 \tag{2}$$

δ^L is the partial derivatives of *Loss* to z^L,

$$\delta^L = \frac{\partial J(W, b, x, y)}{\partial z^L} = \frac{\partial J(W, b, x, y)}{\partial x^L} \odot \sigma'\left(z^L\right) \tag{3}$$

$$\delta^l = \frac{\partial J(W, b, x, y)}{\partial z^{l+1}} \frac{\partial z^{l+1}}{\partial z^l} = \left(\frac{\partial z^{l+1}}{\partial z^l}\right)^T \delta^{l+1} = \left(W^{l+1}\right)^T \delta^{l+1} \odot \sigma'\left(z^l\right) \tag{4}$$

The partial derivatives of W and b are calculated from the losses, and $\odot$ is the Hadamard product.

$$\frac{\partial J(W, b, x, y)}{\partial W^l} = \frac{\partial J(W, b, x, y)}{\partial z^l} \frac{\partial z^l}{\partial w^l} = \delta^l \left(x^{l-1}\right)^T; \ \frac{\partial J(W, b, x, y)}{\partial b^l} = \frac{\partial J(W, b, x, y)}{\partial z^l} \frac{\partial z^l}{\partial b^l} = \delta^l \tag{5}$$

The gradient of the back-propagation of each convolutional layer can be calculated by Equation (5), and finally the convolutional kernel parameters are updated by the following two equations. α is the learning rate.

$$W^{l+1} = W^{l+1} - \alpha \frac{\partial J(W, b, x, y)}{\partial W^{l+1}} \tag{6}$$

$$b^{l+1} = b^{l+1} - \alpha \frac{\partial J(W, b, x, y)}{\partial b^{l+1}} \tag{7}$$

where $l = 1, 2, 3 \ldots, L - 1$, the above steps are repeated to achieve the training of the DAE network.

3.1.3. Model Testing

In the testing process, the test data X are reconstructed by DAE to obtain X', MSE between X and X' is normalized. Setting the threshold value μ, when the MSE value is greater than μ, corresponding samples are judged as anomalies, otherwise, the sample is normal. The calculation equations are as follows.

$$Error(X) = \parallel X - X' \parallel_2^2 \tag{8}$$

$$Predict(X) = MinMax_{0-1}\left\{\parallel X - X' \parallel_2^2\right\} = \begin{cases} 1, & Predict(X) \geq \mu \\ 0, & Predict(X) < \mu \end{cases} \tag{9}$$

3.2. KD-Based STOAD Model

3.2.1. Network Structure Design

KD [78] is a mainstream method applied to model compression, which mainly consists of the Teacher network (T) and Student network (S). The larger T network is usually trained first for the tasks of image classification, target detection, semantic segmentation, etc., and then the smaller S network is trained to learn the generalization ability of the T network. Compared with T-networks, S-networks have fewer parameters and faster computation speed.

Different information will be contained at different depth of the network. For fully extract metallogenic information, the STOAD model is proposed to achieve anomaly detection of layers based on calculating the differences between T encoder and S network. STOAD is divided into two structures: The T network and the S network, where the T network has the same Encoder and Decoder as the DAE, and the S network has the same structure as the Encoder in the T network. Especially, three scales output are designing to express different information at different depth of the network.

In the training process, STOAD model uses all layers to pre-train the T network for reconstruction. The encoder in the T network can be recognized as having the ability to compress the layers, and thus can be seen as having the ability to handle both normal and abnormal data [79]. If the T network is not trained, the anomaly detection can be achieved, but the feature extraction of the T network is not clear, and the interpretability is poor. Sequentially, taking normal data as input, training the S network, calculating the three scale output differences of T encoder and S network using MSE, while h_i, $i = 1, 2, 3$, are the scales represents different depth of the network. Finally, back-propagates to update the parameters of the S network, the S network only has the capacity of handling normal data learned from the encoder of T network.

In the testing process, taking all data as input. Since the S network only learns the ability of dealing with normal data from the encoder of T network, the S network is difficult to maintain the similar multiscale output features with the T network because of a larger MSE. Therefore, the anomaly detection can be achieved by the MSE values (Figure 5). In contrast to DAE, STOAD not only has three outputs, but also uses multiple levels of semantic and detailed information to make full use of the information in the data for anomaly detection.

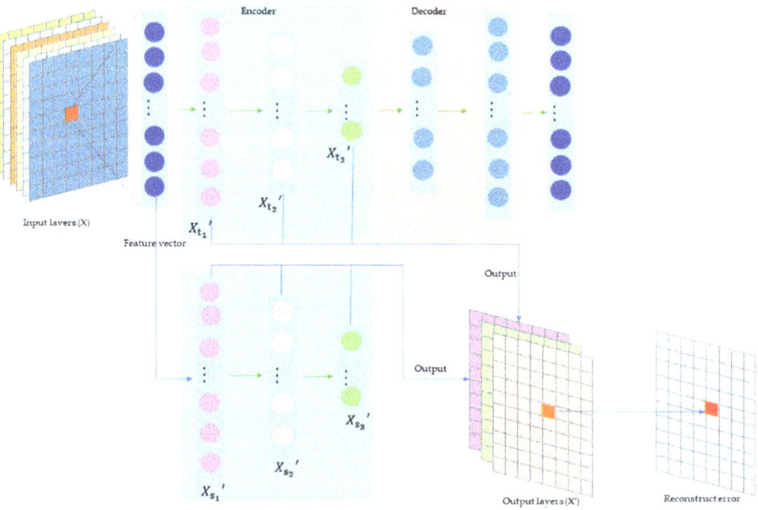

Figure 5. STOAD network structure.

3.2.2. Model Training and Testing

The training loss of STOAD consists of two parts, which are the training loss of the T network and the S network

The overall T-network can be considered as a DAE, and MSE is used as the reconstruction *loss* of the T-network, so that the encoder part of the T-network holds the data compress capability. The training *loss* of the T network can be calculated as follow equation.

$$Loss_{teacher} = \frac{1}{2} \parallel X - X\prime \parallel_2^2 \tag{10}$$

The *loss* of the S network consists of the MSEs of the three scales output of the T network. In order for the S network to learn the normal data compression ability from the T network, let $h_1 = h_2 = h_3 = 1/3$, and the training *loss* of the S network is calculated by Equation (8).

$$Loss_{student} = \sum_{i=1}^{3} \frac{1}{2} h_i \parallel X_{t_i} - X_{s_i} \parallel_2^2 \tag{11}$$

The STOAD model training and testing process is mainly as follows.

1. The essence of T-network is a DAE, and its training process is same with DAE referring to Section 3.1.2.
2. The S-network update is similar to the T-network, mainly in calculating the loss of the multi-scale features output from the encoder of the T-network, and using only normal data for training, thus letting the S-network to only learn the ability to compress normal data. L is the depth of S network.

$$Loss_{student} = J(W, b, x, y) = \sum_{i=1}^{3} \frac{1}{2} h_i \parallel \sigma\left(W_t^{i+1} x^i + b_t^{i+1}\right) - \sigma\left(W_s^{i+1} x^i + b_s^{i+1}\right) \parallel_2^2 \tag{12}$$

δ^L is the partial derivatives of z^L of S network.

$$\delta^L = \frac{\partial J(W, b, x, y)}{\partial z^L} = \sum_{i=1}^{3} \frac{1}{2} h_i \frac{\partial J(W, b, x, y)}{\partial x^L} \odot \sigma'\left(z^L\right) \tag{13}$$

and the δ^l is, i is 1 to $L - 1$:

$$\delta^l = \frac{\partial J(W, b, x, y)}{\partial z^{l+1}} \frac{\partial z^{l+1}}{\partial z^l} = \left(\frac{\partial z^{l+1}}{\partial z^l}\right)^T \delta^{l+1} = \left(W^{l+1}\right)^T \odot \sigma'\left(z^l\right) \delta^{l+1} \tag{14}$$

$$\frac{\partial J(W, b, x, y)}{\partial W^l} = \frac{\partial J(W, b, x, y)}{\partial z^l} \frac{\partial z^l}{\partial w^l} = \delta^l \left(x^{l-1}\right)^T \tag{15}$$

$$\frac{\partial J(W, b, x, y)}{\partial b^l} = \frac{\partial J(W, b, x, y)}{\partial z^l} \frac{\partial z^l}{\partial b^l} = \delta^l \tag{16}$$

Convolutional module is updated by the following two equations. α is the learning rate.

$$W^{l+1} = W^{l+1} - \alpha \frac{\partial J(W, b, x, y)}{\partial W^{l+1}} \tag{17}$$

$$b^{l+1} = b^{l+1} - \alpha \frac{\partial J(W, b, x, y)}{\partial b^{l+1}} \tag{18}$$

where $l = 1, 2, \ldots, L - 1$.

In the testing process, we used T and S networks to calculate the testing dataset, separately. Defining the weights of three scales are $W_{h1} = W_{h2} = W_{h3} = 1/3$, the weighted sum of the output can be calculated by Equation (19), where X_{t_i} and X_{s_i} represent the output of the Encoder part of the T network and the S network, respectively. Predict (x) is the normalized Error (x), selecting a threshold μ, when the Predict (x) is greater than μ the sample point is judged to be an abnormal sample, otherwise it is a normal sample.

$$\text{Error}(x) = \sum_{i=1}^{3} h_i \parallel X_{t_i} - X_{s_i} \parallel_2^2 \tag{19}$$

$$\text{Predict}(x) = MinMax_{0-1}\{\sum_{i=1}^{3} h_i \parallel X_{t_i} - X_{s_i} \parallel_2^2\} = \{_{0,}^{1,} \quad _{Predict(x) < \mu}^{Predict(x) \geq \mu} \tag{20}$$

4. Results

4.1. Geological and Geochemical Quantitative Prediction Model at Depth of Zaozigou Gold Deposit

Before carrying out the deep quantitative mineral prospectivity mapping in the Zaozigou gold deposit, it is necessary to review the geological and geochemical quantitative prediction model established in previous work [80].

Orebodies are strictly controlled by fractures in the Zaozigou gold deposit. The 30 m buffer zone of the fractures can effectively reflect the influence range of fracture, which can be used as a mineral prediction indicator. Factor F4 is an element association of Sb-Hg, which has a close relationship with fractures, and can be used as a favorable indicator for inferring deep fractures. The middle anomaly area of Au extracted by the multiple fractal C-V model can well reflect the spatial distribution of orebodies, which should be used as an important quantitative indicator. Near-ore halo element association B2 is extracted by the knowledge-driven CoDA also can express the location of orebodies well, it should be another mineral prediction indicator. The ratio of front halo to tail halo is an important geochemical parameter for predicting orebodies, B1/B3 is regarded as a prediction indicator accordingly (Table 1; Figure 6). The detailed data analysis and interpretation can refer to the literature [80].

Table 1. Geological and geochemical quantitative mineral resource prediction model at depth of the Zaozigou gold deposit.

Ore-Forming Factor	Description	Prediction Indicator	Variables
Geology	fracture	Influence range of fracture	30 m buffer zone
		Element association of fracture	Hg-Sb (F4)
Geochemistry	Ore-forming element	Geochemical anomaly	Au
	Primary geochemical halo	Element association of near-ore halo	Au-Ag-Cu-Pb-Zn (B2)
		Geochemical parameter (Front halo/tail halo)	As-Sb-Hg (B1)/W-Bi-Co-Mo (B3)

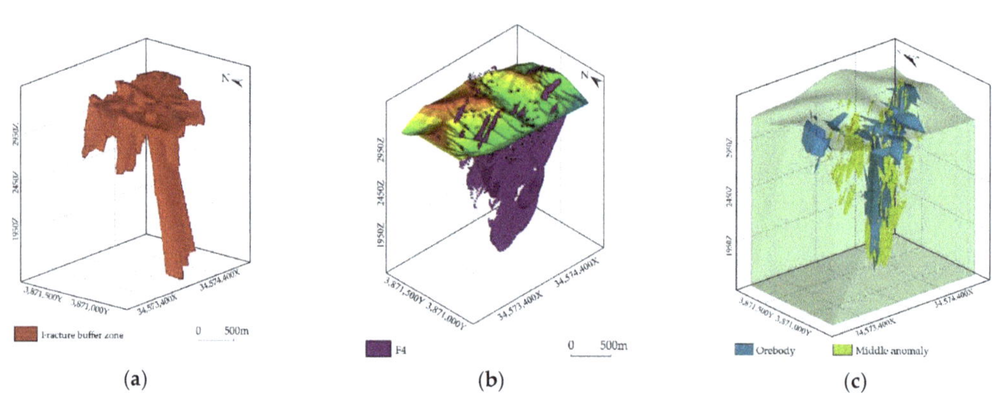

Figure 6. *Cont.*

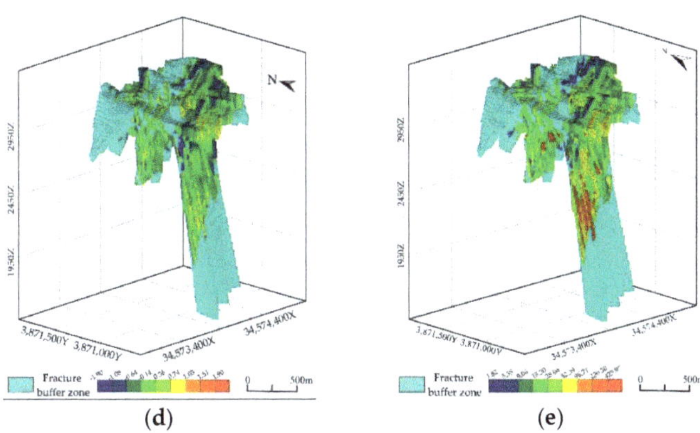

Figure 6. (**a**) Fracture buffer zone; (**b**) Element association of fracture; (**c**) Au anomaly; (**d**) Element association of near-ore halo; (**e**) Geochemical parameter (Front halo/tail halo).

4.2. Training Sample Selection

For mineralization prediction, the target variable (i.e., the label of training samples) is either mineralized or nonmineralized (denoted by 1 and 0, respectively). A total of 39,306 positive samples extracted from known orebodies and 49,686 negative samples extracted from the nonmineralized position confirmed by drillings are used as training dataset. In contrast to mineralization generates concentrated in limited space, the nonmineralization is a widespread phenomenon, and negative samples are selected to be distributed as randomly and uniformly as possible throughout the study area [81] (Figure 7).

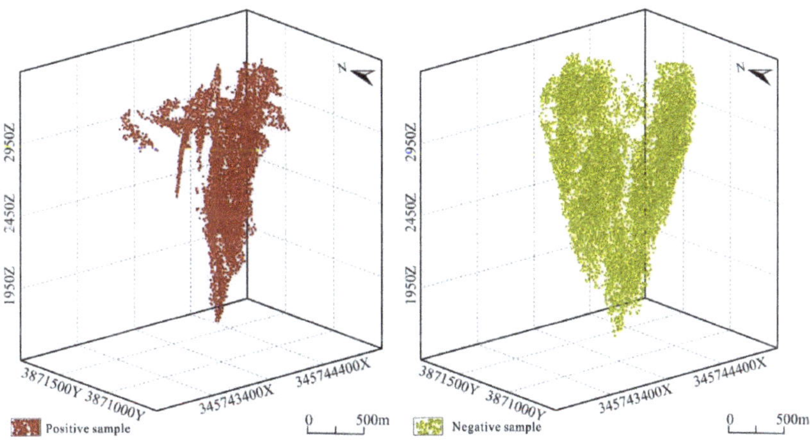

Figure 7. Distribution of positive samples and negative samples.

4.3. DAE-Based 3D MPM

DAE is a widely used unsupervised model in deep learning. In this study, the DAE model is constructed by convolution, and the training process automatically learns to extract the semantic features of the normal evidence layer to compress and reconstruct the corresponding layers. In the testing process, detecting ore-induced anomalies by MSE value, the higher the MSE value is, the greater the loss of reconstruction is, which illustrates the high-probability area of ore-forming.

This DAE model is built using python 3.8, PyTorch 1.71 framework, the optimizer is Adam, the batch size is 16, the learning rate is 0.001, the learning rate decay is 10% per

5 Epochs, the CPU is 12th Gen Intel(R) Core(TM) i7-1270OKF, and the GPU is NVIDIA GeForce RTX 3090.

DAE model uses 70% of the normal data for training, and the testing dataset includes the remaining 30% of the normal samples and the relatively small number of anomaly samples. The DAE model is trained for 500 Epochs and the loss function is plotted as follows. The model relatively converges at an MSE value of about 0.025, but the MSE fluctuates wider (Figure 8). In Figures 4 and 6, the horizontal axis indicates the sampling order and the vertical axis indicates the loss of DAE. The prediction parameters of precision value reaches 93%, Recall value is 88%, and F1-measure is 90%, which demonstrate the high accuracy and reliability of DAE-based MPM. It clearly shows that the high ore-forming probability at depth of Zaozigou gold deposit, which is worth for further investigation (Figure 9).

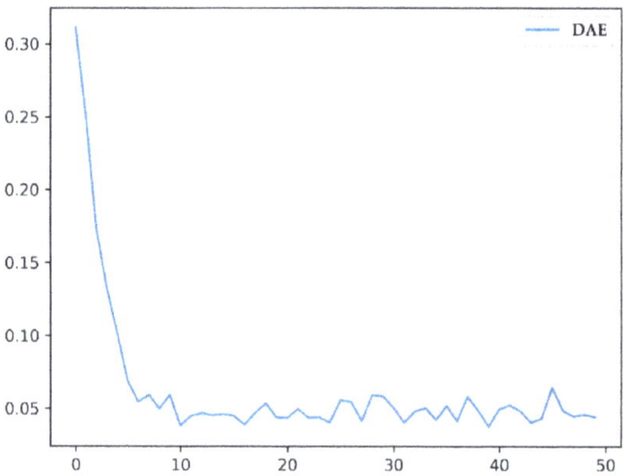

Figure 8. The training loss of DAE model.

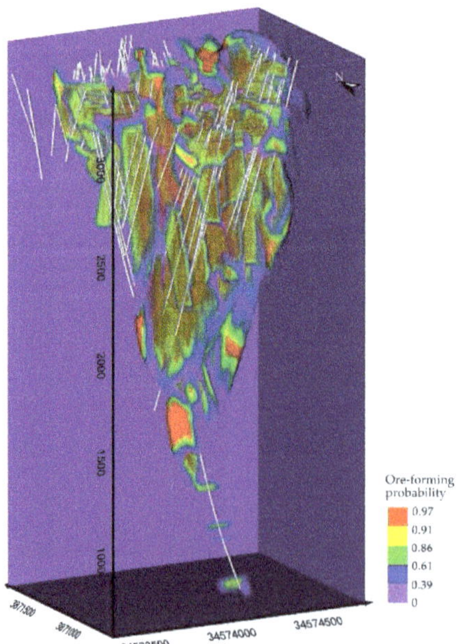

Figure 9. DAE-based 3D MPM.

4.4. STOAD-Based 3D Mineral Prospectivity Mapping

The STOAD model is implemented using python 3.8, and PyTorch 1.71 framework. The T network is mainly composed of two parts: encoder and Decoder. The structure of T network is symmetric for letting the number of parameters of both parts identical, and thus the data compression ability of encoder is approximately equal to the data reconstruction ability of Decoder.

The S network only has the encoder part, which is consistent with the T encoder. ReLU activation function is used for its capability of dealing with nonlinear data.

The same training parameters are used in the T and S networks, with an optimizer of Adam, a batch size of 16, a learning rate of 0.001, and a learning rate decay of 10% per 5 Epochs.

Similarly, STOAD uses 70% of the normal data for training, and the testing dataset includes the remaining 30% of the normal samples and the relatively small number of anomaly samples.

The STOAD model is trained with 500 Epochs, and the Teacher loss function and the Student loss function are plotted in Figure 10, the T network converges at about MSE value of 0.025 (similar to the above DAE), and fluctuates wider. The MSE curve of the S network converges at a lower value of about 0.0125, and distributes more stable than T network. The loss of the T network is the same as DAE indicating the ability to reconstruct all training data, the loss of the S network represents the difference in output between the S and T networks, and the more stable loss of the S network means that the S network can learn better the ability of the T network to reconstruct normal data.

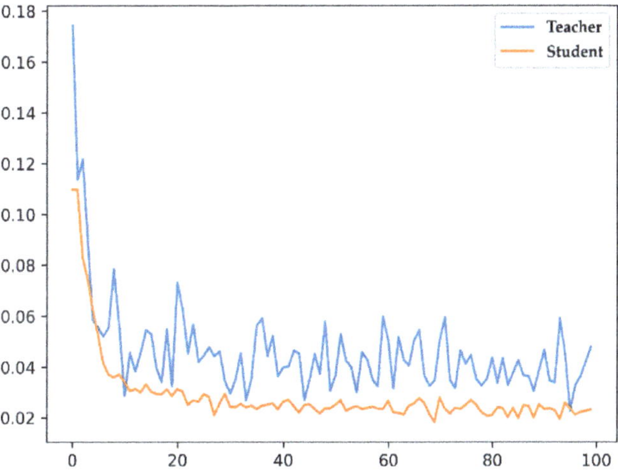

Figure 10. The training loss of STOAD model.

The prediction parameter of Precision value reaches 96%, Recall value is 90%, and F1 value is 93%.

The 3D MPM based on STOAD model shows the high ore-forming probability at depth of Zaozigou gold deposit which has the similar characteristics ore-forming probability distribution at depth. It gives evidence for deep mineral resources prediction (Figure 11).

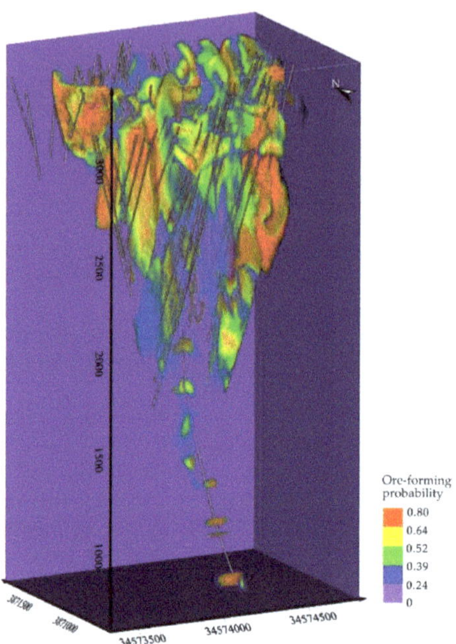

Figure 11. STOAD-based 3D MPM.

5. Discussion

In this study, the deep learning methods of DAE and STOAD model are carried out for 3D MPM in Zaozigou gold deposit.

Compared with the DAE model, STOAD model has a higher precision in detection of ore-induced anomalies, which demonstrates a higher reliability of 3D MPM (Figures 9 and 11).

Meanwhile, the ROCs of the two methods were used to evaluate the models, and the AUC values of the two methods were 0.92 of DAE and 0.97 of STOAD. Observably, the STOAD model expresses a higher performance (Figure 12).

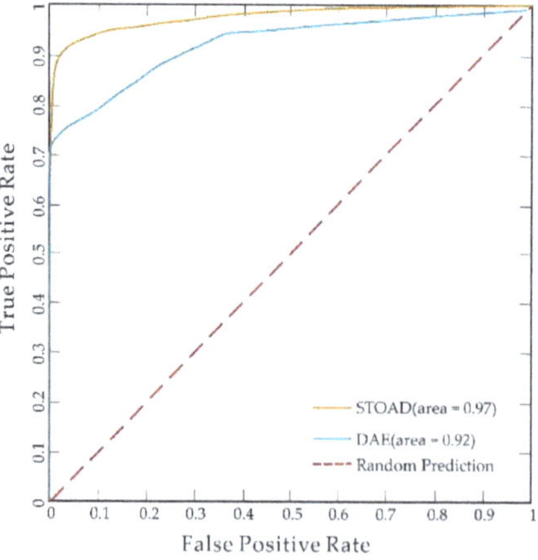

Figure 12. ROC curves of the DAE and STOAD.

From the points of algorithm theory, the STOAD model is proposed based on the KD idea combined with the DAE network, which is an improvement of DAE. So, it is better than the conventional DAE model in terms of algorithm performance.

From the points of application, the results of quantitative mineral resources prediction show that STOAD is more accurate and robust, which is demonstrated by prediction parameters, the loss curve (Figures 8 and 10) and the ROC curve (Figure 12).

Furthermore, a total of 4 methods, two machine learning methods (MaxEnt model and GMM) and two deep learning methods (DAE and STOAD model), have been carried out for 3D MPM (Figure 13). The machine learning-based 3D MPM can refer to the literature [80], which is the previous work of our research team.

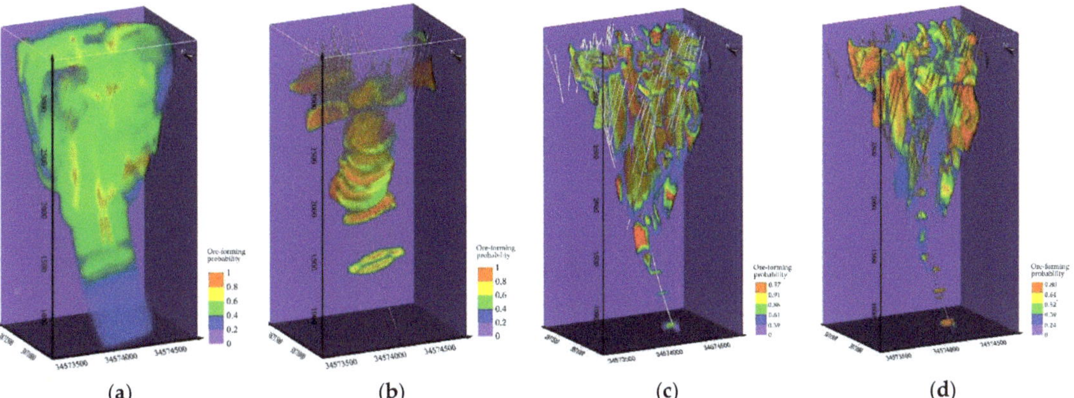

Figure 13. Comparison of the 3D MPM of 4 methods. (**a**) MaxEnt mode. (**b**) GMM. (**c**) DAE model. (**d**) STOAD model.

The prediction results of the four methods exhibit a high correlation with the known orebodies in the shallow part and a slight difference in the deep part (Figure 13). From the comparative analysis of the prediction results and model performance among the four methods, the STOAD takes advantage of mathematical theory and has the highest AUC value (Figure 14). Especially, the high probability area can better reflect the metallogenic regularity at large depth. Therefore, it is decided to delineate a comprehensive mineral exploration targets at large depth of the Zaozigou based on the prediction results of STOAD model and supported by the prediction results of other methods. In order to improve the prediction accuracy, the area with a logical probability greater than 0.75 and less than 0.998 is defined as a medium potential area, and the area with a logical probability greater than 0.998 is defined as a high potential area, based on which three main mineral exploration targets are circled (Figure 15).

Target I is located at elevation 1600~2000 m, belongs to the NE-orientation orebody group, which should be the extension of the Au1 orebody. In this position, we developed the main ore-forming fracture. The Au and Sb concentration in this position, and the ratio of front halo to tail halo has been increasing. Especially, it appears high logical probability calculated by all four methods at this position, indicating the Au1 orebody may extends deeper than the elevation of 1800 m or a concealed orebody exist therein.

Target II is located in the NW-orientation orebody group at the elevation of 2000~2400 m, where the fractures distribute complex, and the tail halo element Co has a significant vertical overlap with the front halo elements Sb and As. The elevated geochemical parameters starting from 2600 m to deep indicate that there may be an extension of ore bodies or new blind orebodies at depth. According to the theory of primary halo zonation, the front halo indicators generally appear at the leading edge of the ore body 100–300 m, and the blind ore body will appear at an altitude of 2400 m deep.

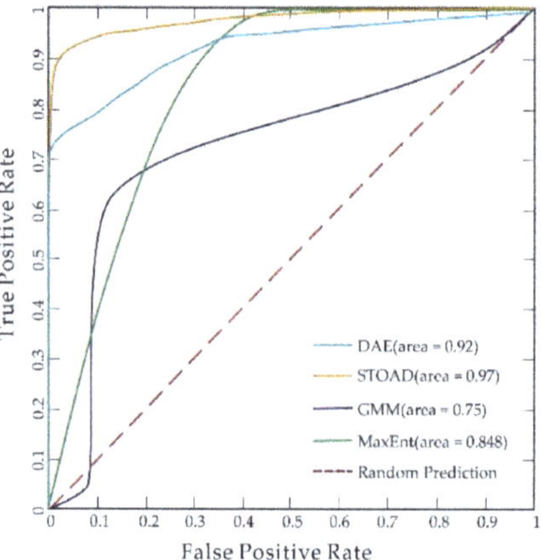

Figure 14. ROC curves of 4 methods, MaxEnt model, GMM, DAE model and STOAD model.

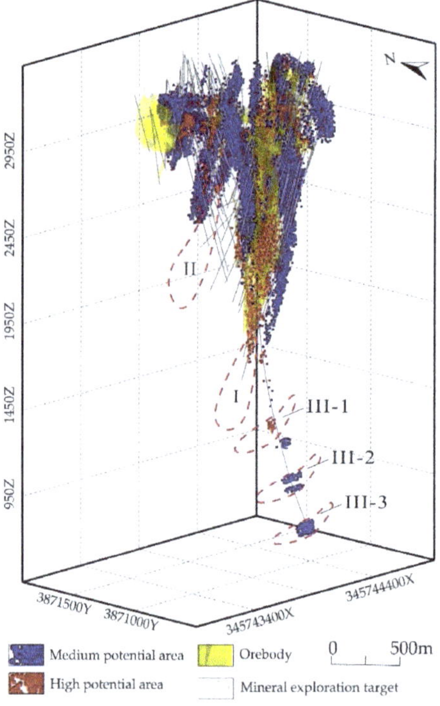

Figure 15. Comprehensive mineral exploration targets.

Target III is located below 1600 m, it can be divided into targets III-1, III-2, and III-3. Target III-1 is located at elevation of about 1300 m, target III-2 located at elevation of about 1000 m., target III-3 located at elevation of 700 m. The deep fracture is the main indicator. It is worth mentioning that the deep drill for scientific research has achieved in 2021, which is the deepest drill of Zaozigou gold deposit, which reaches to the elevation of 700 m (Figure 15). In previous study, there may be an "upper veins and lower layers" distribution feature of orebodies [82]. The shallow stratigraphy is strongly folded, and steeply dipping fracture structures are developed; the orebody is controlled by fractures and vein rocks, while the deep orebody may be mainly controlled by intrusive mass to form laminated ore bodies. This understanding is also reflected in the present prediction targets.

6. Conclusions

To address the scientific problem of quantitative mineralization prediction at large depth, the geological and geochemical quantitative mineral resources prediction model at depth of the Zaozigou gold deposit is constructed based on the quantitative extraction of mineralization signatures, which consists of five quantitative prediction indicators, including 30 m fracture buffer zone, element association (Sb-Hg) of ore-bearing structures, metallogenic element Au, near-ore halo element association Au-Ag-Cu-Pb-Zn, and the ratio of front halo to tail halo (As-Sb-Hg)/(W-Mo-Bi). Then, two machine learning methods (MaxEnt model and GMM) and two deep learning methods (DAE and STOAD model) are carried out for 3D MPM, and mineral exploration targets are delineated at depth (Figure 15).

Because the primary halo of drills is limited by the engineering-controlled boundaries, this study adopts the idea of quantitative expression of ore-controlling factors by element associations and limit the deep inference range within the modeling space to perform deep quantitative mineral prediction. However, the accuracy of the prediction of machine learning and deep learning methods inevitably decreases rapidly with increasing depth. This shortcoming can be supplemented by two aspects of quantitative inversion of geophysical data and numerical simulation of deep mineralization dynamics, which is also a direction worth further research.

In summary, a series of methods are developed in this study of 3D Mineral Prospectivity Mapping of Zaozigou gold deposit, West Qinling, China. The achieved results by machine learning and deep learning methods shows well performance in quantitative mineral resources prediction at depth, which is worth being promoted to similar ore deposits. Moreover, it was concluded that the deep learning-based approach works better with larger amounts of data, and the STOAD proposed in this paper has a higher accuracy in anomaly detection. May the relevant ideas of this study provide a reference for related researchers.

Author Contributions: Ideas, B.L. and Z.Y.; Methodology, Z.Y. and M.X.; software, M.X., Y.W., Y.K., C.L. and G.C.; writing—original draft preparation, Z.Y., B.L., C.L., Y.W., Y.K. and Y.G.; writing—review and editing, Z.Y. and B.L.; visualization, M.X., Y.K., S.Z., H.Z., L.W. and R.T.; supervision, B.L.; funding acquisition, B.L. All authors have read and agreed to the published version of the manuscript.

Funding: This research was funded by the National Key Research and Development Program of China (Grants 2017YFC0601505); the National Natural Science Foundation of China (Grants 41602334; 42072322); the Key Laboratory of Geochemical Exploration, Ministry of Natural Resources (Grant AS2019P02-01); Sichuan Science and Technology Program (Grant 2022NSFSC0510); and the Opening Fund of the Geomathematics Key Laboratory of Sichuan Province (Grant scsxdz2020yb06, scsxdz2021zd04).

Data Availability Statement: Not applicable.

Acknowledgments: The authors thank the anonymous reviewers and the editors for their hard work to this paper. We are grateful to the Development Research Center of China Geological Survey and No. 3 Geological and Mineral Exploration team, Gansu Provincial Bureau of Geology and Mineral Exploration and Development for their data support.

Conflicts of Interest: The authors declare no conflict of interest.

References

1. Frits, A. Principles of probabilistic regional mineral resource estimation. *Earth Sci.* **2011**, *36*, 189–200.
2. Zhou, Y.Z.; Zuo, R.G.; Liu, G.; Yuan, F.; Mao, X.C.; Guo, Y.J.; Xiao, F.; Liao, J.; Liu, Y.P. The Great-leap-forward Development of Mathematical Geoscience During 2010–2019: Big Data and Artificial Intelligence Algorithm Are Changing Mathematical Geoscience. *Bull. Mineral. Petrol. Geochem.* **2021**, *40*, 556–573.
3. Liu, Z.; Li, L.; Fang, X.; Qi, W.; Shen, J.; Zhou, H.; Zhang, Y. Hard-rock tunnel lithology prediction with TBM construction big data using a global-attention-mechanism-based LSTM network. *Automation in Construction.* **2021**, *125*, 103647. [CrossRef]

4. Cheng, Q.M. What are Mathematical Geosciences and its frontiers? *Earth Sci. Front.* **2021**, *28*, 6–25.
5. Zuo, R.G. Deep Learning-Based Mining and Integration of Deep-Level Mineralization Information. *Bull. Mineral. Petrol. Geochem.* **2019**, *38*, 53–60.
6. Zuo, R.G.; Peng, Y.; Li, T.; Xiong, Y.H. Challenges of Geological Prospecting Big Data Mining and Integration Using Deep Learning Algorithms. *Earth Sci.* **2021**, *46*, 350–358.
7. Zhang, Q.; Zhou, Y.Z. Big data will lead to a profound revolution in the field of geological science. *Chin. J. Geol.* **2017**, *52*, 637–648.
8. Zhou, Y.Z.; Li, P.X.; Wang, S.G.; Xiao, F.; Li, J.Z.; Gao, L. Research progress on big data and intelligent modelling of mineral deposits. *Bull. Mineral. Petrol. Geochem.* **2017**, *36*, 327–331.
9. Zhou, Y.Z.; Wang, J.; Zuo, R.G.; Xiao, F.; Shen, W.J.; Wang, S.G. Machine learning, deep learning and Python language in field of geology. *Acta Petrol. Sin.* **2018**, *34*, 3173–3178.
10. Sun, S.; He, Y.X. Multi-label emotion classification for microblog based on CNN feature space. *Adv. Eng. Sci.* **2017**, *49*, 162–169.
11. Simonyan, K.; Zisserman, A. Very deep convolutional networks for large-scale image recognition. *arXiv* **2014**, arXiv:1409.1556.
12. Martens, J.; Sutskever, I. Learning recurrent neural networks with hessian-free optimization. In Proceedings of the 28th International Conference on International Conference on Machine Learning, Bellevue, WD, USA, 28 June 2011–2 July 2011.
13. Bengio, Y.; Lamblin, P.; Popovici, D.; Larochelle, H. Greedy layer-wise training of deep networks. *Adv. Neural Inf. Process. Syst.* **2006**, *19*, 153–160.
14. Sainath, T.N.; Kingsbury, B.; Ramabhadran, B. Auto-encoder bottleneck features using deep belief networks. In Proceedings of the 2012 IEEE International Conference on Acoustics, Speech and Signal Processing (ICASSP), Kyoto, Japan, 25–30 March 2012; pp. 4153–4156.
15. Hinton, G. Deep Belief Nets. In *Encyclopedia of Machine Learning*; Sammut, C., Webb, G.I., Eds.; Springer: Boston, MA, USA, 2011; pp. 267–269. [CrossRef]
16. Ciresan, D.; Giusti, A.; Gambardella, L.; Schmidhuber, J. Deep neural networks segment neuronal membranes in electron microscopy images. *Adv. Neural Inf. Process. Syst.* **2012**, *25*, 2843–2851.
17. Lin, N. *Study on the Metallogenic Prediction Models Based on Remote Sensing Geology and Geochemical Information: A Case Study of Lalingzaohuo Region in Qinghai Province*; Jilin University: Changchun, China, 2015.
18. Yan, G.; Xue, Q.; Xiao, K.; Chen, J.; Miao, J.; Yu, H. An analysis of major problems in geological survey big data. *Geol. Bull. China* **2015**, *34*, 1273–1279.
19. Holtzman, B.K.; Paté, A.; Paisley, J.; Waldhauser, F.; Repetto, D. Machine learning reveals cyclic changes in seismic source spectra in Geysers geothermal field. *Sci. Adv.* **2018**, *4*, o2929. [CrossRef]
20. Rouet Leduc, B.; Hulbert, C.; Lubbers, N.; Barros, K.; Humphreys, C.J.; Johnson, P.A. Machine learning predicts laboratory earthquakes. *Geophys. Res. Lett.* **2017**, *44*, 9276–9282. [CrossRef]
21. Zhou, Y.Z.; Chen, S.; Zhang, Q.; Xiao, F.; Wang, S.G.; Liu, Y.P.; Jiao, S.T. Advances and prospects of big data and mathematical geoscience. *Acta Petrol. Sin.* **2018**, *34*, 255–263.
22. Zuo, R.; Xiong, Y. Big data analytics of identifying geochemical anomalies supported by machine learning methods. *Nat. Resour. Res.* **2018**, *27*, 5–13. [CrossRef]
23. Xiong, Y.; Zuo, R.; Carranza, E.J.M. Mapping mineral prospectivity through big data analytics and a deep learning algorithm. *Ore Geol. Rev.* **2018**, *102*, 811–817. [CrossRef]
24. Zhong, Z.; Sun, A.Y.; Yang, Q.; Ouyang, Q. A deep learning approach to anomaly detection in geological carbon sequestration sites using pressure measurements. *J. Hydrol.* **2019**, *573*, 885–894. [CrossRef]
25. Zuo, R.G. Exploration geochemical data mining and weak geochemical anomalies identification. *Earth Sci. Front.* **2019**, *26*, 67.
26. Li, C.; Jiang, Y.L.; Hu, M.K. Study and application of gravity anomaly separation by cellular neural networks. *Comput. Tech. Geophys. Geochem. Explor.* **2015**, *37*, 16–21.
27. Cai, H.H.; Zhu, W.; Li, Z.X.; Liu, Y.Y.; Li, L.B.; Liu, C. Prediction method of Tungsten-molybdenum prospecting target area based on deep learning. *J. Geo-Inf. Sci.* **2019**, *21*, 928–936.
28. Li, S.; Chen, J.; Xiang, J. Applications of deep convolutional neural networks in prospecting prediction based on two-dimensional geological big data. *Neural Comput. Appl.* **2020**, *32*, 2037–2053. [CrossRef]
29. Li, T.; Zuo, R.; Xiong, Y.; Peng, Y. Random-drop data augmentation of deep convolutional neural network for mineral prospectivity mapping. *Nat. Resour. Res.* **2021**, *30*, 27–38. [CrossRef]
30. Zuo, R. Geodata science-based mineral prospectivity mapping: A review. *Nat. Resour. Res.* **2020**, *29*, 3415–3424. [CrossRef]
31. Zuo, R.; Wang, Z. Effects of random negative training samples on mineral prospectivity mapping. *Nat. Resour. Res.* **2020**, *29*, 3443–3455. [CrossRef]
32. Ran, X.; Xue, L.; Zhang, Y.; Liu, Z.; Sang, X.; He, J. Rock classification from field image patches analyzed using a deep convolutional neural network. *Mathematics* **2019**, *7*, 755. [CrossRef]
33. Sang, X.; Xue, L.; Ran, X.; Li, X.; Liu, J.; Liu, Z. Intelligent high-resolution geological mapping based on SLIC-CNN. *ISPRS Int. J. Geo-Inf.* **2020**, *9*, 99. [CrossRef]
34. Guo, J.; Li, Y.; Jessell, M.W.; Giraud, J.; Li, C.; Wu, L.; Li, F.; Liu, S. 3D geological structure inversion from Noddy-generated magnetic data using deep learning methods. *Comput. Geosci.-UK* **2021**, *149*, 104701. [CrossRef]
35. Rezaei, A.; Hassani, H.; Moarefvand, P.; Golmohammadi, A. Determination of unstable tectonic zones in C–North deposit, Sangan, NE Iran using GPR method: Importance of structural geology. *J. Min. Environ.* **2019**, *10*, 177–195.

36. Rezaei, A.; Hassani, H.; Moarefvand, P.; Golmohammadi, A. Lithological mapping in Sangan region in Northeast Iran using ASTER satellite data and image processing methods. *Geol. Ecol. Landsc.* **2020**, *4*, 59–70. [CrossRef]

37. Liu, Y.P.; Zhu, L.X.; Zhou, Y.Z. Application of Convolutional Neural Network in prospecting prediction of ore deposits: Taking the Zhaojikou Pb-Zn ore deposit in Anhui Province as a case. *Acta Petrol. Sin.* **2018**, *34*, 3217–3224.

38. Liu, Y.P.; Zhu, L.X.; Zhou, Y.Z. Experimental Research on Big Data Mining and Intelligent Prediction of Prospecting Target Area—Application of Convolutional Neural Network Model. *Geotecton. Metallog.* **2020**, *44*, 192–202.

39. Cai, H.H.; Xu, Y.Y.; Li, Z.X.; Cao, H.H.; Feng, Y.X.; Chen, S.Q.; Li, Y.S. division of metallogenic prospective areas based on convolutional neural network model: A case study of the Daqiao gold polymetallic deposit. *Geol. Bull. China* **2019**, *38*, 1999–2009.

40. Li, S.; Chen, J.; Liu, C.; Wang, Y. Mineral prospectivity prediction via convolutional neural networks based on geological big data. *J. Earth Sci.-China* **2021**, *32*, 327–347. [CrossRef]

41. Sun, T.; Li, H.; Wu, K.; Chen, F.; Zhu, Z.; Hu, Z. Data-driven predictive modelling of mineral prospectivity using machine learning and deep learning methods: A case study from southern Jiangxi Province, China. *Minerals* **2020**, *10*, 102. [CrossRef]

42. Zhang, S.H. *Deep Learning for Mineral Prospectivity Mapping of Lala-Type Copper Deposit in the Huili Region, Sichuan*; China University of Geoscience: Beijing, China, 2020.

43. Yang, N.; Zhang, Z.; Yang, J.; Hong, Z.; Shi, J. A convolutional neural network of GoogLeNet applied in mineral prospectivity prediction based on multi-source geoinformation. *Nat. Resour. Res.* **2021**, *30*, 3905–3923. [CrossRef]

44. Zhang, S.; Carranza, E.J.M.; Wei, H.; Xiao, K.; Yang, F.; Xiang, J.; Zhang, S.; Xu, Y. Data-driven mineral prospectivity mapping by joint application of unsupervised convolutional auto-encoder network and supervised convolutional neural network. *Nat. Resour. Res.* **2021**, *30*, 1011–1031. [CrossRef]

45. Rumelhart, D.E.; Hinton, G.E.; Williams, R.J. Learning representations by back-propagating errors. *Nature* **1986**, *323*, 533–536. [CrossRef]

46. Wang, Y.; Yao, H.; Zhao, S. Auto-encoder based dimensionality reduction. *Neurocomputing* **2016**, *184*, 232–242. [CrossRef]

47. Hinton, G.E.; Salakhutdinov, R.R. Reducing the dimensionality of data with neural networks. *Science* **2006**, *313*, 504–507. [CrossRef] [PubMed]

48. Hou, B.; Kou, H.; Jiao, L. Classification of polarimetric SAR images using multilayer autoencoders and superpixels. *IEEE J.-Stars.* **2016**, *9*, 3072–3081. [CrossRef]

49. Lv, F.; Han, M.; Qiu, T. Remote sensing image classification based on ensemble extreme learning machine with stacked autoencoder. *IEEE Access* **2017**, *5*, 9021–9031. [CrossRef]

50. Nayak, R.; Pati, U.C.; Das, S.K. A comprehensive review on deep learning-based methods for video anomaly detection. *Image Vis. Comput.* **2021**, *106*, 104078. [CrossRef]

51. Kingma, D.P.; Welling, M. Auto-encoding variational bayes. *arXiv* **2013**, arXiv:1312.6114.

52. Zhang, S.; Xiao, K.; Carranza, E.J.M.; Yang, F.; Zhao, Z. Integration of auto-encoder network with density-based spatial clustering for geochemical anomaly detection for mineral exploration. *Comput. Geosci.-UK* **2019**, *130*, 43–56. [CrossRef]

53. Xiong, Y.; Zuo, R.; Luo, Z.; Wang, X. A physically constrained variational autoencoder for geochemical pattern recognition. *Math. Geosci.* **2022**, *54*, 783–806. [CrossRef]

54. Luo, Z.; Zuo, R.; Xiong, Y.; Wang, X. Detection of geochemical anomalies related to mineralization using the GANomaly network. *Appl. Geochem.* **2021**, *131*, 105043. [CrossRef]

55. Li, Q.; Jin, S.; Yan, J. Mimicking very efficient network for object detection. In Proceedings of the IEEE Conference on Computer Vision and Pattern Recognition, Honolulu, HI, USA, 21–26 July 2017; pp. 6356–6364.

56. Tung, F.; Mori, G. Similarity-preserving knowledge distillation. In Proceedings of the IEEE/CVF International Conference on Computer Vision, Seoul, Korea; 2019; pp. 1365–1374.

57. Feng, Y.M.; Cao, X.Z.; Zhang, E.P.; Hu, Y.X.; Pan, X.P.; Yang, J.L.; Jia, Q.Z.; Li, W.M. Tectonic Evolution Framework and Nature of The West Qinling Orogenic Belt. *Northwest. Geol. (Xi'an, China)* **2003**, *36*, 1–10. (In Chinese with English Abstract)

58. Wei, L.X. Tectonic Evolution and Mineralization of Zaozigou Gold Deposit, Gansu Province. Master's Thesis, China University of Geosciences, Beijing, China, 2015. (In Chinese with English Abstract)

59. Zeng, J.J.; Li, K.N.; Yan, K.; Wei, L.L.; Huo, X.D.; Zhang, J.P. Tectonic Setting and Provenance characteristics of the Lower Triassic Jiangligou Formation in West Qinling-Constraints from Geochemistry of Clastic Rock and zircon U-Pb Geochronology of Detrital Zircon. *Geol. Rev.* **2021**, *67*, 1–15. (In Chinese with English Abstract)

60. Li, Z.B.; Liu, Z.Y.; Li, R. Geochemical Characteristics and metallogenic Potential Analysis of Daheba Formation in Ta-Ga Area of Gansu Province. *Contrib. Geol. Miner. Resour. Res.* **2021**, *36*, 187–194. (In Chinese with English Abstract)

61. Chen, Y.; Wang, K.J. Geological Features and Ore Prospecting Indicators of Sishangou Silver Deposit. *Gansu. Metal.* **2015**, *37*, 108–111. (In Chinese with English Abstract)

62. Di, P.F. Geochemistry and Ore-Forming Mechanism on Zaozigou gold deposit in Xiahe-Hezuo, West Qinling, China. Ph.D. Thesis, Lanzhou University, Lanzhou, China, 2018. (In Chinese with English Abstract)

63. Li, K.N.; Li, H.R.; Liu, B.C.; Yan, K.; Jia, R.Y.; Wei, L.L. Geochemical characteristics of TTG Dick rock and the Relation with Gold Mineralization in West Qinling Mountain. *Sci. Tech. Engrg.* **2019**, *19*, 52–65. (In Chinese with English Abstract)

64. Kang, S.S. Geological Characteristics and Prospecting Criteria of Nanmougou Copper Deposit, Gansu Province. *Gansu. Metal.* **2018**, *40*, 79–85. (In Chinese with English Abstract)

65. Kang, S.S.; Dou, X.G.; Zhi, C.; Wu, X.L. Geochemical Characteristics and Genetic Analysis of the Namugou Copper Deposit in Sunan County, Gansu. *Gansu. Metal.* **2019**, *41*, 65–72. (In Chinese with English Abstract)

66. Liu, Y. Relationship between Intermediate-acid Dike Rock and Gold Mineralization of the Zaozigou Deposit, Gansu Province. Master's Thesis, Chang'an University, Xi'an, China, 2013. (In Chinese with English Abstract)

67. Hu, J.Q. Mineral Control Factors, Metallogenic Law and Prospecting Direction of Integrated Gold Mine Exploration Area in Shilijba-Yangshan Area of Gansu Province. *Gansu. Sci. Technol.* **2018**, *34*, 27–33. (In Chinese)

68. Zhao, J.Z.; Chen, G.Z.; Liang, Z.L.; Zhao, J.C. Ore-body Geochemical Features of Zaozigou Gold Deposit. *Gansu. Geol.* **2013**, *22*, 38–43. (In Chinese with English Abstract)

69. Lu, J. Study on Characteristics and Ore-host Regularity of Gold Mineral in the Western Qinling Region, Gansu Province. Master's Thesis, China University of Geosciences, Beijing, China, 2016. (In Chinese with English Abstract)

70. Zhang, Y.N.; Liang, Z.L.; Qiu, K.F.; Ma, S.H.; Wang, J.L. Overview on the Metallogenesis of Zaozigou gold deposit in the West Qinling Orogen. *Miner. Explor.* **2020**, *11*, 28–39. (In Chinese with English Abstract)

71. Tang, L.; Lin, C.G.; Cheng, Z.Z.; Jia, R.Y.; Li, H.R.; Li, K.N. 3D Characteristics of Primary Halo and Deep Prospecting Prediction in The Zaozigou Gold Deposit, Hezuo City, Gansu Province. *Geol. Bull. China* **2020**, *39*, 1173–1181. (In Chinese with English Abstract)

72. Chen, G.Z.; Wang, J.L.; Liang, Z.L.; Li, P.B.; Ma, H.S.; Zhang, Y.N. Analysis of Geological Structures in Zaozigou Gold Deposit of Gansu Province. *Gansu. Geol.* **2013**, *22*, 50–57. (In Chinese with English Abstract)

73. Chen, G.Z.; Liang, M.L.; Wang, J.L.; Zhang, Y.N.; Li, P.B. Characteristics and Deep Prediction of Primary Superimposed Halos in The Zaozigou Gold Deposit of Hezuo, Gansu Province. *Geophys. Geochem. Explor.* **2014**, *38*, 268–277. (In Chinese with English Abstract)

74. Jin, D.G.; Liu, B.C.; Chen, Y.Y.; Liang, Z.L. Spatial Distribution of Gold Bodies in Zaozigou Mine of Gansu Province. *Gansu. Geol.* **2015**, *24*, 25–30+41. (In Chinese with English Abstract)

75. Zhu, F.; Wang, G.W. Study on Grade Model of Gansu Zaozigou Gold Mine Based on Geological Statistics. *Acta Mineral. Sin.* **2015**, *35*, 1065–1066. (In Chinese)

76. Chen, G.Z.; Li, L.N.; Zhang, Y.N.; Ma, H.S.; Liang, Z.L.; Wu, X.M. Characteristics of fluid inclusions and deposit formation in Zaozigou gold mine. *J. Jilin Univ. (Earth Sci. Ed.)* **2015**, *45*, 1–2. (In Chinese)

77. Wu, X.M. Study on Geological Characteristics and Metallogenic Regularity of the Gelouang Gold Deposit. Master's Thesis, Lanzhou University, Lanzhou, China, 2018. (In Chinese with English Abstract)

78. Hinton, G.; Vinyals, O.; Dean, J. Distilling the knowledge in a neural network. *arXiv* **2015**, arXiv:1503.02531.

79. Wang, G.; Han, S.; Ding, E.; Huang, D. Student-teacher feature pyramid matching for unsupervised anomaly detection. *arXiv* **2021**, arXiv:2103.04257.

80. Yunhui, K.; Guodong, C.; Bingli, L.; Miao, X.; Zhengbo, Y.; Cheng, L.; Yixiao, W.; Yaxin, G.; Shuai, Z.; Hanyuan, Z.; et al. 3D Mineral Prospectivity Mapping of Zaozigou gold deposit, West Qinling, China: Machine Learning-based mineral prediction. *Minerals* 2022, in press.

81. Carranza, E.J.M. *Geochemical Anomaly and Mineral Prospectivity Mapping in GIS*; Elsevier: Amsterdam, The Netherlands, 2008.

82. Development Research Center of China Geological Survey. The report of mineral resources potential assessment and mineral resources prediction at depth in the Maqu-Hezuo area, Gansu province. 2021; (Unpublished Work).

Article

Combining 3D Geological Modeling and 3D Spectral Modeling for Deep Mineral Exploration in the Zhaoxian Gold Deposit, Shandong Province, China

Bin Li [1], Yongming Peng [2], Xianyong Zhao [2], Xiaoning Liu [1], Gongwen Wang [1,3,4,*], Huiwei Jiang [1], Hao Wang [5] and Zhenliang Yang [6]

1 School of Earth Sciences and Resources, China University of Geosciences (Beijing), Beijing 100083, China
2 Shandong Gold Mining Co., Ltd., Xincheng Gold Mine, Yantai 261438, China
3 MNR Key Laboratory for Exploration Theory & Technology of Critical Mineral Resources, China University of Geosciences (Beijing), Beijing 100083, China
4 Beijing Key Laboratory of Land and Resources Information Research and Development, Beijing 100083, China
5 School of Geography, Southwest University, Chongqing 400715, China
6 Shandong Provincial Engineering Laboratory of Application and Development of Big Data for Deep Gold Exploration, Weihai 264209, China
* Correspondence: gwwang@cugb.edu.cn

Abstract: The Jiaodong Peninsula hosts the main large gold deposits and was the first gold production area in China; multisource and multiscale geoscience datasets are available. The area is the biggest drilling mineral-exploration zone in China. This study used three-dimensional (3D) modeling, geology, and ore body and alteration datasets to extract and synthesize mineralization information and analyze the exploration targeting in the Zhaoxian gold deposit in the northwestern Jiaodong Peninsula. The methodology and results are summarized as follows: The regional Jiaojia fault is the key exploration criterion of the gold deposit. The compression torsion characteristics and concave–convex section zones in the 3D deep environment are the main indicators of mineral exploration using 3D geological and ore-body modeling in the Zhaoxian gold deposit. The hyperspectral detailed measurement, interpretation, and data mining used drill-hole data (>1000 m) to analyze the vectors and trends of the ore body and ore-forming fault and the alteration-zone rocks in the Zhaoxian gold deposit. The short-wave infrared Pos2200 values and illite crystallinity in the alteration zone can be used to identify 3D deep gold mineralization and potential targets for mineral exploration. This research methodology can be globally used for other deep mineral explorations.

Keywords: 3D geological modeling; 3D ore-body modeling; spectral interpretation and 3D modeling; mineral exploration and deep targeting; Zhaoxian gold deposit

Citation: Li, B.; Peng, Y.; Zhao, X.; Liu, X.; Wang, G.; Jiang, H.; Wang, H.; Yang, Z. Combining 3D Geological Modeling and 3D Spectral Modeling for Deep Mineral Exploration in the Zhaoxian Gold Deposit, Shandong Province, China. *Minerals* **2022**, *12*, 1272. https://doi.org/10.3390/min12101272

Academic Editor: José António de Almeida

Received: 3 August 2022
Accepted: 4 October 2022
Published: 9 October 2022

Publisher's Note: MDPI stays neutral with regard to jurisdictional claims in published maps and institutional affiliations.

1. Introduction

As a universal currency, gold has anti-inflation and safe-haven functions. It has always played an important role in the global financial system, especially during financial crises, which highlights its safe-haven function. As such, gold is currently used as a reserve by most governments to maintain economic and material stability. In addition to the official gold reserves, gold is essential for industry, healthcare, and high-tech fields. For China, gold is a credit instrument related to foreign trade and economic cooperation [1]. The Jiaodong Peninsula is the most important gold-producing area in China and the third largest gold-mining area in the world, with proven gold resources exceeding 5000 t [2]. For a long time, researchers have carried out geological, geophysical, and geochemical works in the Jiaodong Gold Mine and accumulated abundant data, which provided an important basis for deep prospecting [2–6]. Since the implementation of the exploration breakthrough strategic action in 2011 [5–7], major exploration breakthroughs have been made in important gold belts such as Sanshandao, Jiaojia, and Zhaoping in Jiaodong [7]. By 2020, the prospecting depth had expanded from 500 m to 4000 m, forming large gold deposits such as Sanshandao, Jiaojia, Linglong, and Denggezhuang. Because of the metallogenic background, the mineralization and production environment are significantly different from those of other types of gold deposits known in the world; this type of deposit is named the Jiaodong-type gold deposit [5]. According to the different mineralization patterns, the gold deposits in this area can be subdivided into Jiaojia-, Linglong-, and Pengjiakuang-type gold deposits [6]. The Zhaoxian gold deposit is located in the northwest of the Jiaojia Gold Mine and is currently the mining area with the largest average exploration depth in China. In-depth analysis of the ore-body geological characteristics of this large gold mine will be of great significance for promoting deep ore prospecting in the Jiaojia ore field [8].

Recently, short-wavelength infrared (SWIR) spectroscopy has been widely applied in mineral exploration. Its wavelength range is between 1300 nm and 2500 nm, which can effectively identify minerals containing hydroxyl groups, amino groups, partial carbonates, and sulfates according to the difference between the reflection absorption of SWIR light and characteristic absorption peaks of the functional groups [9–12]. This method can quickly identify deep minerals and alteration zones in actual exploration and has been successfully applied to porphyry deposits, epithermal deposits, volcanogenic massive sulfide deposits, and some iron oxide–copper–gold (IOCG) deposits [13–16]. With the introduction of computer three-dimensional (3D) modeling technology, geological prospecting prediction has gradually developed from the two-dimensional (2D) plane to 3D space prediction. Three-dimensional geological modeling is a technology that provides 3D visualization through the integration of multisource and multiscale geoscience data [17–19]. Three-dimensional geological modeling allows capturing the geometry of structures at regional and local scales, visualizing the subsurface architecture of deposits and intrusions, and understanding fluid transport processes [20]. Data-driven geographic information system (GIS) based methods for integrating multisource and multisource spatial geoscience datasets have been widely used in mineral-potential mapping for exploration targeting in a data-rich mining area [18].

With the deep mineral-exploration work, SWIR-spectrum 3D alteration mapping has also been applied. For example, Harraden et al. (2013) carried out drill-hole SWIR exploration on the pebble-porphyry Cu–Au–Mo deposit in eastern North America, divided the wall-rock alteration zones in detail on the 3D scale of the mining area, and established the 3D geology and alteration model of the pebble deposit [21]. Chen et al. (2019) [22] performed mineral alteration mapping on the Tonglushan copper–gold deposit and established 3D attribute modeling based on the parameters of chlorite Pos2250 and kaolinite Pos2170, thus establishing altered-mineral-exploration targets in this area. Predecessors have conducted significant research on the altered minerals in the Jiaodong area, but most of them were based on element geochemistry, and no study has combined the SWIR spectrum and 3D modeling in this area. Therefore, this work aimed to use the SWIR spectroscopy measurement and interpretation of the altered-mineral datasets (depth > 1000 m) of the Zhaoxian gold deposit to identify the altered-mineral assemblage and zonation characteristics in the deep part of the mineralization area. By combining the SWIR and 3D geological modeling, the 3D multiparameter model of the altered minerals and the ore body can be quickly established; therefore, it is possible to compare and study the characteristics of alteration-zone minerals in depth and the spatial distribution of the mineralization in one dimension (1D), 2D, and 3D, and then make a 3D comprehensive prediction to optimize the favorable finding area and provide a basis for drilling verification.

2. Deposit Geology and Mineralization

The Zhaoxian gold deposit is located in the Jiaojia ore field of the northwestern Jiaodong Peninsula in China, which is approximately 15 km northeast of Laizhou City (Figure 1) [23]. The gold deposits in the region are controlled by NNE-trending faults, and their gold ore bodies are mostly hosted in the altered rocks within the footwalls of NNE–NE-trending faults. The deep part (1400 m to 2400 m) of the mining zone is bounded by the Jiaojia main fault surface [2,24]. The wall-rock lithology is Linglong granite and metamorphic rock of the Jiaodong Group, and the footwall is ore-bearing altered-rock zones and Linglong-series monzogranite (ca. 160 Ma) and Guojialing-series granodiorite (ca. 130 Ma) [18].

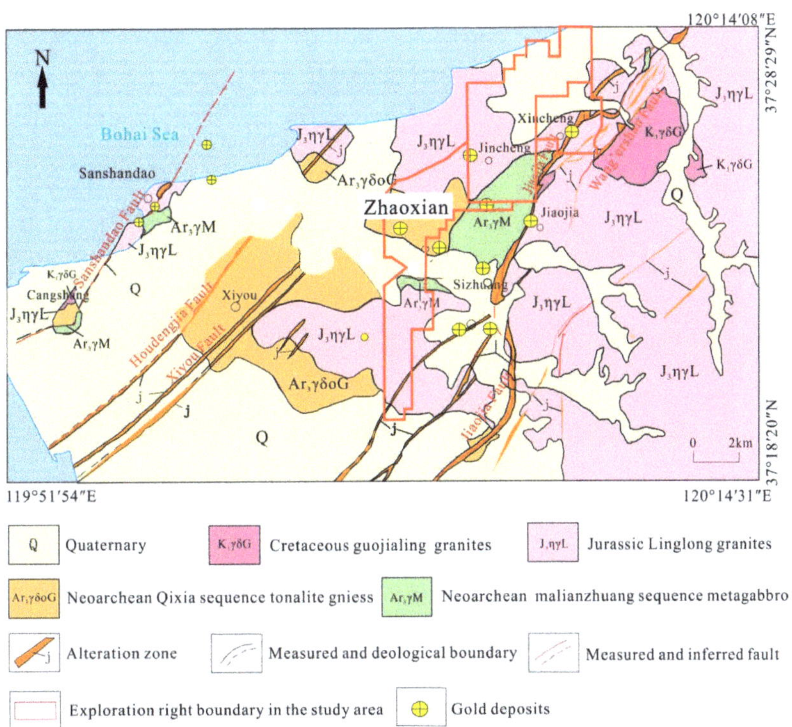

Quaternary — Q
Cretaceous guojialing granites — K₁γδG
Jurassic Linglong granites — J₃ηγL
Neoarchean Qixia sequence tonalite gniess — Ar₃γδoG
Neoarchean malianzhuang sequence metagabbro — Ar₃γM
Alteration zone
Measured and deological boundary
Measured and inferred fault
Exploration right boundary in the study area
Gold deposits

Figure 1. Simplified regional geologic map of the Jiaodong Peninsula (modified by [23]).

The altered rocks in the study area are distributed in strips, which are characterized by the superposition of various alterations and obvious alteration zoning (Figure 2) [3,18]. The altered rocks can be divided into the components of the main regional Jiaojia fault according to the alteration type, degree, and mineral assemblage. The hanging wall of the Jiaojia fault has sericite-granitized cataclastic rock and sericite granite zones, and the footwall has pyrite–sericite–quartz granitized cataclastic rock and pyrite–sericite–quartz granite zones. The gold ore body is mainly in disseminated, veinlet-disseminated, and veinlet forms (Figure 2). Pyrite is the main gold-bearing mineral, and the shallow zone is mostly enriched in pyrite–sericite–quartz cataclastic rocks, whereas the deep zone is mostly enriched in pyrite–sericite–quartz cataclastic rocks and pyrite–sericite–quartz granitized cataclastic rocks. The alteration develops along the wall rock of the fault structures, including potassium feldspar, hematite mineralization (reddenization), pyrite–sericite–quartz alteration, and carbonation, as well as chloritization, kaolinization, etc. Pyrite–sericite–quartz alteration is a general term for sericitization, silicification, and pyritization [8]. Pyrite-sericite-quartz altered rock and potassium-altered (reddenized) rock are usually gold-enrichment zones. The study area has multiperiod and multistage hydrothermal processes, which show the evolution and metasomatism of hydrothermal functions, coupled with multistage tectonic activities. According to the metallogenic relationships among the ore-controlling fault structures, hydrothermal veins, and gold mineralization, previous researchers have divided the hydrothermal metallogenesis into four stages, which are summarized in Table 1 [7,8,25].

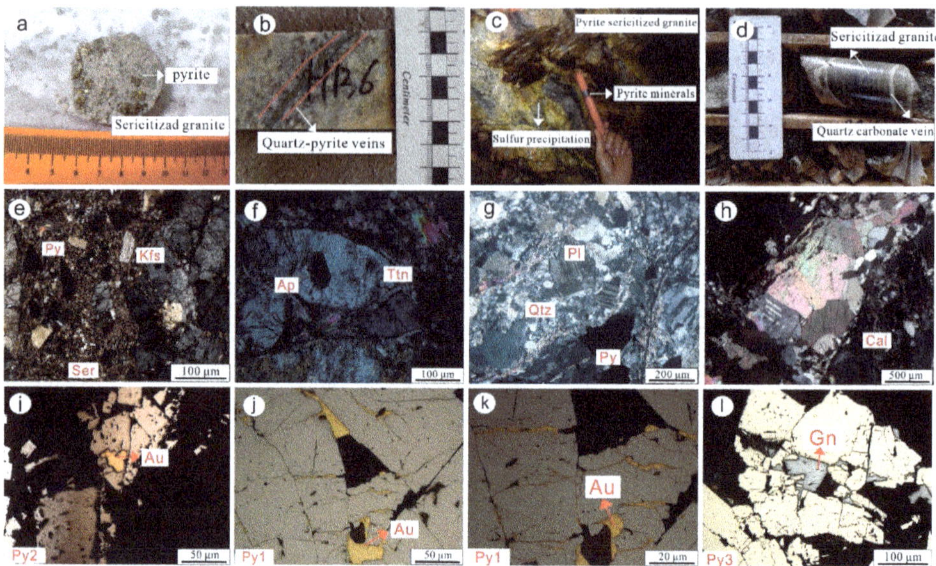

Figure 2. Pyrite-type hand samples and ore-body microscope photos show the occurrence state of transparent minerals and gold (modified by [23]). (**a**)—Type I pyrite in sericitized granites (PyI); (**b**)—Type II pyrite in quartz pyrite veins (PyII); (**c**)—Type III pyrite in quartz sulfide ores (PyIII); (**d**)—Type IV pyrite in pyrite calcite vein ores (PyIV); (**e**–**h**)—transparent mineral forms under polarized light microscopy; (**i**–**l**)—occurrence state of gold in different types of pyrite. Py—pyrite; Ser—sericite; Qtz—quartz; Cal—calcite; Pl—plagioclase; Ap—apatite; Ttn—sphene; Au—natural gold; Gn—galena.

Table 1. Characteristics of hydrothermal metallogenesis and pyrite in different stages.

Metallogenic Stage	Pyrite Distribution	Paragenetic Mineral Assemblages	Pyrite Form	Degree of Mineralization
Pyrite-quartz–sericite stage (I)	Irregular granular or idiomorphic coarse-grained crystals	Pyrite, quartz, sericite	Pyrite quartz vein	Weak
Quartz–pyrite stage (II)	Fine-grained heteromorphic, veinlet, reticulate	Pyrite, sericite, chlorite	Pyrite quartz crushing	Strongest
Quartz–polymetallic sulfide stage (III)	Fine-grained, veinlet, and disseminated	Quartz, pyrite (chalcopyrite), galena, sphalerite, etc.	Quartz–polymetallic sulfide assemblage	Strongest
Quartz–carbonate stage (IV)	Veinlet or reticulate	Quartz, carbonate, and a small amount of pyrite	Intercalation of quartz calcite veins	No

3. Materials and Methods

3.1. 3D Geological Modeling

Three-dimensional geological modeling technology uses the knowledge of geology, geostatistics, space science, and other fields to model the possible extent of geological objects in deep geological bodies based on surface geological data, underground drill-hole data [26,27], etc. Modeling software enables 3D visualization, spatial feature analysis, geological interpretation, and resource evaluation [26,27]. This study used SKUA-GOCAD 18.0 (18.0, Emerson, St. Louis, MI, USA) to establish a 3D geological model of the Zhaoxian gold deposit. In this work, the geological data used included a geological map, several exploration sections, and 19 drill holes of more than 1400 m depths. The four-step 3D geological modeling method is given as follows.

(1) A 3D fault model is created based on 1:2000 exploration-line profile data. The image coordinates are corrected by MapGIS (6.7, Wuhan Zhongdi Information Engineering Co., Ltd., Wuhan, China) to obtain the fault-boundary lines, and then SKUA-GOCAD is used to connect the fault-boundary lines of adjacent profiles and perform surface smoothing.

(2) Based on the exploration-line profile and drill-hole data, the ore-body boundary vectorization is carried out to construct a 3D ore-body model of the developed gold deposit.

(3) According to the geological map, exploration-line profile, and drill-hole data, the formation surface profile is extracted. Based on the workflow of SKUA-GOCAD software, stratum modeling is carried out and the geological significance of curves is given. The block model is established to generate a 3D stratum model.

(4) A kriging spatial interpolation model is established using drill-hole geochemical data. The specific steps are to extract Au grade information first, then establish a grid model and generate ellipsoids using variogram analysis, and, finally, apply kriging for spatial interpolation.

3.2. Spectral Analysis

SWIR spectroscopy is based on the selective absorption characteristics of some specific groups in minerals to SWIR light. The spectral curves can be matched with the standard minerals or mineral assemblages in the spectral library through algorithms (such as The Spectral Assistant in Spectral Geologist (TSG8) software (TSG^{TM} 8, CSIRO, North Ryde, Australia)) to identify minerals (Figure 3) [28,29]. When a sample is illuminated by light in a spectrometer, the molecular bonds in the minerals stretch and bend causing vibrations that result in the adsorption of certain wavelengths of light. In phyllosilicates (including white mica, kaolinite, and montmorillonite), the molecular bonds that cause such vibrations are mainly those in water and hydroxyl groups, including Al–OH, Mg–OH, and Fe–OH [30,31]. Furthermore, a scalar-extraction method has been constructed. This method takes the specific absorption characteristics of the spectrum as the research object (Figure 3), extracts the scalar in the spectral curve, and provides information about the mineral composition and displacement caused by changes in the cationic composition (for example, Tschermak substitution in muscovite) according to the scalar [32,33]. The use of scalars can improve traceability and comparability between different datasets [21].

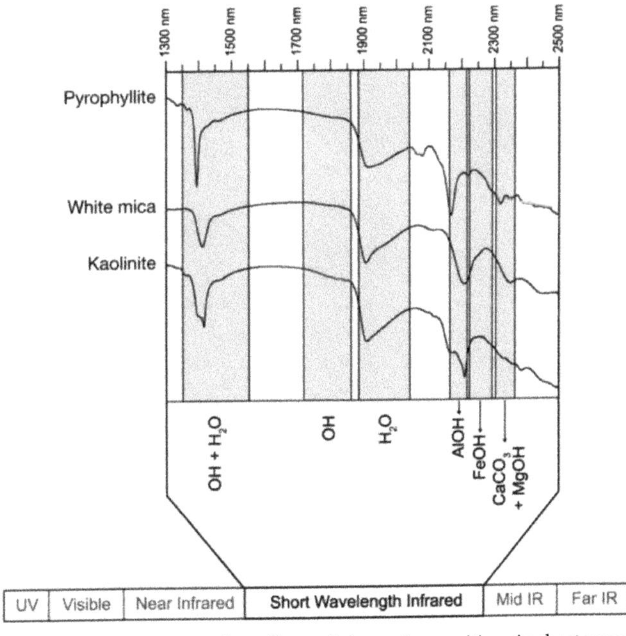

Figure 3. Common spectral profiles and absorption positions in short-wavelength infrared (SWIR) spectrometry (after GMEX, 2008 [30]).

To comprehensively analyze the spectral characteristics of the study area and establish a 3D spectral model, the data involved in this work included five exploration lines in the Zhaoxian gold deposit area and a total of eight drill holes, of which the No. 88 Exploration Line was mainly used for the scalar extraction of layered silicates (Figure 4). Sample locations are roughly equidistant, and intensive sampling was carried out near the mineralization. The sample locations included the surrounding rock, alteration zone, and mineralization center. The spectral analysis of these samples identifies the fluid migration and element migration in the study area.

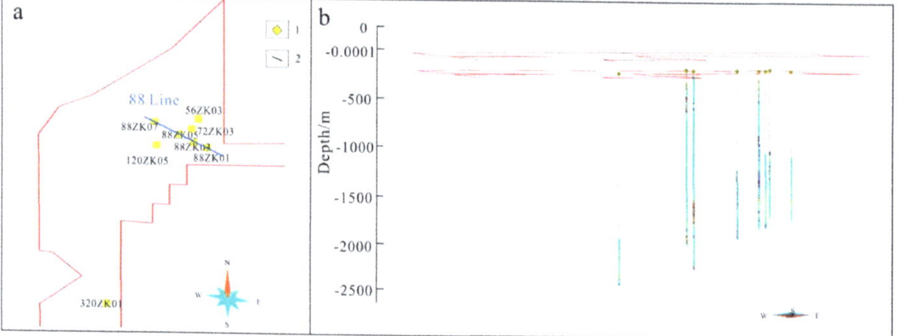

Figure 4. Locations of short-wavelength infrared (SWIR) test holes at the mine site ((**a**)—top view; (**b**)—side view).

The SWIR spectrometer used in the study area is an oreXpress mineral analyzer (oreXpress[TM], Spectral Evolution, Haverhill, MA, USA). The effective test wavelength range of the instrument is 350–2500 nm, which has a good signal-to-noise ratio (SNR) [23]. When the contact probe is used to test the sample, real-time mineral identification is carried out by spectral matching through proprietary EZ-ID spectral data acquisition software (1.4, Spectral Evolution, Haverhill, MA, USA), which has the United States Geological Survey and SpecMIN mineral libraries built in. The sampling bandwidth of the spectrometer is 1 nm, and the minimum scanning speed is 100 milliseconds. When performing spectral testing, the instrument needs to be preheated for 10–30 min, and then an international standard spectral whiteboard is used for calibration. During the testing process, the core was strictly guaranteed to be dry and clean.

Spectral Geologist software (TSG8) (TSG[TM] 8, CSIRO, North Ryde, Australia) was used to process the spectral data. The software matches the waveform of the spectral curve and the position of the absorption peak with the spectral library to determine mineral types and extract spectral information [34]. In the data processing, a Hull quotient removal of the spectral data was carried out. A baseline (Hull) was fitted to each reflectance spectrum. At each wavelength, the reflectance was divided by the corresponding value on the baseline (the Hull quotient) to remove the background effect [32,35]. The absorption characteristics of the spectrum can be enhanced by correcting the baseline [30].

This work adopted the PFIT (A TSG-provided method for extracting scalars to extract more accurate spectral feature parameters)processing method, which is based on the polynomial fitting of the spectral curve after removing the Hull quotient to extract the spectral features [9,10,36]. When analyzing layered silicate minerals in this work, the fourth derivative is calculated after removing the Hull quotient to obtain the fourth-derivative spectrum and extract the scalars. The minimum absorption peak position of Al–OH at 2200 nm and the absorption depths at 2200 nm, 1900 nm, and 2160 nm are extracted. The specific meanings are shown in Table 2.

Table 2. Scalar quantity and extraction used to identify white mica, chlorite, and kaolinite in TSG8 (modified by [37]).

Scalar Name	Mineral Group	Plain Description	Base Algorithm
Al–OH feature depth (2200D)	White mica	Relative depth of the absorption feature near 2200 nm wavelength.	On the spectrum with the hull quotient removed, the fourth-order polynomial fitting is performed near the relative absorption depth (near 2200 nm).
Al–OH feature wavelength (Pos2200)	White mica	Shift of the absorption feature near 2200 nm because of Tschermak's substitution of Al in white mica.	On the spectrum with the hull quotient removed, the fourth-order polynomial fitting is performed near the position of the minimum absorption peak.
Kaolin group crystallinity (2160D)	Kaolin group	The crystallinity order of the kaolinite group minerals can be indicated by 2160D. The larger the relative value of 2160D, the better the crystallinity order.	On the spectrum with the hull quotient removed, the fourth-order polynomial fitting is performed near the relative absorption depth (near 2160 nm).
Fe–OH feature depth (2250D)	Chlorite	Relative absorption depth of absorption feature at 2250 nm wavelength; indicative of Fe–OH mineral abundance.	On the spectrum with the hull quotient removed, the fourth-order polynomial fitting is performed near the relative absorption depth (near 2250 nm).
Fe–OH feature wavelength (W2250)	Chlorite	Estimation of Mg/(Mg+Fe) in chlorite, where the wavelength position is caused by Mg, Fe, or relative Al, Fe^{3+}, or Ca content.	The minimum wavelength of 2250 nm absorption of the continuous medium is removed near 2250 nm, which is determined by four-band polynomial fitting around the band with the lowest reflectivity.
Mg–OH feature depth (W2350)	Chlorite	Depth of 2350 nm feature, evident in white mica, chlorite, and carbonate; used to separate white mica from Al-smectites, when Al–OH feature is present.	On the spectrum with the hull quotient removed, the fourth-order polynomial fitting is performed near the position of the minimum absorption peak (near 2350 nm).

4. Results

4.1. 3D Geological Model

The 3D geological model used for alteration modeling in the Zhaoxian mining area includes a 3D fault model and a 3D ore-body and grade-interpolation model. A 3D structural alteration-zone model is constructed to constrain the spatial range of the alteration-zone mineral parameters and the Au grade-interpolation modeling.

4.1.1. 3D Fault Modeling

Using SKUA-GOCAD software, 3D ore-body and fault models of the Xincheng, Zhaoxian, Sizhuang, and Wang'ershan deposits were established from drill holes and geological sections, which can reflect the spatial location and geometry of the fault structures in the region, as well as the spatial correspondence between them and the main ore bodies. The 3D fault and ore-body model shown in Figure 5a was established by the location of the fault clay in the drill holes. The main fault surface is continuous and stable, extending from west to east; the fault controls the Xincheng, Zhaoxian, and Sizhuang gold deposits. The secondary faults in the footwall of the main fault zone, such as the Sanshandao and Wang'ershan faults, are relatively developed, and the occurrence changes of the main and secondary faults are in a gentle wave shape in the strike and tendency. The concave or convex zones in Figure 5a,b are the intrusion body boundaries of the Linglong-series monzogranite and the Guojialing-series granodiorite at depth. Additionally, the regional structure controls the distribution of gold ore bodies, keeping them close to the footwall with good continuity (Figure 5b).

4.1.2. 3D Ore-Body and Au Grade-Interpolation Modeling

The 3D gold-deposit ore-body modeling of the area uses the traditional explicit modeling method, including roughly determining the ore-body range using the ore-body boundary line determined from mining engineering and drill-hole datasets (Figures 5a and 6). For this work, additional gold grade geochemistry datasets were acquired in the drill holes [18]. The constructed ore-body model is shown in Figure 6. The ore body has inclined veins and stable shape, and its occurrence is consistent with the fault (Figure 5b), most of which are distributed in the pyrite–sericite–quartz alteration zone (Figure 6). There is a close spatial and temporal relationship between the ore body and pyrite–sericite–quartz alteration zone.

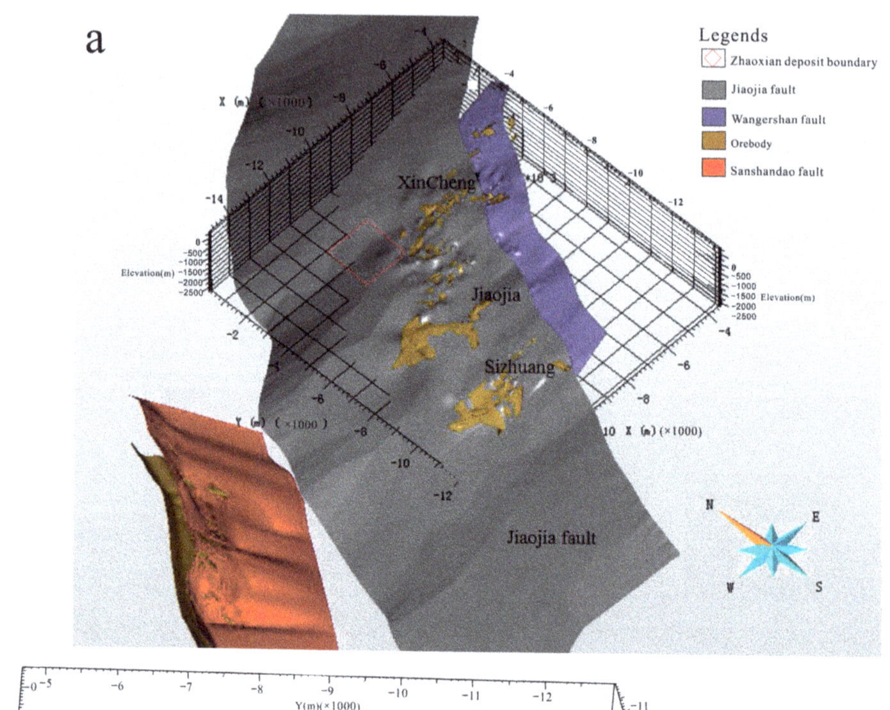
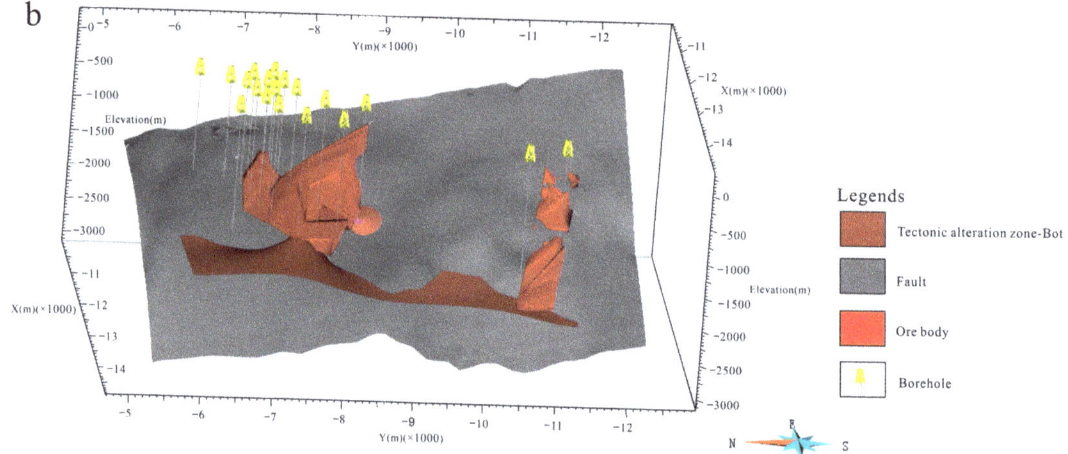

Figure 5. (**a**)—Northwestern Jiaodong Peninsula three-dimensional (3D) fault and ore-body model; (**b**)—3D ore-body and fault model of Zhaoxian gold deposit.

To intuitively display the spatial distribution behavior of the gold grades in the study area, this study used the drill-hole datasets to perform 3D explicit modeling. Through the geostatistical analysis of the gold grade data at different depths in the drill hole and calculations of the variograms in different directions, each ore block is given a search radius ellipsoid. This process estimates the average-grade space allocation of each ore block (Figure 7a). The 3D grade-interpolation model is shown in Figure 7b. The 3D grade-interpolation model is constrained by the alteration zone. From the interpolation results, there are numerous high-grade ore areas within the alteration zone, and within the constrained range, the grade gradually increases from shallow to deep.

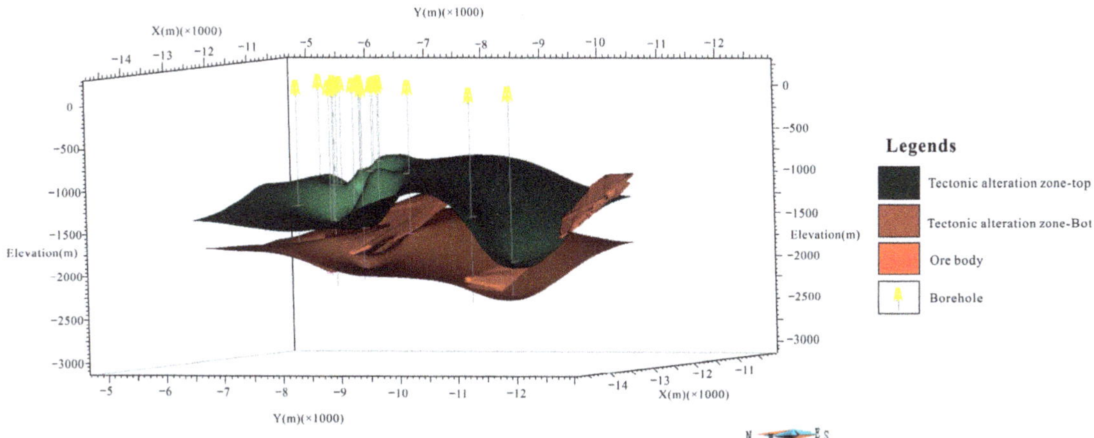

Figure 6. Three-dimensional (3D) ore-body modeling of Zhaoxian gold deposit.

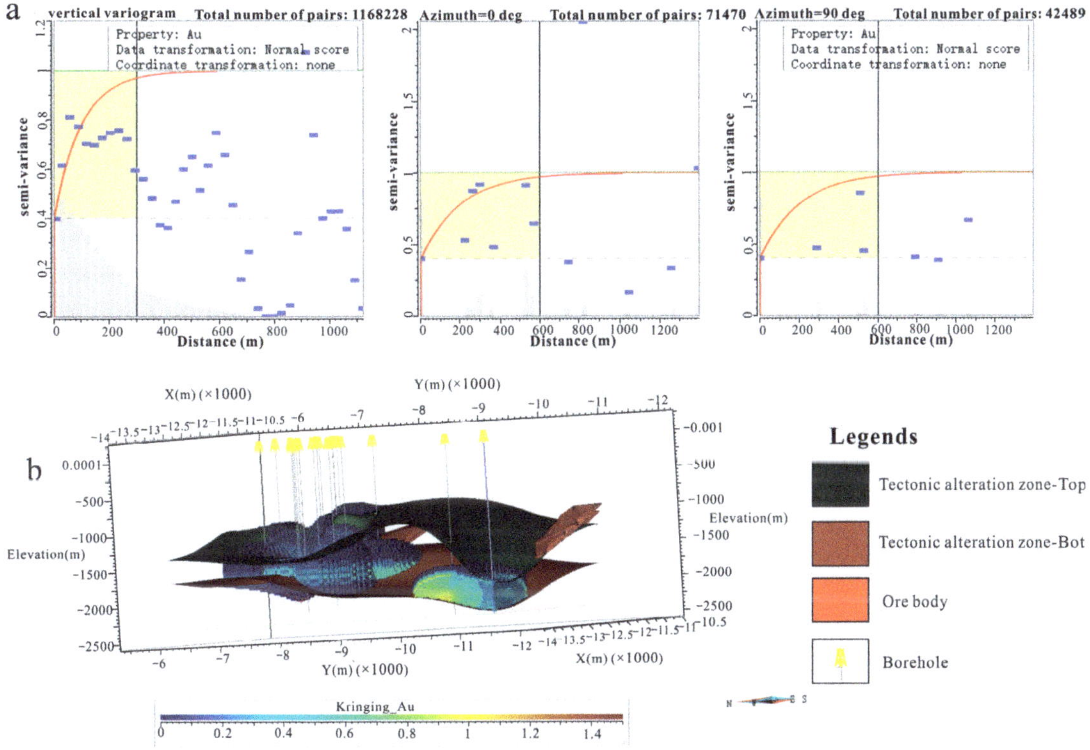

Figure 7. Three-dimensional (3D) fault and grade-interpolation modeling of Zhaoxian gold deposit. (**a**)—Variograms in 3D environment in Jiaojia gold belt; (**b**)—3D grade-interpolation model of Zhaoxian gold deposit.

4.1.3. 3D Alteration-Zone Modeling

Here, the alteration-zone model is constructed based on the pyrite–sericite–quartz alteration lithology that is symmetrically distributed on the hanging and lower walls of the fault [38]. The outline of the unit is extracted from the 1:2000 exploration-line profile.

The structural morphology and spatial distribution of the altered zone can be described by the altered structure model. The final 3D alteration-zone model is shown in Figure 8. The 3D model can better demonstrate the spatial correlations among the fracture structure, ore body, and alteration zone. The gold ore is mainly distributed in the footwall of the fault, and the morphology and distribution are controlled by the fracture. The range of the alteration zone is larger than the mineralization range, and its distribution is in the form of a belt, intersected by the main fracture surface.

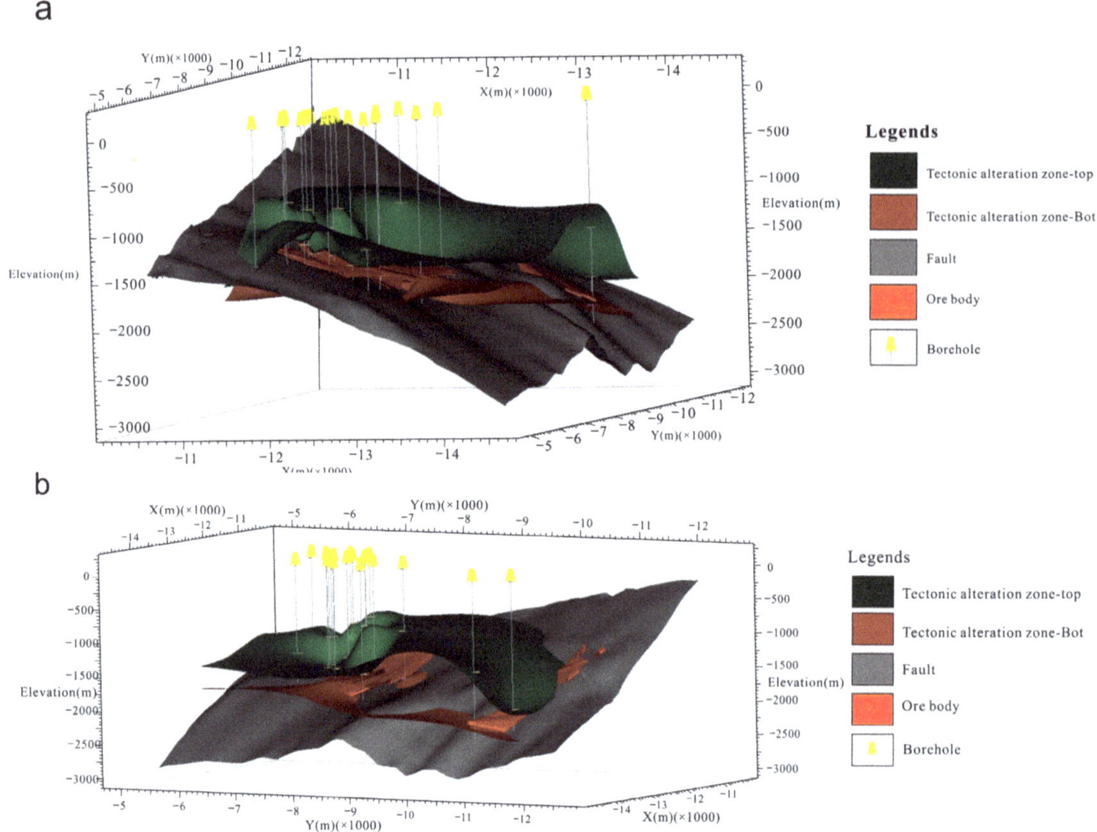

Figure 8. Three-dimensional tectonic alteration-zone and ore-body modeling.(**a**)—front view; (**b**)—side view.

4.2. Interpretation Based on SWIR Spectra

SWIR spectroscopy results identified the main alteration-mineral groups in the eight holes collected in the Zhaoxian gold deposit as white mica, carbonate (siderite and magnesite), kaolinite, smectite, and chlorite, with most minerals co-occurring with white mica. According to the spectral interpretation, the white mica of the Zhaoxian deposit mainly consists of muscovite, paragonite, phengite, muscovitic illlite, paragonitic illlite, and phengitic illlite; muscovitic illlite, for example, is an intermediate product of the conversion of muscovite to illite (the latter three are classified in this work as muscovite, paragonite, and phengite and not separately discussed).

Kaolinite is the most widely distributed clay mineral, an Al-rich silicate that forms through acidic alteration. The crystallinity of kaolinite can be identified by SWIR spectroscopy [34]. The mineral has typical double absorption peaks near 2160 nm and 2206 nm, the intensities of which depend on the type of kaolinite mineral and its crystallinity. Therefore, the influence of kaolinite on Pos2200 needs to be considered when a mixture of white mica and kaolinite is present [16,39,40]. In the Zhaoxian gold deposit, the kaolinite is mainly combined with white mica as a second major constituent, with only a small proportion being found as a single mineral.

In the Zhaoxian main ore-body zone, the chlorite is mainly found in deep granite in small quantities and always associated with muscovite and carbonate. Chlorite has diagnostic Fe–OH and Mg–OH absorptions centered at 2250 and 2350 nm. Because the Mg–OH in chlorite may be affected by the presence of carbonate, Fe–OH absorption is commonly used to determine the composition in the chlorite.

4.3. Alteration Features and Zonation

The ore bodies in the study area are controlled by the Jiaojia Fault and are mainly distributed in fractured, altered rocks [8]. Drill-hole cores are mainly concentrated in the northern part of the Zhaoxian gold deposit, and most of the ore bodies are present. The distribution of ore bodies along Line 88 is continuous, and their thicknesses are stable. Therefore, this study mainly focused on the Line 88 drill holes to study the cross-section of the ore body and host rock and to illustrate the spatial distribution of SWIR alteration-zone minerals.

4.3.1. No. 88 Exploration Line

The No. 88 Exploration Line includes drill holes 88ZK01, 88ZK03, and 88ZK05 and crosses the hanging wall and footwall of the main fracture. The hanging wall of the main fracture is composed of medium-grained monzogranite and sericite-quartz altered rock. The ore bodies developed in the footwall of the fault zone, including most of the I, II-1, and II-2 ore bodies. Both ends of the ore bodies are pinched out, and their occurrence is controlled by the Jiaojia fault. From the center of the ore body to the outside, there is a symmetrical distribution of pyrite–sericite–quartz altered rock, pyrite-sericitized granitic altered rock, sericite–quartz altered rock,, and medium-grained monzonitic granites (Figure 9). The ore body is relatively simple, with strong continuity, and has good metallogenic conditions [8,23]. Ore body I is present in the pyrite–sericite–quartz cataclastic rocks and sericite-quartz granitic cataclastic rocks. Ore body II mainly exists in sericite-quartz granitic cataclastic rocks. All the drill holes penetrated the intact alteration zone.

Drill-hole 88ZK01 is an oblique drill with a test depth of 905.46 m to 1594.99 m. SWIR spectroscopy indicates that it contains diorite and porphyrite interlayers. The lithology histogram can be described as follows. Five main alteration-zone mineral groups were extracted by SWIR spectroscopy, mainly white mica, kaolinite, smectite, carbonate, and other-Al–OH minerals; white mica was the main alteration-zone mineral (Figure 10). The mineralization is concentrated at approximately 1300 m to 1469 m. The type of sericite is muscovite, and its distribution basically corresponds to the mineralization. Kaolinite is less developed, and the muscovite–carbonate alteration zone is also developed. The distribution of alteration-zone minerals near the mineralization can be divided into five zones. The shallow depth of 1070 m is alteration-zone V, and the corresponding lithology is monzogranite. The main minerals are Al-rich sericite, carbonate, and a small quantity of montmorillonite, among which the carbonate is well-developed. The mineral assemblage can be preliminarily classified as the quartz–carbonate mineralization stage [23]. Alteration-zone IV extends from 1070 m to 1300 m, where large quantities of Al-rich sericite–kaolinite alteration-zone minerals are developed, and montmorillonite and carbonate are generally developed. This layer has a unique kaolinite mineral with good crystallinity (kaolinite-wx). The 1469–1546.2 m deep mineralization is alteration-zone II, and the mineral assemblage is sericite–carbonate and a small quantity of low-crystallinity kaolinite (kaolinite-px). The Pos2200 value of sericite in this zone is lower than it is in zone IV. Chlorite is developed only in the monzonitic granite formation of alteration-zone I at 1546.2 m, and the mineral assemblage is sericite–carbonate–chlorite. Generally, sericite is widely distributed in drill-hole 88ZK01, and there are obvious high Pos2200 values near the mineralization, which is developed in the pyrite–sericite–quartz cataclastic rock. The content of montmorillonite is low, there is no obvious regularity, and the carbonate alteration is relatively continuous (Figure 10).

Drill-hole 88ZK03 was vertically drilled and spectroscopically tested at depths of 870.39 m to 1679.77 m. There are gabbro interlayers near the mineralization, and potassic granite gneiss is developed in the deep part. There are tectonic cataclastic rocks near the Jiaojia fault zone. The lithology histogram is described as follows (Figure 11). The zoning of the mineral assemblages is the same as that of 88ZK01, where the shallow 870.39–1123.32 m area is alteration-zone V, and the corresponding lithology is monzogranite and sericite–quartz granite, mainly consisting of Al-rich muscovite and carbonate. Alteration-zone IV is located near the fault, with depths of 1123.32–1345.09 m, and the pyrite sericitization is obvious. The mineralization is roughly concentrated at 1345.0–1500 m in alteration-zone II, and strong pyrite–sericite–quartz and sericite alterations are developed in the cataclastic rock (Figure 11). A large quantity of phengite appears in this layer, and kaolinite is rare. Alteration-zone I, which is developed in potassic granitic gneiss at depths of 1621–1679.77 m, contains unique chlorite minerals with relative contents higher than those of 88ZK01.

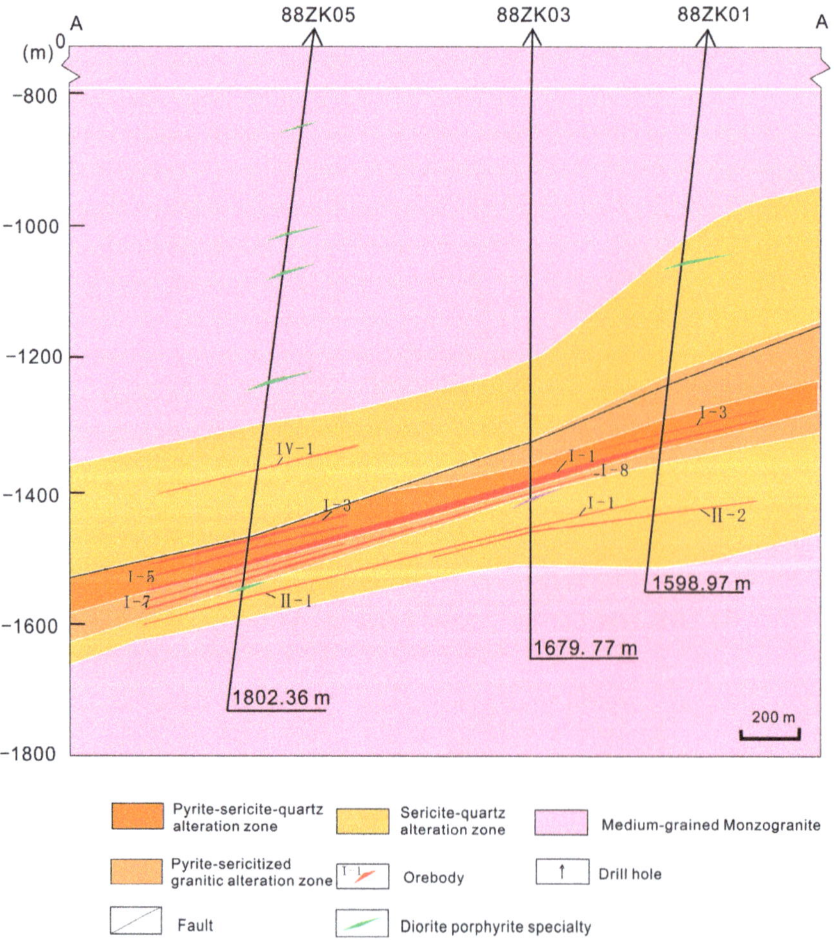

Figure 9. Cross-section of No. 88 Exploration Line of Zhaoxian gold deposit.

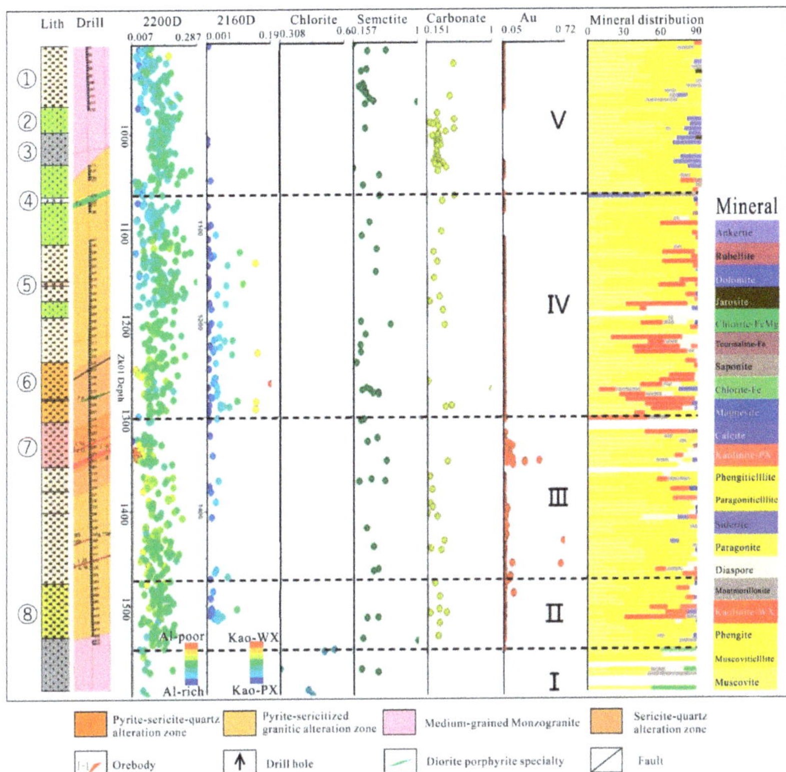

Figure 10. Distribution of alteration-zone minerals in drill-hole 88ZK01. ①—Sericite–quartz granitized cataclastic rock; ②—sericitized granite; ③—biotite monzogranite; ④—white diorite porphyrite; ⑤—sericite–quartz cataclastic rock; ⑥—pyrite–sericite–quartz granitic cataclastic rock; ⑦—pyrite–sericite–quartz cataclastic rock; ⑧—potassic sericite–quartz granitic cataclastic rock. 2200D indicates relative abundance of sericite minerals; 2160D indicates relative abundance of kaolinite; I—sericite–carbonate–chlorite alteration-zone; II—sericite–carbonate–kaolinite-px alteration-zone; III—phengite–carbonate–kaolinite-wx alteration-zone; IV—phengite–carbonate alteration-zone; V—sericite–carbonate alteration-zone.

The sampling depth of drill-hole 88ZK05 is deeper than that in the previous two drill holes, ranging from 1055.05 m to 1800 m. The drill hole passes through many ore veins, which pinch out on both sides. Compared with the previous two drill holes, the shallow area lacks pyrite–sericite–quartz granitoid cataclastic rock. The lithologic histogram for drill-hole 88ZK05 can be described as follows (Figure 12). Owing to the increased acquisition depth, there was no alteration-zone V. Alteration-zone I is located in the granitic cataclastic rocks and sericite–quartz cataclastic rocks at 1774.5–1800 m. Compared with the other two drill holes, the content of montmorillonite in this alteration zone is relatively high. Alteration-zone II developed at 1599–1774.5 m. Alteration-zone III is the mineralized zone. Unlike the other drilling holes, there is a small quantity of kaolinite near the mineralization. Alteration-zone IV is shallower than 1475 m. The 88ZK05 drill hole is quite different from the previous two drill holes owing to the mineralized area and multiple ore bodies. A large quantity of sericite and carbonate are continuously distributed in the drill hole, with high relative contents, including abundant kaolinite-wx (Figure 12).

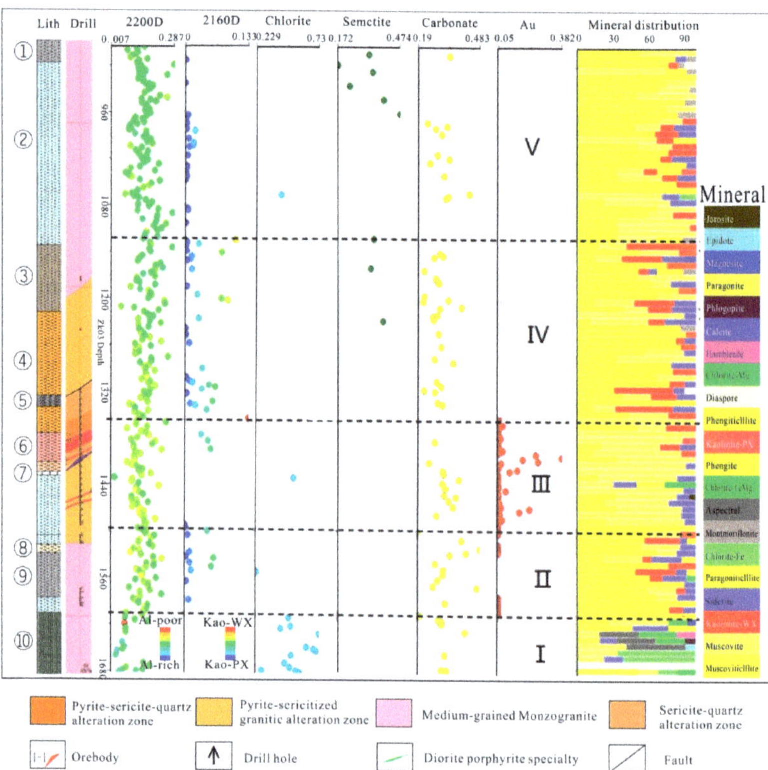

Figure 11. Distribution of alteration-zone minerals in drill-hole 88ZK03. ①—Biotite monzonitic granite; ②—sericite monzogranite; ③—pyrite–sericite–quartz granite; ④—pyrite–sericite–quartz granitic cataclastic rock; ⑤—tectonic cataclastic rock; ⑥—lithified cataclastic rock; ⑦—gabbro; ⑧—sericite–quartz cataclastic rock; ⑨—potassic granitic gneiss. 2200D indicates relative abundance of sericite-group minerals; ⑩—2160D indicates relative abundance of kaolinite; I—sericite–carbonate–chlorite alteration-zone; II—sericite–carbonate–kaolinite-px alteration-zone; III—phengite–carbonate–kaolinite-wx alteration-zone; IV—phengite–carbonate alteration-zone; V—sericite–carbonate alteration-zone.

4.3.2. Spatial Distribution of Alteration-Zone Minerals in the Section of No. 88 Exploration Line

The No. 88 Exploration Line has obvious alteration zoning, clearly recording the mineralization and alteration characteristics of the deposit area, and the deep mineralization is typical. The SWIR results can be used to divide the alteration-zone mineralization into five alteration zones. Alteration-zone I, which is located in the monzogranite strata, consists of sericite–chlorite and only appears in the deep mineralization. Additionally, with increasing depth, the relative content of chlorite also increases. Alteration-zone II is a sericite–carbonate zone, containing a small quantity of kaolinite-px, usually near the intersection of the pyrite–sericite–quartz granitized cataclastic rock and deep potassic granite zone. Alteration-zone III is a mineralized zone with abundant pyritic sericite; its lithology is mainly pyrite–sericite–quartz cataclastic rock and pyrite–sericite–quartz granitic cataclastic rock. The alteration-zone minerals are phengite–carbonate, with a few other minerals. Alteration-zone IV is the sericite–kaolinite-wx–carbonate zone, which is located in the hanging wall of the fault zone, and is characterized by the development of unique kaolinite-wx. Alteration-zone V is the sericite–carbonate zone, containing small quantities of kaolinite and montmorillonite. Generally, the minerals have obvious zonal distribution, in which phengite is mostly developed near the mineralization (Figures 10–12).

Kaolinite-wx develops on the hanging wall of the fault zone, and the position near the ore body contains almost no kaolinite. Moreover, chlorite only exists in the deep granite plutons far from the mineralization.

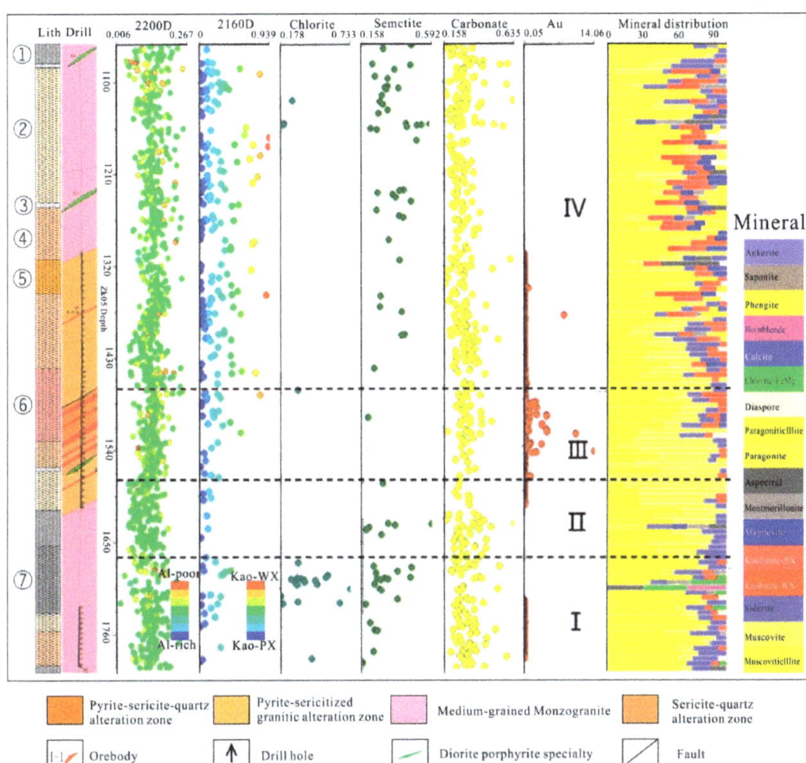

Figure 12. Distribution of alteration-zone minerals in drill-hole 88ZK05. ①—Biotite monzonitic granite; ②—sericite–quartz granitic cataclastic rock; ③—diorite porphyrite; ④—sericite–quartz cataclastic rock; ⑤—pyrite–sericite–quartz granitic cataclastic rock; ⑥—pyrite–sericite–quartz cataclastic rock; ⑦—biotite-bearing monzonitic granitic cataclastic rock. 2200D indicates relative abundance of sericite-group minerals; 2160D indicates relative abundance of kaolinite; I—sericite–carbonate–chlorite alteration-zone; II—sericite–carbonate–kaolinite-px alteration-zone; III—phengite–carbonate–kaolinite-wx alteration-zone; IV—phengite–carbonate alteration-zone.

4.4. Spectral Characteristics of Sericite

Sericitization is the main alteration type of the Zhaoxian deposit; sericite is distributed in the center of the alteration zone and is closely related to gold mineralization [3]. Sericite is commonly used to describe fine-grained white mica (muscovite, phengite, and/or illite) developed in hydrothermally altered rocks [28]. Muscovite is a dioctahedral layered silicate. When the octahedral coordination cation undergoes Tschermak substitution, the octahedral aluminum is replaced by other cations (such as iron and magnesium) to form three common end members: muscovite, paragonite, and phengite [28]. Tschermak displacement is common in phengite solid solutions and involves coupled displacement between tetrahedral and octahedral layers (($Al \leftrightarrow Si$)tet = (Al)$\leftrightarrow$(Fe^{2+},Mg)oct) [9]. When the temperature and pressure are changed, the Al in the octahedral position is replaced by other cations, and the ratio of Si to Al becomes greater than 3; that is, Al-poor muscovite (phengite) is formed [41].

The muscovite group has a diagnostic Al–OH absorption signature centered at 2200 nm, which is associated with the vibrations of the octahedral coordination atoms, the wavelength positions of which provide information about the mineral composition and shifts owing to changes in the cationic composition [21,31,41]. A useful parameter is the wavelength of the Al–OH band (Table 2), which increases with a decrease in Al^{vi}. Because montmorillonite has similar SWIR spectral characteristics (1900 nm and 2200 nm) to muscovite, illite crystallinity values (IC=2200D/1900D) are also widely used to evaluate muscovite and montmorillonite crystallinity [21,37]. Other octahedral silicate minerals, such as kaolinite and montmorillonite, also have absorption features near 2200 nm that overlap with many muscovite features, even though they can be identified and separated on the basis of other specific spectral properties.

4.4.1. Al–OH Feature Wavelength Pos2200

The substitution of octahedral Al by other cations such as Fe and Mg is the main reason for the change in muscovite composition. Therefore, the wavelength of the Al–OH band can be used to spectroscopically quantify the octahedral cation composition (Al^{vi}) of muscovite, reflect compositional changes, and infer fluid characteristics [31,40,42]. Sericite exhibits strong absorption near 2200 nm, which is called Pos2200. When Tschermak substitution occurs, the composition of cations in the octahedron changes; usually, Pos2200 > 2205 nm is Al-poor muscovite with high Fe and Mg content [41]. Furthermore, because kaolinite and montmorillonite also have absorption peaks near 2200 nm, when the muscovite-group minerals are mixed with these minerals, the absorption peak position of Al–OH will slightly change.

The sericite minerals identified by SWIR in the eight drill holes in the Zhaoxian mining area mainly include muscovite, paragonite, phengite, and illite. The minimum absorption peak position (Pos2200) of Al–OH is concentrated around 2200 nm, approximately at 2185–2220 nm with a normal distribution, indicating that the dolomite minerals in the study area are distinct (Figure 13). It can be seen from Figure 10 that Pos2200 > 2205 near the ore body is Al-poor muscovite. In the deep part of the mineralization, the Pos2200 value decreases, which reflects the significant correlation between mineralization, depth, and Al–OH wavelength. For the convenience of distinction and description, this study used Pos2205 as the division between muscovite and phengite.

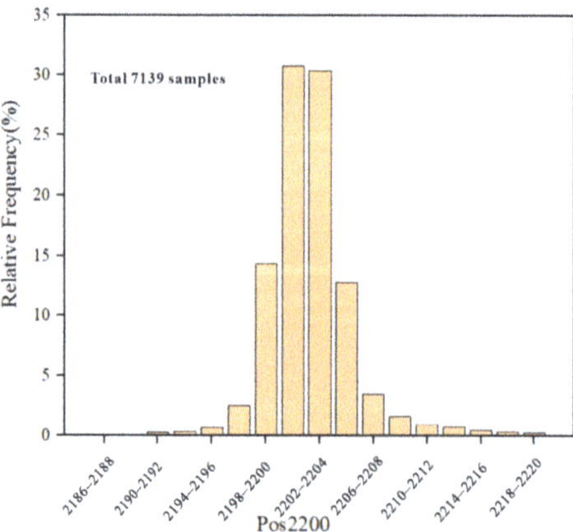

Figure 13. Relative frequencies of Al–OH Pos2200 in white mica.

The Pos2200 values of the four selected drill holes passing through the main ore bodies are all in a normal distribution, which roughly shows that the Pos2200 wavelength of the shallow sericite is relatively short, generally less than 2204, indicating Al-rich muscovite. With the increase in depth, the Pos2200 value of sericite gradually increased to approximately 2208. According to Figure 13, the Pos2200 values of the selected drill holes passing through the main ore body at 800–1500 m depths have almost a normal distribution; as the depth increases, the Pos2200 values also increase (Figure 14). Conversely, because of the depth of 88ZK05, the Pos2200 value actually decreases in the deepest part. The study of the offset of Pos2200 reflects that the gold mineralization corresponds to high values of Pos2200. This phenomenon is more obvious in the contour profile, and the high value area of Pos2200 is basically consistent with the mineralization trend (Figure 15).

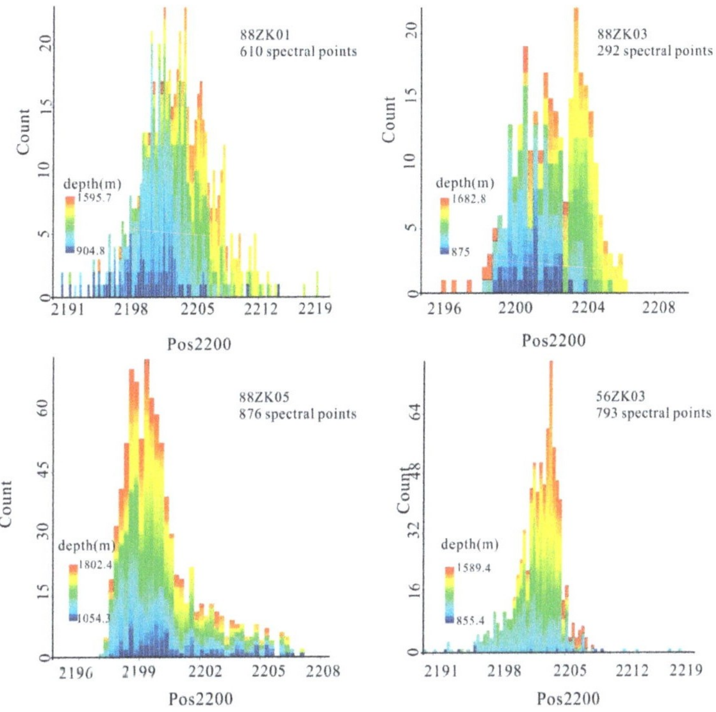

Figure 14. Wavelength relationship histogram of sericite Al–OH Pos2200 and depth.

4.4.2. Al–OH Feature Depth 2200D

The quantity of infrared radiation absorbed is a function of the quantity of absorbing material in the sample. However, the absorption intensity is also affected to varying degrees by the physical conditions of the sample, such as the particle size and orientation of the absorbing minerals [15]. As a first approximation, the intensity of the Al–OH band was taken as an indication of the relative abundance of muscovite by assuming similar sample conditions for all pulverized core samples [31]. The intensity of the Al–OH band is the relative absorption depth of layered silicate minerals such as muscovite, kaolinite, and montmorillonite around 2200 nm, which is a fourth-order polynomial fitted between 2180 nm and 2220 nm of the Hull quotient. The derivative is followed by an extraction of relative depth, which indicates muscovite abundance. According to the contour profile, the 2200D in the No. 88 Exploration Line profile has a slight change in the vertical direction, and the correlation with the mineralization is not high, but the relative content of muscovite gradually increases from 88ZK01 to 88ZK05 in the horizontal direction (Figure 16).

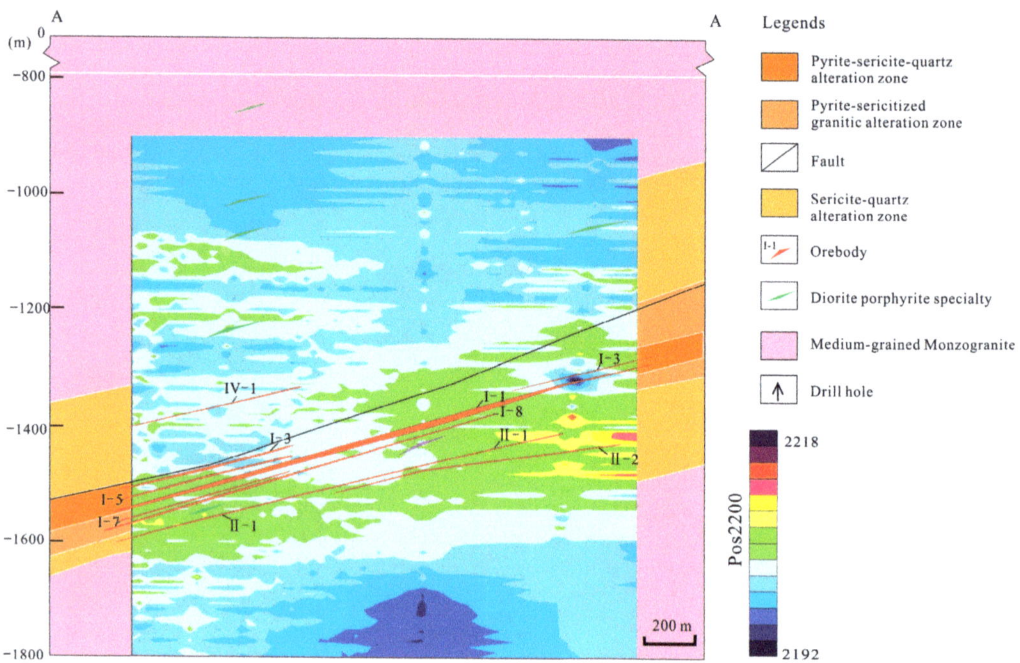

Figure 15. No. 88 Exploration Line diagram of two-dimensional spatial variation of characteristic absorption peak position of muscovite Al–OH Pos2200.

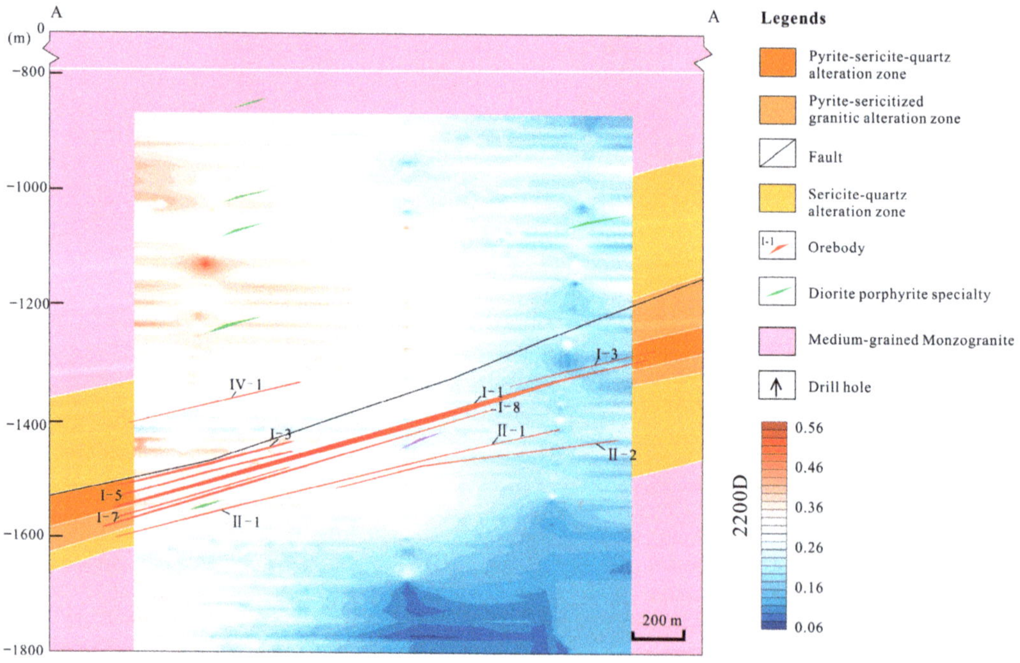

Figure 16. No. 88 Exploration Line diagram of two-dimensional spatial variation of characteristic absorption depth of muscovite Al–OH 2200D.

4.4.3. Illite Crystallinity SWIR-IC Value

The illite crystallinity (SWIR-IC) value refers to the ratio of the absorption depth of sericite at 2200 nm to the absorption depth at 1900 nm. Numerous experiments have been conducted by predecessors. Compared with X-ray diffraction illite crystallinity (XRD-IC), the two have a good negative correlation [21]. Therefore, SWIR-IC, such as XRD-IC, can indicate the crystallinity of sericite minerals. The characteristic absorption peak at 1900 nm may indicate the water absorption of sericite. When the fluid temperature is relatively high, the muscovite minerals contain less crystalline water and have a high SWIR-IC value, so the degree of change in crystallinity can be used to reflect variations in mineral formation temperature [21,37,43].

The SWIR-IC values in the study area varied from 0.2 to 3.8 with a majority concentrated between 0.5 and 1.5 (Figure 17). The crystallinities of the muscovite-group minerals in the drill holes were extracted and combined with the spatial locations of the drill holes to make a 2D contour map. As shown in Figure 16, the SWIR-IC values are relatively high near the ore body in the vertical direction, gradually decrease from the center of the mineralization to the sides, and are lower in the deeper areas of the mineralization. In the horizontal direction (Figure 17), the distribution of high IC values is generally consistent with the mineralization strike, thus confirming that the center of mineralization can be indicated by the IC values.

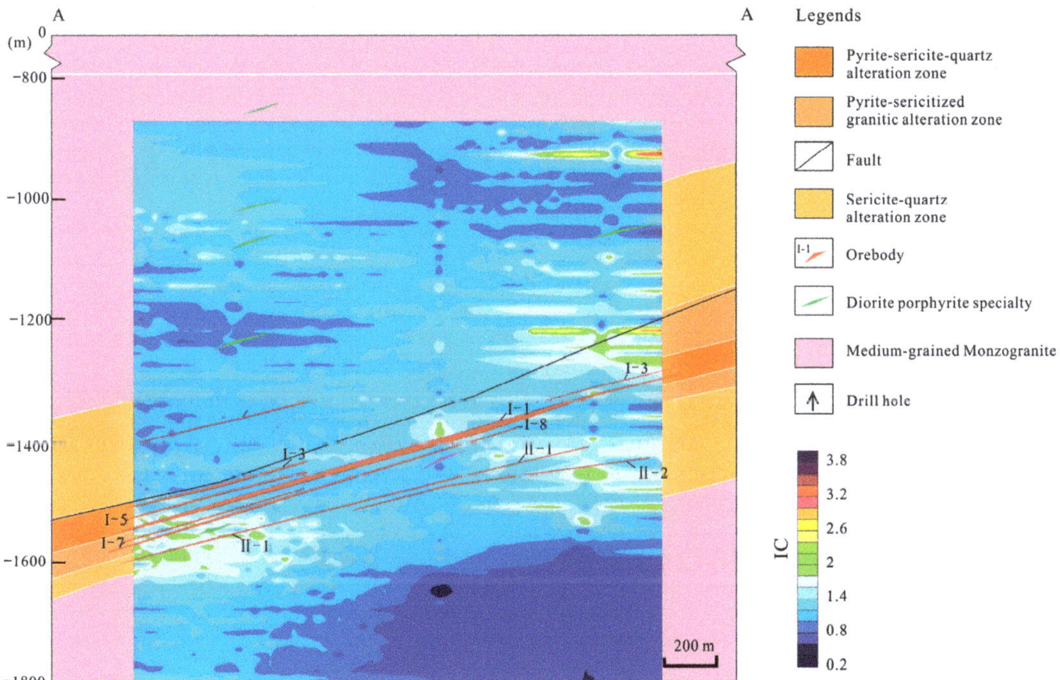

Figure 17. No. 88 Exploration Line diagram of two-dimensional spatial change in short-wave infrared illite crystallinity.

5. Discussion

5.1. Metallogenic Center Indication

The purpose of auxiliary exploration can be achieved by tracking the migration of ore-forming fluid and change in fluid composition through alteration zoning [21]. By using the SWIR-spectrum test, the alteration type in the rock core can be intuitively interpreted at the micron scale, and the boundary of the alteration zone can be refined through the spectral characteristic parameters and mineral abundance changes [21]. Based on spectral (hyperspectral) imaging, determining the best stratigraphic combination and mineralogical alteration type related to mineralization is carried out to determine the mining probability with similar stratigraphic and structural conditions [44,45].

At the deposit scale, the common spatial zonation pattern in the Jiaojia gold deposit advances toward the hydrothermal center. According to the SWIR results, by establishing 3D alteration-zoning models (Figures 18 and 19), the spatial distribution of mineral assemblages can be determined [46,47]. The model shows the location and geometry of the five alteration zones in the area, clearly reveals their spatial distributions, and confirms the 2D alteration-zoning model (Figure 18). The alteration zoning in this area is obvious, and the distribution is relatively stable. The phengite–carbonate alteration zone intersects with the ore body (Figure 19), which agrees well with, and can be a good indication of, the spatial distribution of the mineralization center. The alteration-zone minerals in the surrounding rocks are controlled by the original rock composition and structural zoning. The pyrite–sericite–quartz alteration zone in the footwall of the fault is close to the main fault plane. The pyrite–sericite–quartz alteration is strong, and the main alteration-zone minerals are developed in the center of the ore body. There were large quantities of phengite and carbonate minerals, followed by montmorillonite and a small quantity of kaolinite. Sericite and carbonate rocks are mainly developed in the monzogranite, in the quartz–carbonate stage, and are basically not mineralized. A small quantity of chlorite is developed in the deep granite. In addition to sericite, kaolinite has the most obvious spectral characteristics. The sericite–quartz granitized cataclastic rock on the hanging wall of the main fault surface is rich in kaolinite-wx, indicating an acidic and water-rich environment. Kaolinite was basically undeveloped in the mineralized section. The petrological and SWIR spectral characteristics show that the distribution of these minerals is related to mineralization temperature changes. To sum up, the presence of alteration-zone minerals such as phengite, kaolinite, and chlorite can be used as a direct indicator of the deep ore body of the Zhaoxian gold deposit.

In addition to refining the distribution of layered silicate mineralogy, the Al–OH datasets can be used to track the fluid path that forms the clay alteration and mineralization center [21]. Temperature and pH are the most important factors affecting the hydrothermal system [40]. Taking muscovite and phengite as examples, the increase in the pH value helps to replace Al^{vi} with Fe^{2+} or Mg^{2+}, so that the Si/Al^{vi} ratio increases and Pos2200 moves toward the long-wave direction. Therefore, muscovite is formed at a lower pH, whereas phengite is formed at a relatively high pH. According to previous studies, the crystallinity of layered silicate minerals can be used to infer the relative temperature change [9,29]. The greater the crystallinity, the higher the formation temperature [40]. Based on the results of the above analysis of the Al–OH spectral characteristics of sericite in the 1D and 2D space (Section 4.4), the Pos2200 and SWIR-IC values of Al–OH are strongly correlated (Figure 20) and closely associated with mineralization. As the spectral test data for the drill holes are relatively continuous, the 3D modeling for these two parameters can be established by referring to the method of grade modeling in Section 4.1.2. Figures 21 and 22 are constructed by ordinary kriging interpolation in SKUA-GOCAD; the ranges of Pos2200 and SWIR-IC interpolation are constrained in the 3D space by the tectonic alteration zone, clearly reflecting that both Pos2200 and SWIR-IC have high values within the tectonic alteration zone and coincide with the ore body. The range of high values for Pos2200 is wider than the range of high values for SWIR-IC. Therefore, Pos2200 and illite crystallinity can be used as vectors and exploration tools to delineate hydrothermal centers.

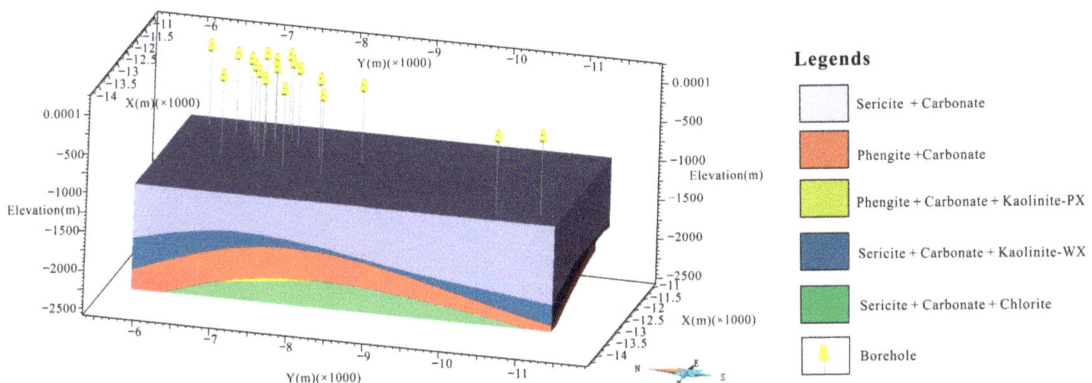

Figure 18. Three-dimensional modeling of alteration zones in Zhaoxian gold deposit.

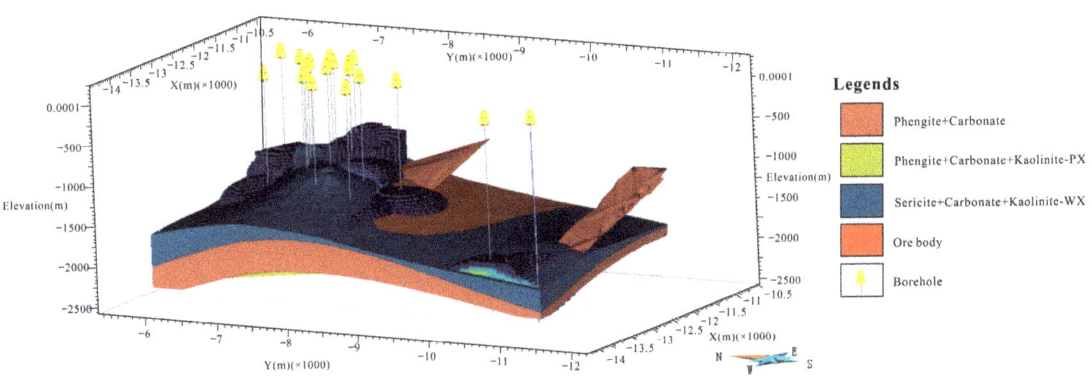

Figure 19. Relationship between alteration zone and mineralization.

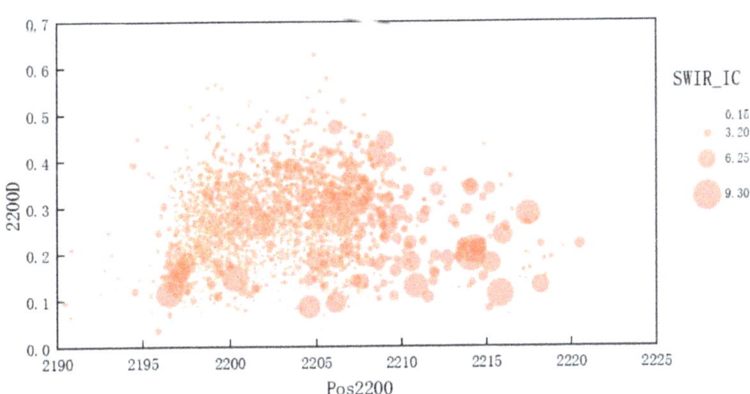

Figure 20. Relationship between characteristic parameters of muscovite (2200D represents relative content of muscovite).

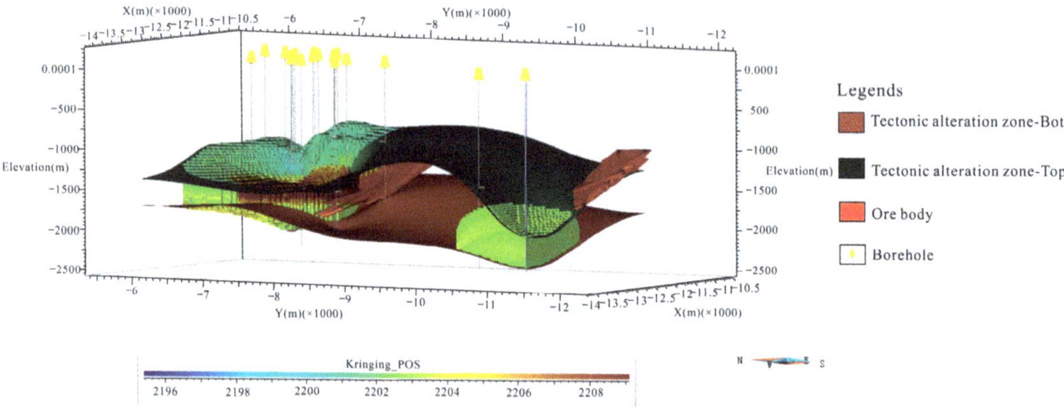

Figure 21. Changes in Pos2200 values in alteration zone.

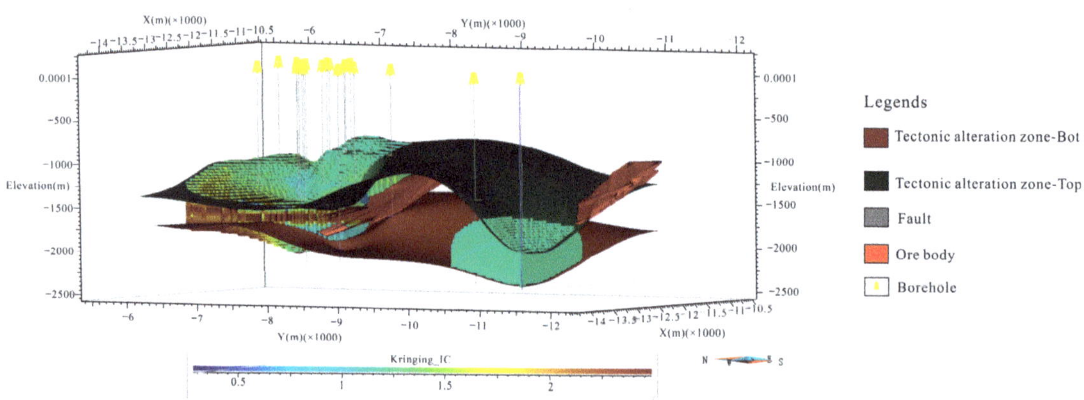

Figure 22. Changes in short-wave infrared illite crystallinity values in alteration zone.

5.2. Mineral-Exploration Indications

The 3D alteration-zoning model (Figure 19) shows the location and geometric shape of each alteration zone and the gold mineralization and ore body [21,48]. The model clearly displays the spatial relationships among different alteration types and the ore bodies and fault structures and confirms the relative positions of alteration zones. The alteration zone is relatively continuous, and the phengite zone and underlying kaolinite zone are basically consistent with the ore body enrichment [34,48]. Based on the regional spreading of 3D fracture structures and ore bodies, the main Jiaojia fault was shown to be the key to conducting mineral exploration in the area. By developing a 3D model, it is possible to visualize the geological characteristics of the hanging wall and footwall of the ore body's occurrence stratum and their variations along the strike [49]. Understanding the spatial range of alteration-zone mineral assemblages and the relationship between the ore body and fault is of great significance for exploration around the deposit and is conducive to developing the mining plan [49]. In addition, based on the 3D geological model, the predecessors proposed a new method of building 3D mineral prospect with a convolutional neural network (CNN), thus reducing the uncertainty of exploration targets [50].

When the quantity of alteration-zone mineral data in the study area is sparse, the mineralized center can be roughly determined through the IC value, which is associated with the sulfide ore-forming stage in the gold deposit, and then the characteristic IC parameters of muscovite can be introduced to delineate the high-temperature core zone of the gold deposit [16]. In the Zhaoxian gold deposit, the exploration targets can be accurately located along with datasets such as the distribution of Al–OH absorption characteristics of sericite (Pos2200 and IC) with pyritization alteration, sericite–kaolinite and phengite alteration zones, the change in kaolinite crystallinity, hand specimen observation and microscopic identification, and geochemical data.

The results of alteration zoning provide an important reference for metallurgical test design. An accurate and comprehensive alteration combination diagram is crucial for optimizing the beneficiation process. The study of gold migration in the Zhaoxian gold deposit showed that each type of alteration accounts for different proportions of chalcopyrite and pyrite [25]. With the support of SWIR analysis, the alteration model established for the Zhaoxian gold deposit describes the precipitation characteristics of gold in different occurrence states, which is essential for the possible grinding and concentration processes in future prefeasibility work [21]. Additionally, because different layered silicate minerals have different flotation reactions, the layered silicate mineral assemblages identified by scalar extraction can help to understand the clay mineral content changes, thus helping to reduce the impact of clay minerals on flotation and bringing obvious economic benefits to mineral processing [51].

6. Conclusions

The analyses of the SWIR spectra show that muscovite, carbonate, kaolinite, montmorillonite, and chlorite are the main alteration-zone minerals in the Zhaoxian gold deposit, among which phengite is closely related to gold mineralization. The mineral assemblages in the study area have obvious zonality, and the changes between zones are gradual. Alteration-zone I consists of sericite–carbonate–chlorite and developed in deep granite. Phengite–carbonate zones are mainly distributed in alteration-zones II and III. Strong pyrite–sericite–quartz alteration is developed in alteration-zone III, and kaolinite is not developed, whereas alteration-zone II contains a small quantity of kaolinite-px. Sericite–carbonate–kaolinite-wx is developed in sericite–quartz granitized cataclastic rock (alteration-zone IV). Alteration-zone V (sericite–carbonate) is located in the monzogranite.

The 3D geological and spectral scalar models show that the study area has obvious tectonic control features, with spatial correlation between alteration zones, fracture structures, and ore bodies. The mineralization zone is a semiopen shear space and provides sufficient space for the water–rock metasomatic reaction of the fluid. The compressive torsional features and concave–convex section zone (depth 3000 m) are the main signs of 3D geological ore-body modeling in the Zhaoxian gold deposit for ore prospecting. A large quantity of high-crystallinity phengite was found near the mineralization and in the deep area, which indicates that a large quantity of gold in the fracture was precipitated in a hydrothermal environment with gradually lower temperatures and relatively higher pH.

The SWIR spectral features indicate that phengite is the proximal alteration-zone mineral to gold mineralization in a strong pyritic sericitization zone. Phengite with relatively high Pos2200 values (>2205 nm) as well as relatively high SWIR-IC values can correspond well to gold mineralization. Both Pos2200 and SWIR-IC can be used as mineralization indicators for the study area.

Author Contributions: Conceptualization, G.W.; methodology, B.L., X.L. and H.W.; software, B.L., X.L. and H.J.; validation, Y.P., X.Z. and H.W.; formal analysis, Y.P., X.Z., Z.Y. and H.W.; investigation, B.L. and G.W.; project administration, G.W. and X.L.; data curation, B.L., X.L. and G.W.; writing—original draft preparation, B.L.; writing—review and editing, G.W. All authors have read and agreed to the published version of the manuscript.

Funding: This research was funded by the open project of the Shandong Provincial Engineering Laboratory of Application and Development of Big Data for Deep Gold Exploration (Grant No. SDK202210) and National Key Research and Development Program of China (Grant No. 2022YFC2903604).

Data Availability Statement: The data presented in this study are available on request from the corresponding author. The data are not publicly available due to the confidentiality of some of the data.

Acknowledgments: The authors thank the Shandong Xincheng Gold Mining Co. Ltd. for its cooperation in field work. We are grateful to Ling Zuo for her constructive comments and discussion. We also thank Xuewei Shao for the improvements to the pictures.

Conflicts of Interest: The authors declare no conflict of interest.

References

1. Li, P.; Li, J.; Ma, Z. Preference of China s gold resources overseas investment. *China Min. Mag.* **2021**, *30*, 23–29. (In Chinese)
2. Deng, J.; Yang, L.-Q.; Groves, D.I.; Zhang, L.; Qiu, K.-F.; Wang, Q.-F. An integrated mineral system model for the gold deposits of the giant Jiaodong province, eastern China. *Earth-Sci. Rev.* **2020**, *208*, 103274. [CrossRef]
3. Yang, L.; Deng, J.; Wang, Z.; Guo, L.; Li, R.; Groves, D.I.; Danyushevsky, L.V.; Zhang, C.; Zheng, X.; Zhao, H. Relationships between gold and pyrite at the Xincheng gold deposit, Jiaodong Peninsula, China; implications for gold source and deposition in a brittle epizonal environment. *Econ. Geol. Bull. Soc. Econ. Geol.* **2016**, *111*, 105–126. [CrossRef]
4. Deng, J.; Wang, J.P.; Peng, R.M.; Liu, J.; Yang, L. Researches on deep ore prospecting. *Miner. Depos.* **2004**, *23*, 142–149.
5. Yang, L.; Deng, J.; Wang, Z.L.; Liang, Z.; Guo, L.N.; Song, M.C.; Zheng, X.L. Mesozoic gold metallogenic system of the Jiaodong gold province, eastern China. *Acta Petrol. Sin.* **2014**, *30*, 2447–2467.
6. Deng, J.; Wang, C.; Bagas, L.; Carranza, E.J.M.; Lu, Y. Cretaceous-Cenozoic tectonic history of the Jiaojia Fault and gold mineralization in the Jiaodong Peninsula, China: Constraints from zircon U-Pb, illite K-Ar, and apatite fission track thermochronometry. *Miner. Depos.* **2015**, *50*, 987–1006. [CrossRef]
7. Yu, X.; Yang, D.; Li, D.; Shan, W.; Xiong, Y.; Chi, N.; Liu, P.; Yu, L. Mineralization characteristic and geological significance in 3000 m depth of Jiaojia gold metallogenic belt, Jiaodong Peninsula. *Acta Petrol. Sin.* **2019**, *35*, 2893–2910.
8. Zhu, D.; Zhang, W.; Wang, Y.; Tian, J.; Liu, H.; Hou, J.; Gao, H.L. Characteristics of ore bodies and prospecting potential of Zhaoxian gold deposit in Laizhou City of Shandong Province. *Shandong Land Resour.* **2018**, *34*, 14–19.
9. Duke, E.F. Near infrared spectra of muscovite, Tschermak substitution, and metamorphic reaction progress: Implications for remote sensing. *Geology* **1994**, *22*, 621–624. [CrossRef]
10. Green, D.; Schodlok, M. Characterisation of carbonate minerals from hyperspectral TIR scanning using features at 14,000 and 11,300 nm. *Aust. J. Earth Sci.* **2016**, *63*, 951–957.
11. Laukamp, C.; Cudahy, T.; Thomas, M.; Jones, M.; Cleverley, J.S.; Oliver, N.H. Hydrothermal mineral alteration patterns in the Mount Isa Inlier revealed by airborne hyperspectral data. *Aust. J. Earth Sci.* **2011**, *58*, 917–936. [CrossRef]
12. Laukamp, C.; Salama, W.; Gonzalez-Alvarez, I. Proximal and remote spectroscopic characterisation of regolith in the Albany-Fraser Orogen (Western Australia). *Ore Geol. Rev.* **2016**, *73*, 540–554. [CrossRef]
13. Thompson, A.J.; Hauff, P.L.; Robitaille, A.J. Alteration mapping in exploration: Application of short-wave infrared (SWIR) spectroscopy. *SEG Discov.* **1999**, *39*, 1–27. [CrossRef]
14. Jones, S.; Herrmann, W.; Gemmell, J.B. Short-wavelength infrared spectral characteristics of the RW horizon: Implications for exploration in the Myra Falls volcanic-hosted massive sulfide camp, Vancouver Island, British Columbia, Canada. *Econ. Geol.* **2005**, *100*, 273–294. [CrossRef]
15. Yang, K.; Browne, P.R.L.; Huntington, J.F.; Walshe, J.L. Characterising the hydrothermal alteration of the Broadlands–Ohaaki geothermal system, New Zealand, using short-wave infrared spectroscopy. *J. Volcanol. Geotherm. Res.* **2001**, *106*, 53–65. [CrossRef]
16. Laakso, K.; Peter, J.M.; Rivard, B.; White, H.P. Short-wave infrared spectral and geochemical characteristics of hydrothermal alteration at the Archean Izok Lake Zn-Cu-Pb-Ag volcanogenic massive sulfide deposit, Nunavut, Canada: Application in exploration target vectoring. *Econ. Geol.* **2016**, *111*, 1223–1239. [CrossRef]
17. de Kemp, E.A.; Schetselaar, E.M.; Hillier, M.J.; Lydon, J.W.; Ransom, P.W. Assessing the workflow for regional-scale 3D geologic modeling: An example from the Sullivan time horizon, Purcell Anticlinorium East Kootenay region, southeastern British Columbia. *Interpretation* **2016**, *4*, SM33–SM50. [CrossRef]
18. Wang, G.; Zhang, Z.; Li, R.; Li, J.; Sha, D.; Zeng, Q.; Pang, Z.; Li, D.; Huang, L. Resource prediction and assessment based on 3D/4D big data modeling and deep integration in key ore districts of North China. *Sci. China Earth Sci.* **2021**, *64*, 1590–1606. [CrossRef]
19. Malehmir, A.; Thunehed, H.; Tryggvason, A. The Paleoproterozoic Kristineberg mining area, northern Sweden: Results from integrated 3D geophysical and geologic modeling, and implications for targeting ore deposits3D geologic modeling. *Geophysics* **2009**, *74*, B9–B22. [CrossRef]
20. Laudadio, A.B.; Schetselaar, E.; Mungall, J.; Houlé, M. 3D modeling of the Esker intrusive complex, Ring of Fire intrusive suite, McFaulds Lake greenstone belt, Superior Province: Implications for mineral exploration. *Ore Geol. Rev.* **2022**, *145*, 104886. [CrossRef]
21. Harraden, C.L.; McNulty, B.; Gregory, M.J.; Lang, J.R. Shortwave infrared spectral analysis of hydrothermal alteration associated with the Pebble porphyry copper-gold-molybdenum deposit, Iliamna, Alaska. *Econ. Geol.* **2013**, *108*, 483–494. [CrossRef]
22. Chen, H.; Zhang, S.; Chu, G.; Zhang, Y.; Cheng, J.; Tian, J.; Han, J. The short wave infrared (SWIR) spectral characteristics of alteration minerals and applications for ore exploration in the typical skarn-porphyry deposits, Edong ore district, eastern China. *Acta Petrol. Sin.* **2019**, *35*, 3629–3643.

23. Shao, X.; Peng, Y.; Wang, G.; Zhao, X.; Tang, J.; Huang, L.; Liu, X.; Zhao, X. Application of SWIR, XRF and thermoelectricity analysis of pyrite in deep prospecting in the Xincheng gold orefield, Jiaodong peninsula. *Earth Sci. Front.* **2021**, *28*, 236.

24. Song, M.C.; Song, Y.X.; Ding, Z.J.; Wei, X.F.; Sun, S.L.; Song, G.Z.; Zhang, J.J.; Zhang, P.J.; Wang, Y.G. The discovery of the Jiaojia and the Sanshandao giant gold deposits in Jiaodong Peninsula and discussion on the relevant issues. *Geotecton. Metallog.* **2019**, *43*, 92–110.

25. Song, Y.; Song, Y.X.; Cui, S.X.; Jiang, H.L.; Yuan, W.H.; Wang, H.J. Characteristic comparison between shallow and deep-seated gold ore bodies in Jiaojia superlarge gold deposit, northwestern Shandong peninsula. *Miner. Depos.* **2011**, *30*, 923–932.

26. Wang, G.; Zhu, Y.; Zhang, S.; Yan, C.; Song, Y.; Ma, Z.; Hong, D.; Chen, T. 3D geological modeling based on gravitational and magnetic data inversion in the Luanchuan ore region, Henan Province, China. *J. Appl. Geophys.* **2012**, *80*, 1–11. [CrossRef]

27. Zeqiri, R. Geostatistical analysis of the nickel source in Gllavica mine, Kosovo. *Min. Miner. Depos.* **2020**, *14*, 53–58. [CrossRef]

28. Zhou, Y.; Li, L.; Yang, K.; Xing, G.; Xiao, W.; Zhang, H.; Xiu, L.; Yao, Z.; Xie, Z. Hydrothermal alteration characteristics of the Chating Cu-Au deposit in Xuancheng City, Anhui Province, China: Significance of sericite alteration for Cu-Au exploration. *Ore Geol. Rev.* **2020**, *127*, 103844. [CrossRef]

29. Zuo, L.; Wang, G.; Carranza, E.J.M.; Zhai, D.; Pang, Z.; Cao, K.; Mou, N.; Huang, L. Short-Wavelength Infrared Spectral Analysis and 3D Vector Modeling for Deep Exploration in the Weilasituo Magmatic–Hydrothermal Li–Sn Polymetallic Deposit, Inner Mongolia, NE China. *Nat. Resour. Res.* **2022**. [CrossRef]

30. Pontual, S.; Merry, N.; Gamson, P. Spectral Interpretation-Field Manual. In *GMEX. Spectral Analysis Guides for Mineral Exploration*; AusSpec International Pty: Kwinana, WA, Australia, 2008.

31. Yang, K.; Huntington, J.; Gemmell, J.B.; Scott, K. Variations in composition and abundance of white mica in the hydrothermal alteration system at Hellyer, Tasmania, as revealed by infrared reflectance spectroscopy. *J. Geochem. Explor.* **2011**, *108*, 143–156. [CrossRef]

32. Clark, R.N.; King, T.V.V.; Klejwa, M.; Swayze, G.A.; Vergo, N. High spectral resolution reflectance spectroscopy of minerals. *J. Geophys. Res. Solid Earth* **1990**, *95*, 12653–12680.

33. Gaillard, N.; Williams-Jones, A.E.; Clark, J.R.; Lypaczewski, P.; Salvi, S.; Perrouty, S.; Piette-Lauzière, N.; Guilmette, C.; Linnen, R.L. Mica composition as a vector to gold mineralization: Deciphering hydrothermal and metamorphic effects in the Malartic district, Quebec. *Ore Geol. Rev.* **2018**, *95*, 789–820. [CrossRef]

34. Haest, M.; Cudahy, T.; Laukamp, C.; Gregory, S. Quantitative mineralogy from infrared spectroscopic data; I, Validation of mineral abundance and composition scripts at the Rocklea Channel iron deposit in Western Australia. *Econ. Geol.* **2012**, *107*, 209–228. [CrossRef]

35. Clark, R.N. Spectral properties of mixtures of montmorillonite and dark carbon grains: Implications for remote sensing minerals containing chemically and physically adsorbed water. *J. Geophys. Res. Solid Earth* **1983**, *88*, 635–644. [CrossRef]

36. Doublier, M.P.; Roache, T.; Potel, S.; Laukamp, C. Short-wavelength infrared spectroscopy of chlorite can be used to determine very low metamorphic grades. *Eur. J. Mineral.* **2012**, *24*, 891–902. [CrossRef]

37. Lampinen, H.M.; Laukamp, C.; Occhipinti, S.A.; Hardy, L. Mineral footprints of the Paleoproterozoic sediment-hosted Abra Pb-Zn Cu-Au deposit Capricorn Orogen, Western Australia. *Ore Geol. Rev.* **2019**, *104*, 436–461. [CrossRef]

38. Hosseini, S.T.; Asgharl, O.; Emery, X. An enhanced direct sampling (DS) approach to model the geological domain with locally varying proportions: Application to Golgohar iron ore mine, Iran. *Ore Geol. Rev.* **2021**, *139*, 104452. [CrossRef]

39. Zhang, S.; Chu, G.; Cheng, J.; Zhang, Y.; Tian, J.; Li, J.; Sun, S., Wei, K, Short wavelength infrared (SWIR) spectroscopy of phyllosilicate minerals from the Tonglushan Cu-Au-Fe deposit, Eastern China: New exploration indicators for concealed skarn orebodies. *Ore Geol. Rev.* **2020**, *122*, 103516.

40. Halley, S.; Dilles, J.H.; Tosdal, R.M. Footprints: Hydrothermal Alteration and Geochemical Dispersion Around Porphyry Copper Deposits. *SEG Discov.* **2015**, *100*, 1–17. [CrossRef]

41. Wang, R.; Cudahy, T.; Laukamp, C.; Walshe, J.L.; Bath, A.; Mei, Y.; Young, C.; Roache, T.J.; Jenkins, A.; Roberts, M.; et al. White Mica as a Hyperspectral Tool in Exploration for the Sunrise Dam and Kanowna Belle Gold Deposits, Western Australia. *Econ. Geol.* **2017**, *112*, 1153–1176. [CrossRef]

42. Yang, K.; Huntington, J.F.; Browne, P.R.; Ma, C. An infrared spectral reflectance study of hydrothermal alteration minerals from the Te Mihi sector of the Wairakei geothermal system, New Zealand. *Geothermics* **2000**, *29*, 377–392.

43. Haest, M.; Cudahy, T.; Rodger, A.; Laukamp, C.; Martens, E.; Caccetta, M. Unmixing the effects of vegetation in airborne hyperspectral mineral maps over the Rock lea Dome iron-rich palaeochannel system (Western Australia). *Remote Sens. Environ.* **2013**, *129*, 17–31. [CrossRef]

44. Sun, L.; Khan, S.; Shabestari, P. Integrated Hyperspectral and Geochemical Study of Sediment-Hosted Disseminated Gold at the Goldstrike District, Utah. *Remote Sens.* **2019**, *11*, 1987. [CrossRef]

45. Khan, S.D.; Okyay, Ü.; Ahmad, L.; Shah, M.T. Characterization of Gold Mineralization in Northern Pakistan Using Imaging Spectroscopy. *Photogramm. Eng. Remote Sens.* **2018**, *84*, 425–434. [CrossRef]

46. Wang, G.; Zhang, S.; Yan, C.; Pang, Z.; Wang, H.; Feng, Z.; Dong, H.; Cheng, H.; He, Y.; Li, R.; et al. Resource-environmental joint forecasting in the Luanchuan mining district, China through big data mining and 3D/4D modeling. *Earth Sci. Front.* **2021**, *28*, 139.

47. Wang, G.; Li, R.; Carranza, E.J.; Zhang, S.; Yan, C.; Zhu, Y.; Qu, J.; Hong, D.; Song, Y.; Han, J.; et al. 3D geological modeling for prediction of subsurface Mo targets in the Luanchuan district, China. *Ore Geol. Rev.* **2015**, *71*, 592–610. [CrossRef]

48. Haest, M.; Cudahy, T.; Laukamp, C.; Gregory, S. Quantitative mineralogy from infrared spectroscopic data. II. Three-dimensional mineralogical characterization of the Rocklea channel iron deposit, Western Australia. *Econ. Geol.* **2012**, *107*, 229–249. [CrossRef]
49. Petlovanyi, M.V.; Ruskykh, V.V. Peculiarities of the underground mining of high-grade iron ores in anomalous geological conditions. *J. Geol. Geogr. Geoecol.* **2019**, *28*, 706–716. [CrossRef]
50. Deng, H.; Zheng, Y.; Chen, J.; Yu, S.; Xiao, K.; Mao, X. Learning 3D mineral prospectivity from 3D geological models using convolutional neural networks: Application to a structure-controlled hydrothermal gold deposit. *Comput. Geosci.* **2022**, *161*, 105074. [CrossRef]
51. Petlovanyi, M.; Ruskykh, V.; Zubko, S.; Medianyk, V. Dependence of the mined ores quality on the geological structure and properties of the hanging wall rocks. *E3S Web Conf.* **2020**, *201*, 01027.

Article

3D Geophysical Predictive Modeling by Spectral Feature Subset Selection in Mineral Exploration

Bahman Abbassi [1,*], Li-Zhen Cheng [1], Michel Jébrak [2] and Daniel Lemire [3]

[1] Institut de Recherche en Mines et en Environnement (IRME), Université du Québec en Abitibi-Témiscamingue (UQAT), Rouyn-Noranda, QC J9X 5E4, Canada
[2] Département de la Science de la Terre et de l'Atmosphère, Université du Québec à Montréal (UQAM), Montreal, QC H2X 1L4, Canada
[3] Département de Science et Technologie, Université TÉLUQ, Montreal, QC G1K 9H5, Canada
* Correspondence: bahman.abbassi@uqat.ca

Abstract: Several technical challenges are related to data collection, inverse modeling, model fusion, and integrated interpretations in the exploration of geophysics. A fundamental problem in integrated geophysical interpretation is the proper geological understanding of multiple inverted physical property images. Tackling this problem requires high-dimensional techniques for extracting geological information from modeled physical property images. In this study, we developed a 3D statistical tool to extract geological features from inverted physical property models based on a synergy between independent component analysis and continuous wavelet transform. An automated interpretation of multiple 3D geophysical images is also presented through a hybrid spectral feature subset selection (SFSS) algorithm based on a generalized supervised neural network algorithm to rebuild limited geological targets from 3D geophysical images. Our self-proposed algorithm is tested on an Au/Ag epithermal system in British Columbia (Canada), where layered volcano-sedimentary sequences, particularly felsic volcanic rocks, are associated with mineralization. Geophysical images of the epithermal system were obtained from 3D cooperative inversion of aeromagnetic, direct current resistivity, and induced polarization data sets. The recovered cooperative susceptibilities allowed locating a magnetite destructive zone associated with porphyritic intrusions and felsic volcanoes (Au host rocks). The practical implementation of the SFSS algorithm in the study area shows that the proposed spectral learning scheme can efficiently learn the lithotypes and Au grade patterns and makes predictions based on 3D physical property inputs. The SFSS also minimizes the number of extracted spectral features and tries to pick the best representative features for each target learning case. This approach allows interpreters to understand the relevant and irrelevant spectral features in addition to the 3D predictive models. Compared to conventional 3D interpolation methods, the 3D lithology and Au grade models recovered with SFSS add predictive value to the geological understanding of the deposit in places without access to prior geological and borehole information.

Keywords: independent component analysis; 3D modeling; spectral feature subset selection

Citation: Abbassi, B.; Cheng, L.-Z.; Jébrak, M.; Lemire, D. 3D Geophysical Predictive Modeling by Spectral Feature Subset Selection in Mineral Exploration. *Minerals* **2022**, *12*, 1296. https://doi.org/10.3390/min12101296

Academic Editor: José António de Almeida

Received: 7 September 2022
Accepted: 11 October 2022
Published: 14 October 2022

Publisher's Note: MDPI stays neutral with regard to jurisdictional claims in published maps and institutional affiliations.

1. Introduction

A fundamental problem in integrated geophysical interpretation is the proper geological understanding of multiple inverted physical property images, which demands high dimensional techniques for extracting geological information from modeled physical property images. The study of seismic attributes is an example of high-dimensional pattern recognition that seeks to extract relevant geological features from broadband seismic data [1–7]. Non-seismic data interpretations in mineral exploration involve a similar classification problem, where extracting 3D geological information from inverted potential field data is a computational challenge.

Several techniques are available for feature extraction in the high dimensional space, and this computational challenge can be seen as a dimensionality increase problem that consequently leads to an overload of information [8–10]. For example, the spectral decomposition of geophysical images provides a robust way for feature extraction in the frequency domain [11–13]. However, the underlying patterns inside the high-dimensional images are only related to certain frequencies, and most decomposed spectra are redundant. The question is which frequency or frequency ranges are geologically relevant. This computational challenge can be seen as a dimensionality reduction problem, in which we try to extract the best representative components of high-dimensional images to facilitate visual interpretations and machine learning optimization [14,15].

Conventionally, high redundancy in the spectral domain is treated by incorporating principal component analysis (PCA) to extract the principal components of multiple images [14,16]. PCA is an unsupervised algorithm that linearly transforms the multivariate datasets into new features called Principal Components (PCs) by maximizing the variance of the input data [14,16]. However, the uncorrelatedness of principal components is a weaker form of independence; therefore, PCA is not a good choice for separating overlapped features inside multiple physical property images [17]. Alternatively, independent component analysis (ICA), precisely the negentropy maximization approach, has been proposed as a robust tool for separating background lithotypes from alteration events [17,18]. Unlike PCA, ICA performs beyond the Gaussian assumption and incorporates higher-order statistics to find maximally independent latent features [14,16].

As expected, the spectra of the ICs still contain certain hidden overlapped features in different frequencies. A similar problem has been addressed by Honório et al. [19] to visualize seismic broadband data in different frequencies. They used ICA to maximize the non-gaussianity of the decomposed seismic frequency volume and stacked the resulting ICs in three red, green, and blue (RGB) channels. The unified RGB image delineated a hidden sedimentary channel system [19]. However, this approach lacks automation and involves manually selecting and stacking representative ICs of the decomposed seismic data. The extraction of spectral features provides more detailed input features for the machine learning algorithm; however, the existence of redundant features decreases the performance of the learning process. Finding a way to detect these redundancies and avoid them can improve machine learning predictions. Moreover, it is challenging to decide how much dimensionality reduction is necessary to represent the most prominent spectral features to machine learning algorithms.

Feature subset selection (FSS) provides a powerful tool for adaptive dimensionality reduction and feature learning [15,20–22]. SFSS reduces the size of a large set of input features to a new set of relatively small and representative input features that improve the accuracy of the artificial neural network (ANN) prediction, either by decreasing the learning speed and model complexity or by increasing generalization capacity and classification accuracy [23]. Therefore, unlike conventional statistical dimensionality reduction methods that are blind source separation algorithms, there is a criterion for the selection of the best features that can be expressed as an adaptation of the learning process in the light of the optimization of both neural synaptic weights and the number of selected input features [21–23].

This study's spectral feature subset selection (SFSS) procedure is seen as a multi-objective optimization problem in which both the ANN weights and the number of the selected features are updated in different iterations. Several heuristic algorithms are known to solve these multi-objective optimization problems [24–26]. We use a bi-objective genetic algorithm (GA) optimization to regularize the ANN weights and find the most adaptable combination of input spectral features in the supervised machine learning procedure. The advantage of this approach to the conventional representative learning algorithms is that it can direct the learning of ANN in a way that only a relevant set of input features are allowed to be used during the learning process.

The main objective of this study is to present a robust algorithm to train ANN with automatically selected spectral features to predict geological targets such as mafic to felsic 3D lithological variations and 3D Au-grade distribution. The predicted outputs of the

SFSS procedure are the estimated geological targets and selected sets of spectral features related to each geological target. The study also aims to show that the selected features and predicted targets also provide a unique way for interpreters to see which spatial and spectral features are prominent in relation to the geological targets.

2. Materials and Methods

We developed an algorithm that automatically selects the best representative components based on multi-objective machine learning optimization. Our self-proposed 3D SFSS algorithm works for dimensionality reduction and separating high-dimensional spectral overlapped features. However, SFSS implementation in this study demands several pre-processing stages, including 3D inversion of geophysical data and 3D feature extraction methods. The accuracy of inverted physical property images is critically important in the subsequent feature extraction and selection procedures. 3D inversion algorithms are dedicated to tackling this problem [27–30]. Even if the geophysical inversion provides the most accurate physical property distributions, each physical property image provides only partial information about subsurface geology. This phenomenon can be seen as a linear mixing process, and ICA can help to reconstruct the hidden geological features from multiple physical property images [17,18]. The advantage of this method is that it can separate the host geology from alteration overprints in multidimensional physical property space [17].

Subsequently, spectral decomposition reveals many latent features that are not properly visible in the spatial domain. However, the statistical interdependence of the decomposed images is a major difficulty in extracting and selecting geological information. The large volumes of decomposed spectra also limit the spectral decomposition's efficiency and demand effective dimensionality reduction methods to prevent information overload. Our self-proposed SFSS algorithm solves these high-dimensional problems in five stages (Figure 1):

1. 3D inversion of geophysical data.
2. Separation of physical properties through ICA (spatial feature extraction).
3. Spectral decomposition of ICs through a continuous wavelet transform (CWT).
4. Dimensionality reduction and separation of raw spectral features through ICA (spectral feature extraction).
5   SFSS based on a supervised feature selection algorithm optimized by a multi-objective GA optimization (spatiospectral feature selection).

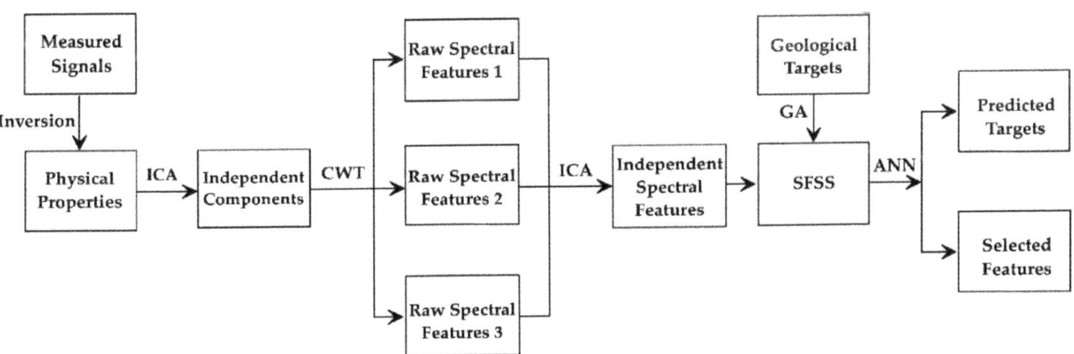

Figure 1. Workflow of SFSS for 3D predictive modeling.

2.1. Geological and Geophysical Settings

We test our method on an Au/Ag epithermal deposit known as the Newton deposit in British Colombia, Canada [28–30]. Regionally, the Late Cretaceous volcanic sequence is overlain by Miocene–Pliocene Chilcotin Group flood basalts and Quaternary glacial deposits, which are variably eroded to expose the older rocks. Quaternary glacial tills cover most of the Newton property on a deposit scale. Consequently, deposit scale geological information has primarily been obtained from borehole samples [27–30]. A bedrock geology map of the property is compiled from the mapping of limited outcrops, drill cores, and cross-section interpretations (Figure 2a).

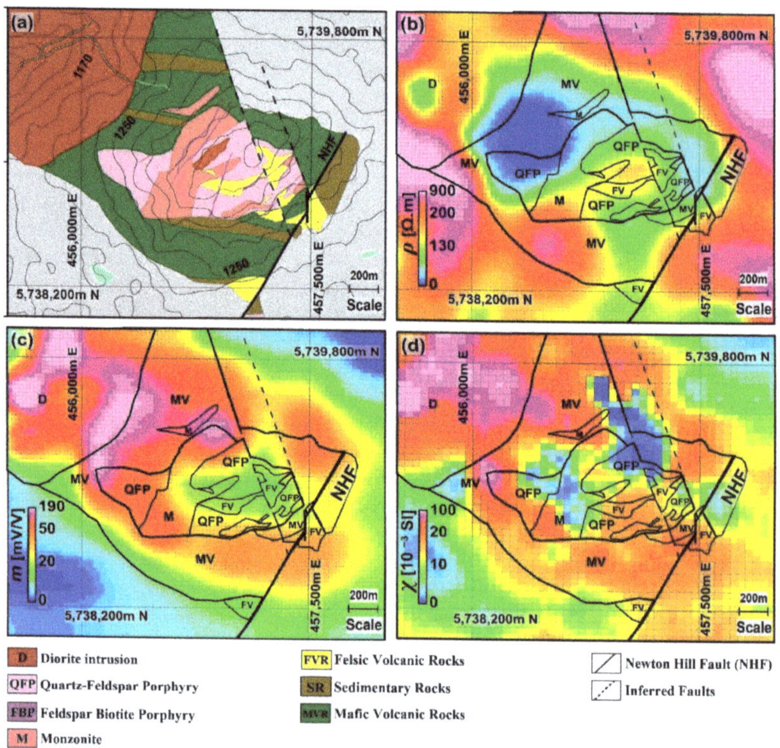

Figure 2. Geological and geophysical settings of the deposit: (**a**) Bedrock geology map at the deposit scale (sliced at an elevation of 1000 m). The light gray parts are undefined regions due to the lack of borehole information in the periphery of the deposit. Contours indicate topography. (**b**) Electrical resistivity model. (**c**) IP chargeability model. (**d**) Magnetic susceptibility model. The bedrock geology is overlaid on all geophysical models [27].

Three layered volcano-sedimentary sequences overlain from bottom to top are mafic volcanic rocks (MVR), sedimentary rocks (SR), and felsic volcanic rocks (FVR). The layered rocks are intruded by felsic to intermediate porphyritic intrusions, including the monzonite (M), quartz-feldspar porphyry (QFP), feldspar biotite porphyry (FBP), and the younger dioritic intrusion (D). The epithermal mineralization associated with the Newton deposit is at a depth of ~50 m to ~600 m. Au/Ag mineralization is mainly hosted within the felsic volcanic sequence [27–30]. The deposit is offset by the Newton Hill Fault (NHF), displacing geology and mineralization by ~300m of normal dip-slip movement.

Geophysical images of Newton deposit were obtained from 3D inversion of direct current (DC) resistivity, induced polarization (IP), and magnetic data sets [27,31]. In 2010, an 85 line-kilometer DC/IP survey was conducted in the Newton Hill area using a "pole-dipole" electrode configuration [27,31]. The DC/IP survey lines were placed at 200m intervals in the east–west direction in the hydrothermal alteration zone. The dipole length was set to 100 m and 200 m, with a maximum of ten times dipolar separations. Apparent chargeability (mV/V) was measured by recording the voltage drop after current interruption. The total magnetic field data were collected from a helicopter flying at an average altitude of 155 m above the ground. The magnetic data were collected along N–S flight lines spaced 200 m apart, and E–W tie lines were flown approximately every 2000 m to 500 m. A total of 7071 line-kilometers of magnetic data were collected, covering an area of 1293 km^2. The magnetometer sampling rate was 0.1 s, yielding a measurement interval of approximately 10 m along each profile, depending on the helicopter speed [27,31].

The interrelationships between physical properties enabled the cooperative inversion of multiple datasets. The chargeability image in this study was used to constrain the magnetic inversion. The constrained susceptibilities set aside the less consistent equivalent solutions and narrowed down to a final solution that is well matched with the IP-DC resistivity inversion results, which improves deposit scale susceptibility imaging [27,31]. The value of this approach is that it does not necessarily require prior information for physical properties coupling. The total number of cells for forward DC/IP calculation is 288,000, and the size of each cell is 25 m × 50 m in the horizontal plane. The model cells are also set to grow exponentially with depth in 20 intervals up to 600 m. The unit cell size for cooperative susceptibility inversion is also set to X = 25 m, Y = 25, and Z = 10 m, with a total of 613,800 cells.

The recovered cooperative susceptibilities allowed to locate a magnetite destructive zone associated with porphyritic intrusions and felsic volcanoes (Au host rocks). Although, in some places, the distinction between magnetic highs of diorite and mafic volcanic rocks and magnetic lows of felsic volcanic rocks (Au host rocks) and porphyritic intrusions is still challenging. The result of the cooperative magnetic- IP-DC resistivity inversion is also shown in Figure 2. The high magnetic cap imaged from the constrained inversion is related to a mixture of the mafic volcanic rocks (MV) and the younger dioritic intrusions (D). A low magnetic zone (LMZ) was positioned deeper in earlier unconstrained models, but the cooperative inversion has replaced it upward (about 500 m) and imaged right under the high magnetic cap (magnetic susceptibilities > 0.04 SI). Several flanks of this LMZ locally reduce the magnetic susceptibilities to 0.015 SI. This structure is probably a response of the intermediate-to-felsic porphyritic intrusions (FIP) and interbeds of felsic volcanic rocks (FV) within the upper high magnetic cap [27,31].

2.2. Feature Extraction

The feature extraction in this study is based on implementing several blind source separation methods, including PCA, ICA, and CWT, to recover latent features from a set of highly mixed images. Consider a mixing model where every image (g) is a mix of several hidden features (f) with different contributions to constructing the observed image, determined by a set of mixing weights *A*, in the form of a linear mixing model [18]:

$$g = Af \tag{1}$$

To recover the hidden features (f), one needs to find a separation matrix (*W*) and unmixes the observed images:

$$f = Wg = A^{-1}g \tag{2}$$

The separation matrix (*W*) can be estimated as an optimization problem by minimization of a cost function. By making a few assumptions about the statistical measures of the data sets, the feature extraction process iteratively reduces the effect of the mixing on the observed images. In PCA, the second-order statistical measure, i.e., variance, is maximized for image separation, and the outcome is linearly separated uncorrelated images. However, observed images with the nonlinear form of correlation (dependency) pose a significant difficulty to PCA-based separation methods. We can solve this problem by maximization of higher-order statistical measures (non-gaussianity) to separate the images into nonlinearly uncorrelated images through ICA.

The ICA starts with a PCA as a preprocessing step that removes the mean of the input images (g). The principal components (y) are related to the centered images (g_c) through a weight matrix D:

$$y = Dg_c \tag{3}$$

The matrix D can be estimated through variance maximization of the principal components (y). Finally, increasing the non-gaussianity of the principal components (y) produces the ICs of the images. The problem is to find a rotation matrix that, during the multiplication with the principal components, produces the least Gaussian outputs [16,18].

The Fast-ICA algorithm, through negentropy maximization and the Hyvärinen fixed-point method, was used to obtain the rotation matrix that incorporates higher-order statistics to recover the latent independent sources [16,18]. The entropy (H) of an image (g) is defined as [16,18]:

$$H(g) = - \int p(g) log p(g) dg \tag{4}$$

where $p(g)$ is the probability density of the image g. Entropy is a measure of randomness. The more unpredictable and unstructured the variable is, the larger its entropy. Theoretically, the Gaussian variable possesses the largest entropy. The negentropy (J) of an image is the normalized differential entropy of that image:

$$J(g) = H\left(g_{gausss}\right) - H(g) \tag{5}$$

where $H(g)$ is the entropy of the image, $H\left(g_{gausss}\right)$ is the entropy of a Gaussian random image of the same covariance matrix, and $neg(g) \geq 0$ is always non-negative and zero when the image has a pure Gaussian distribution. The negentropy maximization is based on bringing an objective function to an approximated maximum value [14,16–18,27].

The CWT in two dimensions (horizontal planes) was used for feature extraction in the frequency domain. Since wavelets are localized in space and have finite durations, the sharp changes in images are efficiently detectable by 2D wavelet decomposition, which provides a unique way for spectral feature extraction [19,32]. The output of wavelet decomposition effectively reflects the sharp changes in images, making it an ideal tool for feature extraction [19,32]. The continuous wavelet transform of an image $I(x, y)$ is defined as a decomposition of that image by translation and dilation of a mother wavelet $\psi(x, y)$. The resulting wavelet coefficients (C_s) are then given by

$$C_S = (b_1, b_2, a) = \frac{1}{\sqrt{|a|}} \iint I(x,y)\psi^*(\frac{x - b_1}{a}, \frac{y - b_2}{a}) dx dy \tag{6}$$

where b_1 and b_2 control the spatial translation, $a > 1$ denotes the scale, and ψ^* is the complex conjugate of the mother wavelet $\psi(x, y)$. The mother wavelet shifts and scales in multiple directions and produces numerous features. In this study, taking advantage of the nICA statistical properties, we can keep the most geologically pertinent information within the spectral decomposed volumes [18].

2.3. Learning Spectral Features

Spatial and spectral feature extraction provides inputs for predictive modeling through supervised machine learning algorithms. Training machine learning models directly with raw data sets often yield unsatisfactory results. The feature extraction process identifies the most discriminating characteristics in raw images, which a machine learning algorithm can more easily consume. However, the curse of dimensionality prevents efficient machine learning when extracted features are too large [8–10]. Anyway, dimensionality reduction can improve spectral learning performance through feature extraction and feature subset selection methods. The main difference is that feature extraction combines the original features and creates a set of new features, while feature selection selects a subset of the original features [15,21–23,26]. We propose a hybrid algorithm that includes several phases of preprocessing, spatial feature extraction, spectral feature extraction, and machine learning for geophysical predictive modeling (Figure 3). In this section, we first outline a basis for SRL through multilayer perceptron (MLP), and then we describe our proposed SFSS algorithm for 3D geophysical predictive modeling, feature selection, and dimensionality reduction.

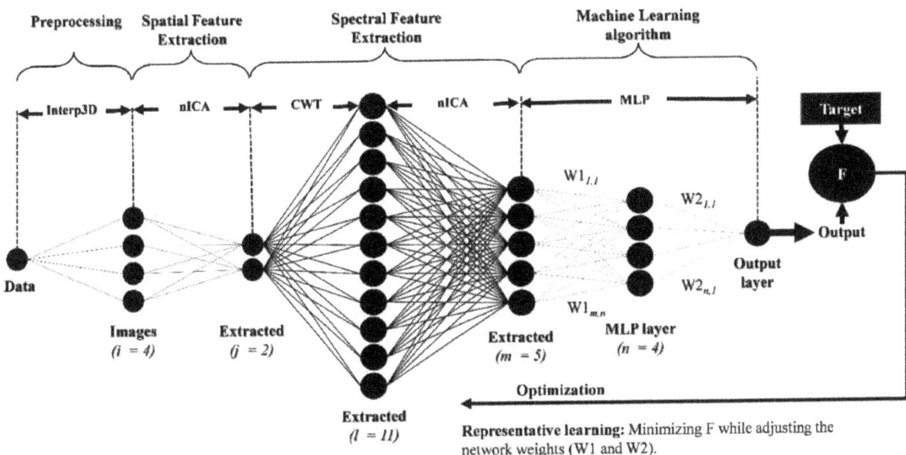

Figure 3. Schematic view of SRL for predictive modeling. The algorithm consists of four sub-modules: The preprocessing prepares the data sets by 3D inversion, interpolation, and filtering methods. The spatial feature extraction with nICA separates the image overlaps in 3D. Spectral feature extraction with CWT-nICA extracts the wavelet decomposed features. Furthermore, the machine learning algorithm with MLP adjusts network weights (W1 and W2) to learn patterns inside the extracted features based on the sample targets.

The basic building block of MLP is the Perceptron, a mathematical analog of the biological neuron, first described by Rosenblatt [33]. In this model, the integrated weights and biases of input vectors (x) are activated by a sigmoid function (activation function) to produce the output (y). After setting the initial weights randomly, the network is ready to train. Then, the output will be compared to the target in an iterative process, adjusting the weights of the ANN. This process can be performed by defining an Objective Function (F) and minimizing it over N training samples. F is defined as the least mean square (LMS) of the output vector (y_i) and the desired output or the target vector (t_i):

$$F = \frac{1}{N} \sum_{i=1}^{N} (t_i - y_i)^2 \tag{7}$$

During the back-propagation process, the optimization updates the synaptic weights to make output closer to the specified target.

Despite successful applications of ANNs in nearly all branches of science and engineering, we still face many problems and potential pitfalls. A potential pitfall is an over-fitting problem, in which designing too many neurons and too many iterations gives rise to noisy outputs. On the other side, under-fitting happens when a too-simple network is created, sometimes leading to under-estimated values. Preprocessing of original data is also an important criterion. Random noises in inputs or targets can propagate through networks and disturb the outputs. For example, surficial noises in geophysical images can lead to underestimating the results. Smoothing the inputs and targets can reduce spurious effects but also eliminate valuable attributes in the images, leading to overestimation in the final models. An optimum smoothness filtering strategy can be used to avoid over-fitting and under-fitting effects [34–36].

Choosing an efficient optimization algorithm is also critical because a suitable optimization algorithm can make a considerable difference in final estimations. The Levenberg–Marquardt optimization technique [37,38] is a fast and robust technique for minimizing the performance function. Because of the local minima problem inherent in nonlinear optimization procedures, finding a global optimal or better local solution is the goal of an effective learning process. Several novel heuristic stochastic techniques are available to solve this complex hyper-dimensional optimization problem [24,35,38–42].

In this study, we use GA optimization to minimize the ANN cost function and the number of input features simultaneously. The process is called feature subset selection and begins with initializing the input parameters we wish to optimize. These considerations are codified in an objective or cost function with different weights depending on the survivability of individuals. The best-fitted individuals meet the specifications of the objective function and survive, while the rest of the population is extinct [24,41,42]. The optimal solution emerges from parallel processing among the population with a complex fitness landscape. The whole idea is moving the population away from local optima that classical hill-climbing techniques usually might be trapped in. This ability makes GA an extremely robust global optimization technique that can resolve the traditional problem of local pockets existing in older optimization routines.

During the SFSS procedure, the spectral extracted features are fed to the MLP network, and the learning process involves adjusting the network weights and the number of the selected features while minimizing the bi-objective function F (Figure 4). In this study, we used a fast implementation of the Non-dominated Sorting Genetic Algorithm (NSGA) for bi-objective optimization. The NSGA algorithm [43] optimizes a global bi-objective cost function in the general form of:

$$E_{\text{Global}} = f\left(E_{\text{ANN}}, n_f\right) \tag{8}$$

where f is the global cost function operator. The bi-objective GA simultaneously minimizes the number of spectral features (n_f) and ANN cost (E_{ANN}).

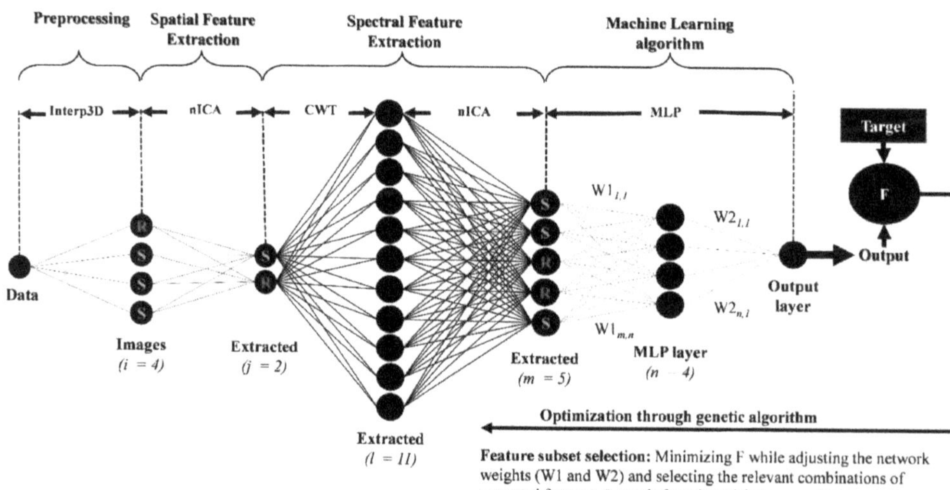

Figure 4. Schematic view of spectral feature subset selection for predictive modeling. The algorithm consists of four main sub-routines like the SRL algorithm, except that the MLP is integrated with NSGA-II to adjust the network weights and the selection of inputs simultaneously.

The solution of multi-objective optimization problems gives rise to a set of Pareto-optimal solutions instead of a single solution. Conventional optimization methods convert multi-objective optimization problems to single-objective optimization problems by emphasizing one Pareto-optimal solution at a time. The algorithm has to be iterated to obtain a different solution at each run to find multiple solutions. Over the past decades, evolutionary algorithms have been suggested to find multiple Pareto-optimal solutions in one single simulation run, emphasizing moving toward the actual Pareto-optimal region. This study used an improved version of NSGA called NSGA-II [44,45]. NSGA-II can be detailed in the following steps:

Step 1: Population initialization

NSGA-II randomly initializes a population P_t of size N. An offspring population Q_t is generated using the genetic operators (tournament selection, crossover, and mutation), and P_t and Q_t are combined to create a population R_t of size $2N$.

Step 2: Fast non-dominated sorting procedure

The algorithm sorts the combined population R_t according to non-domination to obtain different non-dominated fronts F_i.

Step 3: Crowding distance estimation

To maintain the population diversity in each non-dominated front, the crowding distance is computed; that is, the average distance between two points on either side of a particular solution. Then, repeat the process with other objective functions.

Step 4: Binary tournament selection

Individuals are selected using a binary tournament selection with a crowded comparison operator to create a mating pool: The best solutions in the combined population are those belonging to the lower-level non-dominated set F_1. When the two solutions have an equal non-domination level, the solution with a higher crowding distance is preferred to preserve diversity.

Step 5: Crossover and mutation

The algorithm creates an offspring population from the mating pool by applying the simulated binary crossover operator and the polynomial mutation operator.

Step 6: Recombination and selection

The last stage of the NSGA-II involves combining the current population with the offspring and then selecting the best solutions from this combined population to create a new population for the next generation. The process is iterated for the desired generations to achieve the best multi-objective solution in the last generation.

3. Results and Discussion

We compiled several MATLAB functions and scripts to predict geological targets in 3D, such as 3D lithotype estimation and 3D Au grade estimation. We used unevenly distributed borehole datasets, including the lithotype data (Figure 5a) and the Au concentrations in g/t (Figure 5d). Borehole datasets show that the hydrothermal alteration has destroyed magnetite and replaced it with pyrite in the felsic volcanic rocks that are the host rocks of epithermal Au/Ag mineralization [27–30]. Therefore, the high-Au grades are accompanied by low-magnetic anomalies of the felsic rocks. This correlation made it possible to separate the high magnetics of the mafic volcanic rocks and dioritic intrusions from low-magnetic felsic volcanic rock and porphyritic intrusions based on the 3D cooperatively recovered magnetic susceptibility model.

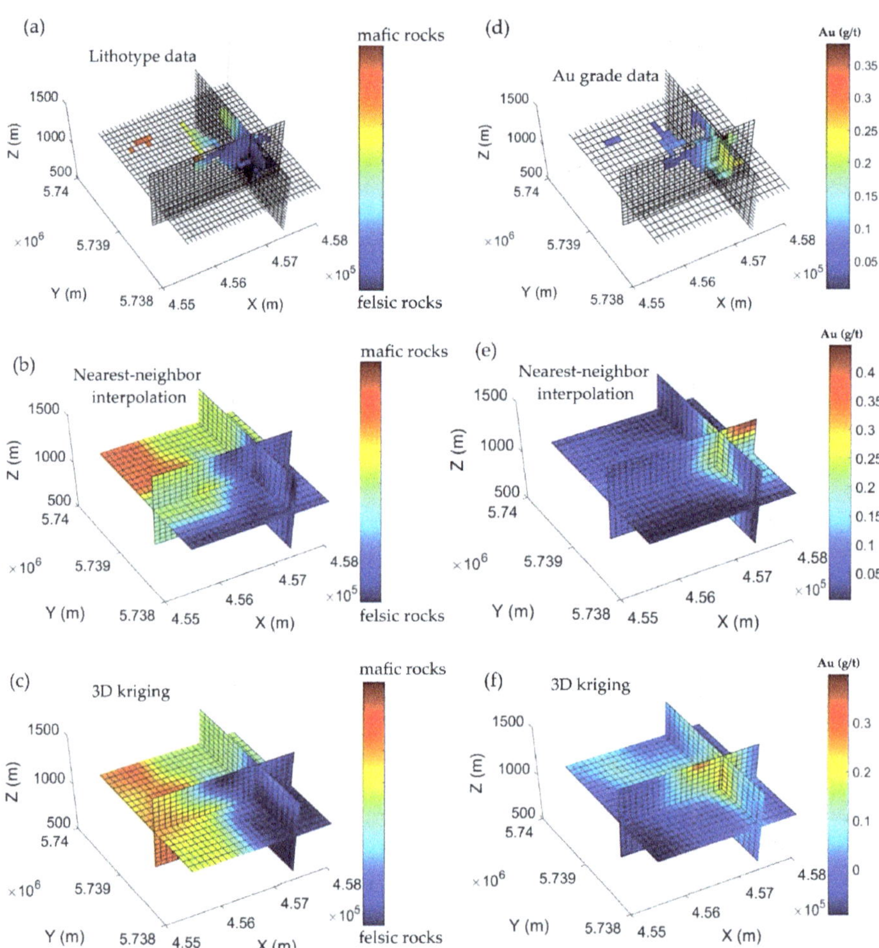

Figure 5. Interpolation of geological targets with conventional methods. (**a**) Lithotype data (mafic/felsic). (**b**) Nearest-neighbor interpolation of lithotypes. (**c**) Kriging of lithotypes. (**d**) Au grade data. (**e**) Nearest-neighbor interpolation of Au grades. (**f**) Kriging of Au grades [27].

Consequently, we assigned three lithotype codes to quantify the lithological variations: Code 1 for felsic volcanic rocks, code 2 for mafic volcanic rocks, and code 3 for dioritic intrusions. The lithotype codes are matched and sorted according to their magnetic properties, from low magnetics of felsic volcanic rocks and intrusions to high magnetics of mafic volcanic rocks and diorites. The datasets are interpolated with the nearest neighbor (direct gridding) and kriging methods for comparison. Figure 5 shows that traditional interpolation methods cannot add any new geological information in places without borehole data. This study aims to solve this 3D interpolation problem by reconstructing lithological and Au grade patterns from 3D geophysical images.

The 3D physical property images are derived from cooperative inversion [27,31] of magnetic and DC/IP datasets (Figure 2b–d). A preliminary Fast-ICA separates the latent features inside the multiple physical property images in the form of three negentropy-maximized ICs (spatial feature extraction). Then, three ICs are decomposed to form a set of raw spectral features through a continuous wavelet transform (CWT).

We calculated the 2D wavelet coefficients with a gaussian mother wavelet for three scales in 72 directions (every 5°). We used only three scales because we observed that for more than three scales, the changes in low-frequency features (high scales) would disappear, and there will not be any difference in the raw spectral features for scales more than three. The 3D spectral decomposition is iterated for the three initial ICs and has produced 648 ($3 \times 72 \times 3$) features containing several frequency-dependent raw features that need further extraction. Figure 6 demonstrates the raw spectral features sliced at an elevation of 1000 m. X and Y UTM coordinates are removed to save space for display.

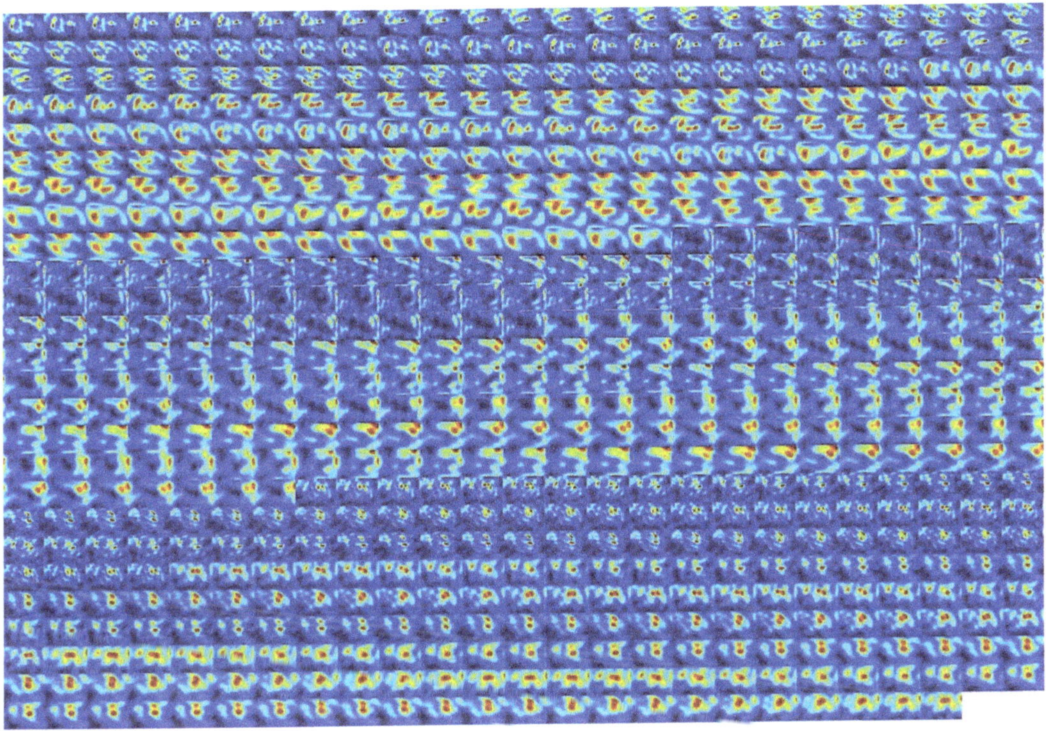

Figure 6. CWT with a gaussian mother wavelet produced raw 3D spectral features with three scales in 72 directions (0°, 5°, 10°, . . . , 355°). The extracted features are sliced at an elevation of 1000 m.

Fast-ICA, through negentropy maximization, separates the 648 raw features to produce the 3D independent spectral inputs necessary for feature selection. The Fast-ICA algorithm can also reduce the dimensionality of the raw features. The best strategy to reduce the dimensionality of features depends on two factors: computer hardware resources and the stability of the SFSS output. Low reductions result in high computational costs, and severe reductions lead to loss of valuable information and poor predictions. There is always an optimum way to moderately reduce the dimensionality of features through several testing and running of the algorithm by the user. In cases with no hardware limitation, this trial-and-error strategy is unnecessary. In this study, we reduced the dimensionality of the raw spectral features by 35 percent to 227 extracted spectral features (Figure 7).

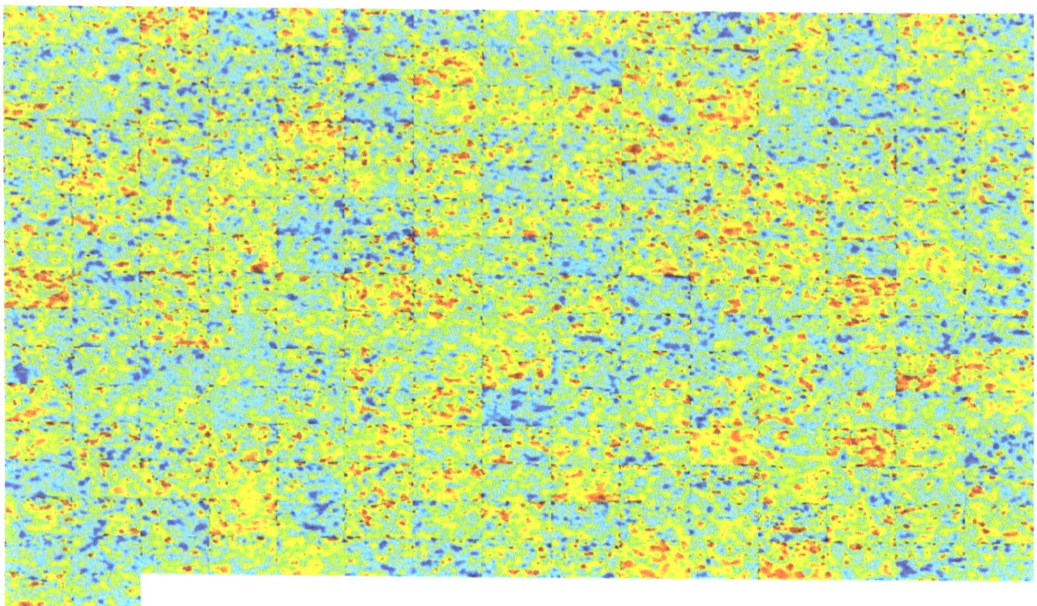

Figure 7. Total 227 number of separated 3D spectral features after Fast-ICA dimensionality reduction for lithological feature selection. The extracted features are sliced at an elevation of 1000 m.

We ran the SFSS algorithm two times in parallel to predict lithotypes and Au grades. For feature selection, we used an MLP network with independent spectral features as inputs, 20 neurons in the hidden layer, a maximum of 20 iterations, and one output. Of the borehole data, 70% are used for training and the rest for validation (10%) and testing (20%). The GA is iterated in 20 generations with 20 populations, 70% crossover, 40% of mutation, and a mutation rate of 0.05.

For the lithotypes model reconstruction, evaluation results are compared for both SRL optimized with the Levenberg–Marquardt method and the SFSS algorithm optimized with GA (Figure 8a). As can be seen, the SFSS algorithm gives better validation and test results (98% and 91% R-squares, relatively) compared to the poor results of the conventional SRL method (with 90% and 27% R-squares, relatively).

For Au grade prediction, the same network parameters are used, and the Fast-ICA dimensionality reduction has reduced the number of independent spectral features to 227 (Figure 7). Evaluation results are compared for both SRL optimized with the Levenberg–Marquardt method and the SFSS algorithm optimized with GA (Figure 8b). As can be seen, the SFSS algorithm gives much better validation and test results (with 97% and 88% R-squares, relatively) compared to the poor results of the conventional SRL method (with 95% and 56% R-squares, relatively).

The error landscapes of the lithotype and Au grade prediction by SFSS are also shown in Figure 9. The best results come after the last generation of GA and for the least global error indicated by the dark blue color. For lithotype prediction, the SFSS algorithm selects 200 out of 227 features for spectral learning (Figure 10). As can be seen, the SFSS can recover a much better lithological model compatible with prior geological information. The selected features for lithotype prediction are also shown in Figure 11. In this case, SFSS detects 27 redundant features in the prediction of lithological variations that helps to facilitate the machine learning predictions.

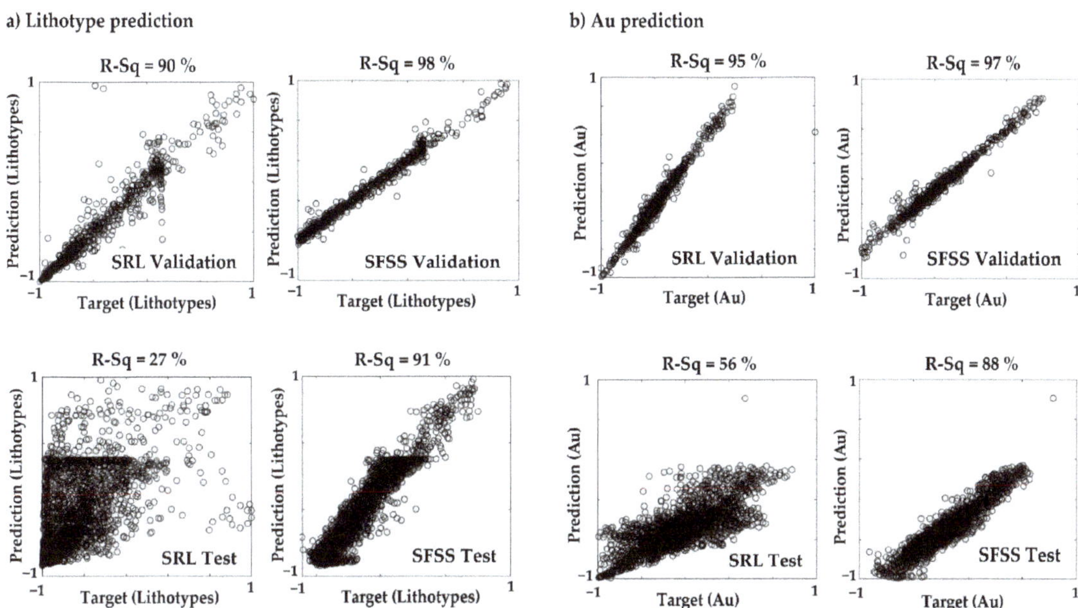

Figure 8. Evaluation of lithotype (**a**) and Au (**b**) predictive modeling for SRL and SFSS algorithms. R-sq denotes the R-squared values.

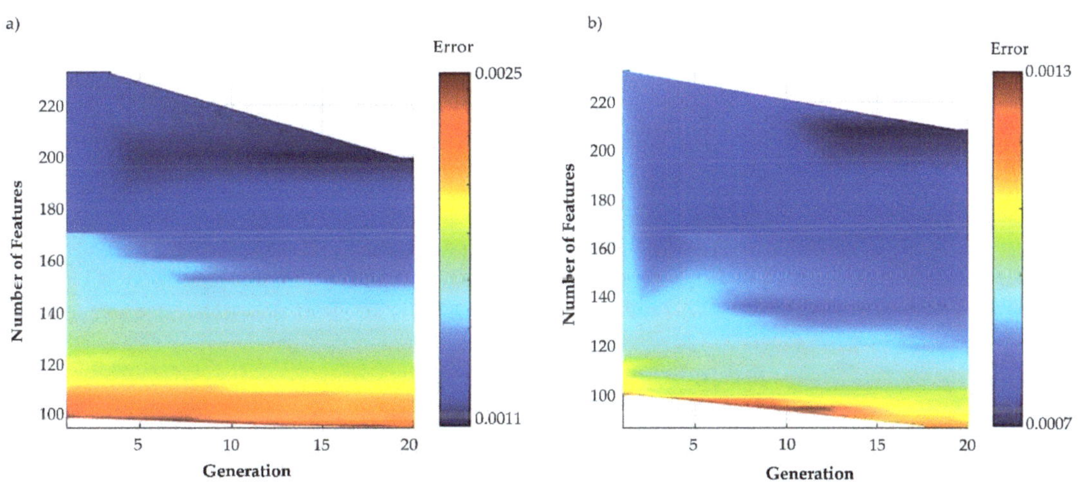

Figure 9. Bi-objective error landscape of SFSS algorithm during lithotype (**a**) and Au (**b**) estimation.

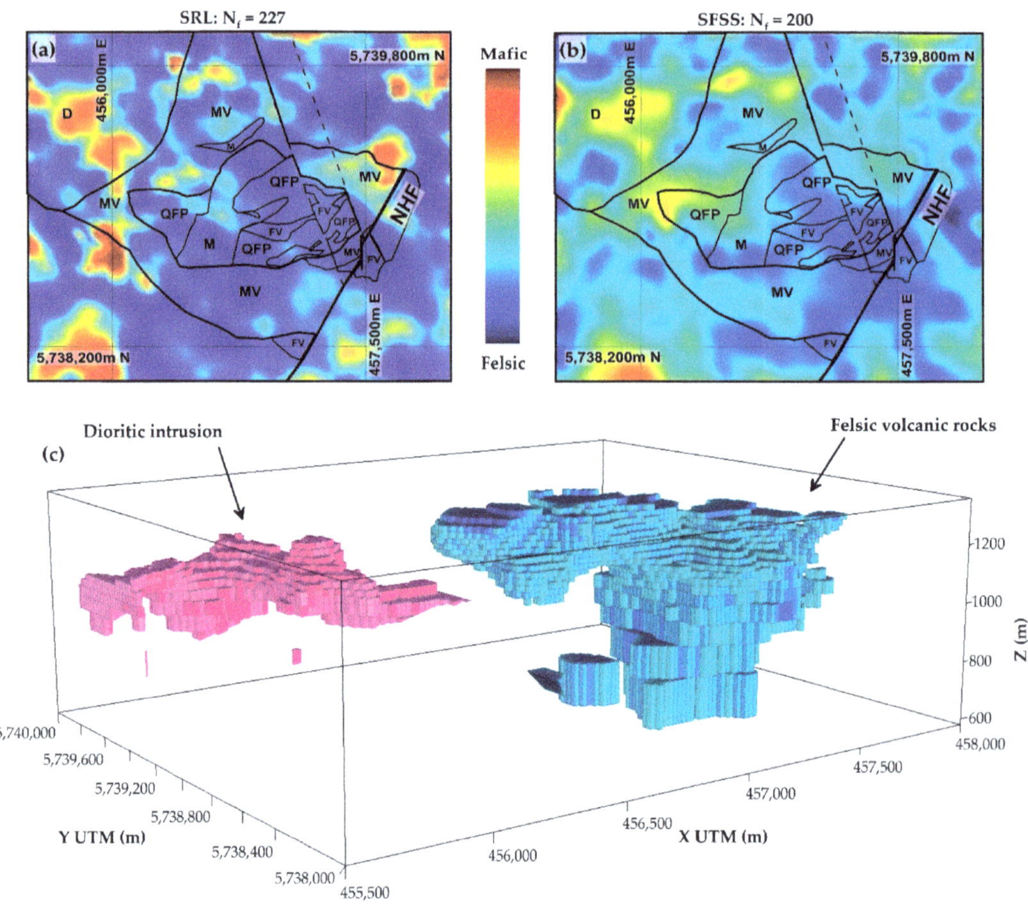

Figure 10. Results of SRL lithotype predictions (**a**) versus SFSS lithotype predictions (**b**,**c**). SFSS resulted in the better reconstruction of lithotypes with a smaller number of input spectral features (Nf = 200). Results are sliced horizontally at the elevation of 1000 m.

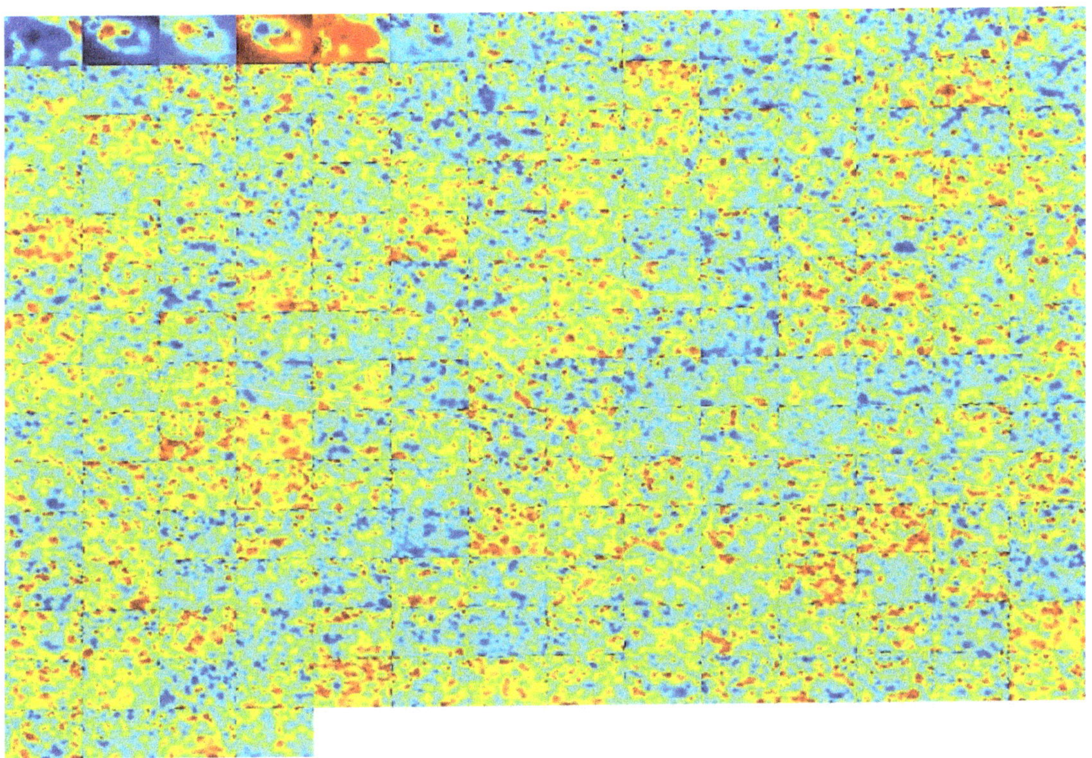

Figure 11. A total of 200 selected features were used for lithotype prediction. The selected features are sliced at an elevation of 1000 m.

The error landscape of the Au grade prediction by SFSS is also shown in Figure 10b. For the last generation, the SFSS algorithm selects 209 out of 227 features for spectral learning (Figure 12). The SFSS can recover a much better Au grade model than conventional SRL estimations. The selected features are also shown in Figure 13. In this case, SFSS detects 18 extremely redundant features in the prediction of Au concentrations that help to facilitate the machine learning predictions.

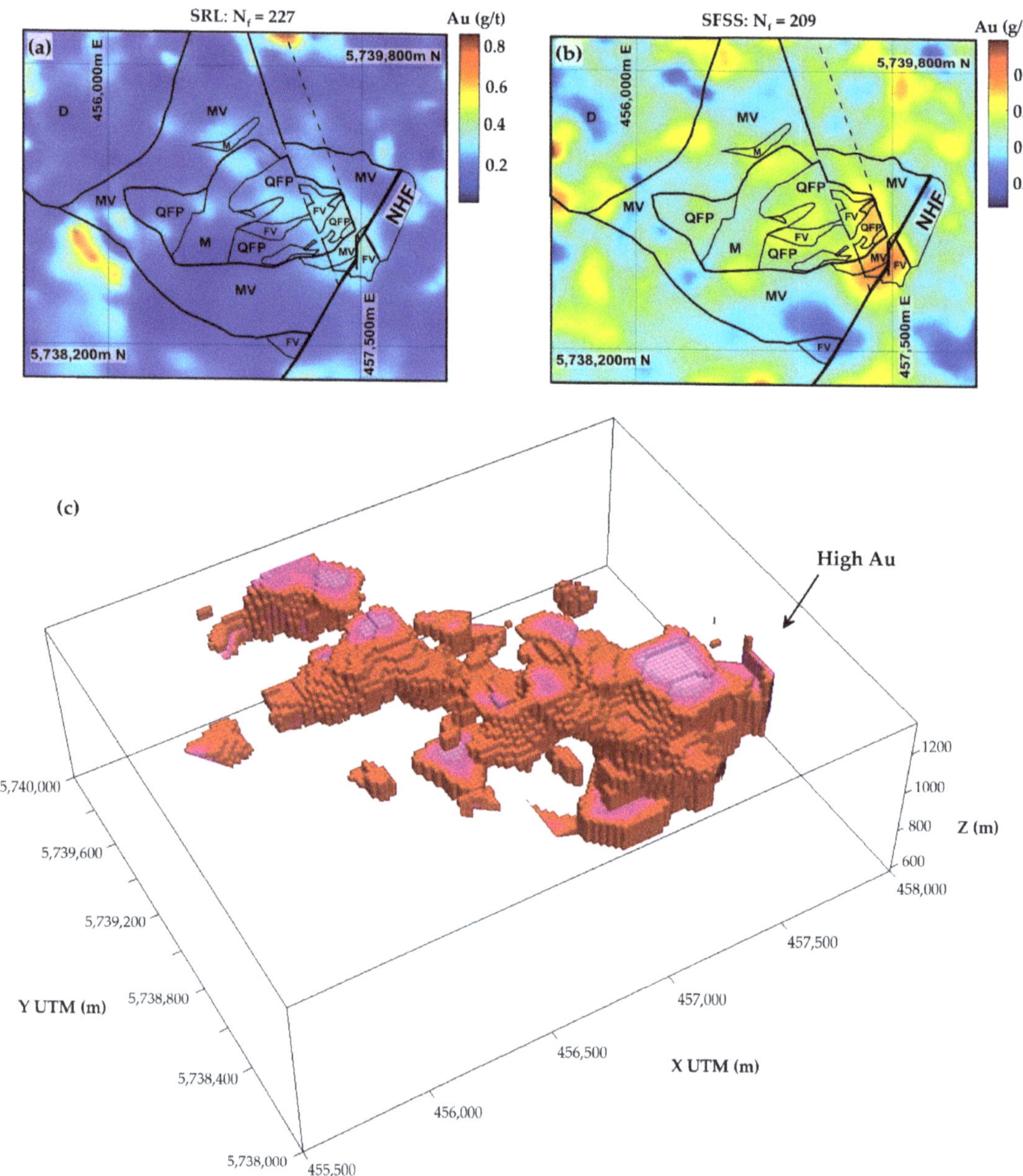

Figure 12. Au grade predictions: Results of 3D SRL predictions (**a**) versus 3D SFSS predictions (**b,c**). SFSS resulted in the better reconstruction of Au-grade distributions with a smaller number of input spectral features (N_f). Results are sliced horizontally at the elevation of 1000 m.

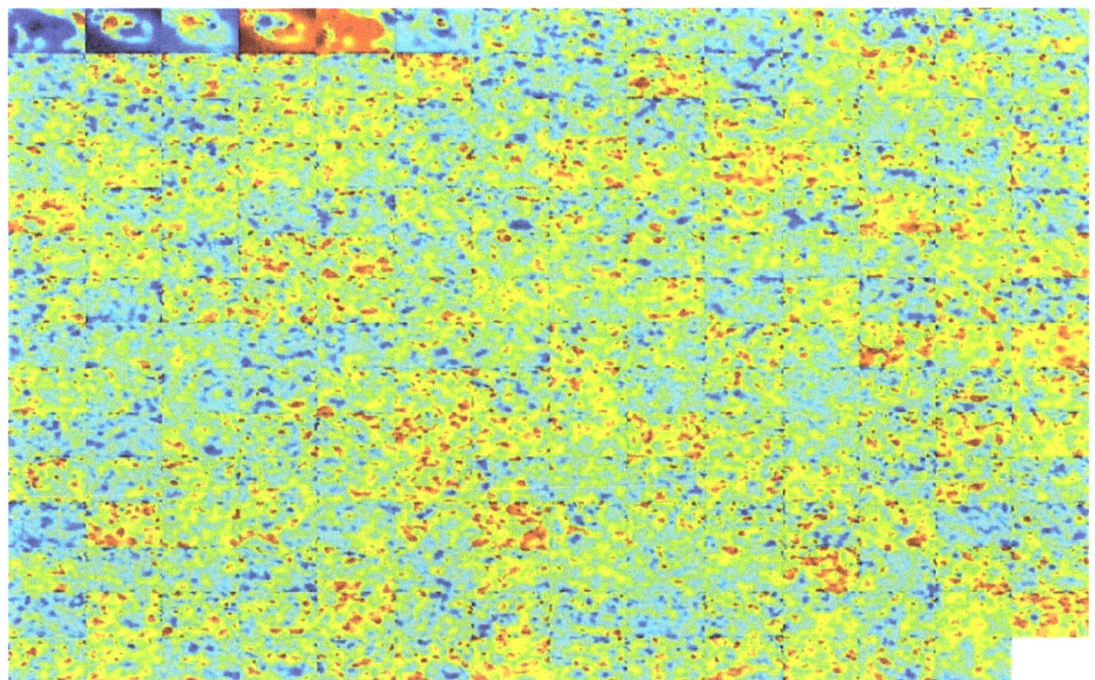

Figure 13. A total of 209 selected features were used for Au prediction. The selected features are sliced at an elevation of 1000 m.

4. Conclusions

We propose a spectral feature selection algorithm for supervised learning of geological patterns to perform 3D predictive modeling from 3D inverted physical properties. Our work is based on the synergy between the ICA and CWT feature extraction methods and multi-objective machine learning optimization through GA for feature selection. The spectral feature extraction method provided the inputs of the feature selection algorithm, and we show that our self-proposed SFSS algorithm can pick the relevant spectral features necessary for 3D predictive modeling of the targets. The practical implementation of the SFSS algorithm is also evaluated for an epithermal Au/Ag deposit in British Columbia, Canada. The results show that the spectral learning scheme proposed can efficiently learn geological patterns to make predictions based on 3D physical property inputs. The SFSS also minimizes the number of extracted spectral features and tries to pick the best representative geophysical features for each target learning case. This automated dimensionality reduction strategy also gives interpreters a precise predictive model and an understanding of the relevant and irrelevant selected geological features at the end, which adds value to the interpretability of the machine learning process. Although tested on Newton's epithermal deposit, the proposed feature selection approach should be applicable to similar mineral deposits. Other 3D geophysical imageries acquired with inversion of seismic, gravity, and electromagnetic data sets could be considered as well to enhance the accuracy of the feature extraction and the subsequent feature selection. Future research will be focused on the automatic fine-tuning of SFSS hyperparameters and adapting the 3D SFSS algorithm to provide a practical tool for integrated 3D imaging of mineral deposits.

Author Contributions: Conceptualization, B.A. and L.-Z.C.; methodology, B.A.; software, B.A.; validation, B.A. and L.-Z.C.; formal analysis, B.A.; investigation, B.A.; resources, B.A.; data curation, B.A.; writing—original draft preparation, B.A.; writing—review and editing, B.A., L.-Z.C., M.J., and D.L.; visualization, B.A.; supervision, L.-Z.C.; project administration, L.-Z.C.; funding acquisition, L.-Z.C. All authors have read and agreed to the published version of the manuscript.

Funding: This study was supported by grants from the Natural Sciences and Engineering Research Council of Canada (Grant No. 43505) and Fonds de recherche du Québec—Nature et technologies (Grant No. 133896).

Data Availability Statement: Not applicable.

Conflicts of Interest: The authors declare no conflict of interest.

References

1. Babikir, I.; Elsaadany, M.; Sajid, M.; Laudon, C. Evaluation of principal component analysis for reducing seismic attributes dimensions: Implication for supervised seismic facies classification of a fluvial reservoir from the Malay Basin, offshore Malaysia. *J. Pet. Sci. Eng.* **2022**, *217*, 110911. [CrossRef]
2. Brown, A.R. Seismic attributes and their classification. *Lead. Edge* **1996**, *15*, 1090. [CrossRef]
3. Chen, Q.; Sidney, S. Seismic attribute technology for reservoir forecasting and monitoring. *Lead. Edge* **1997**, *16*, 425–840. [CrossRef]
4. Chopra, S.; Marfurt, K.J. Emerging and future trends in seismic attributes. *Lead. Edge* **2008**, *27*, 281–440. [CrossRef]
5. Chopra, S.; Pruden, D. Multi-Attribute Seismic Analysis on AVO derived parameters—A case study. *CSEG Rec.* **2003**, *28*, 998–1002.
6. Hall, M. White Magic: Calibrating Seismic Attributes. *Views News Geosci. Technol.* **2006**. Available online: https://agilescientific. com/blog/2016/1/25/white-magic-calibrating-seismic-attributes (accessed on 13 October 2022).
7. Lindseth, R. Seismic Attributes—Some recollections. *Recorder* **2005**, *30*. Available online: https://csegrecorder.com/articles/ view/seismic-attributes-some-recollections (accessed on 13 October 2022).
8. Fernández-Martínez, J.L.; Fernández-Muñiz, Z. The curse of dimensionality in inverse problems. *J. Comput. Appl. Math.* **2020**, *369*, 112571. [CrossRef]
9. Murtagh, F.; Starck, J.-L.; Berry, M.W. Overcoming the Curse of Dimensionality in Clustering by Means of the Wavelet Transform. *Comput. J.* **2000**, *43*, 107–120. [CrossRef]
10. Salimi, A.; Ziaii, M.; Amiri, A.; Hosseinjani Zadeh, M.; Karimpouli, S.; Moradkhani, M. Using a Feature Subset Selection method and Support Vector Machine to address curse of dimensionality and redundancy in Hyperion hyperspectral data classification. *Egypt. J. Remote Sens. Space Sci.* **2018**, *21*, 27–36. [CrossRef]
11. Castagna, J.P.; Sun, S. Comparison of spectral decomposition methods. *First Break.* **2006**, *24*, 5. [CrossRef]
12. Sinha, S.; Routh, P.S.; Anno, P.D.; Castagna, J.P. Spectral decomposition of seismic data with continuous-wavelet transform. *Geophysics* **2005**, *70*, P19–P25. [CrossRef]
13. Tian, L. Seismic spectral decomposition using short-time fractional Fourier transform spectrograms. *J. Appl. Geophys.* **2021**, *192*, 104400. [CrossRef]
14. Anowar, F.; Sadaoui, S.; Selim, B. Conceptual and empirical comparison of dimensionality reduction algorithms (PCA, KPCA, LDA, MDS, SVD, LLE, ISOMAP, LE, ICA, t-SNE). *Comput. Sci. Rev.* **2021**, *40*, 100378. [CrossRef]
15. Liu, H.; Motoda, H. *Feature Selection for Knowledge Discovery and Data Mining*; Springer: Berlin/Heidelberg, Germany, 2013; Volume 454.
16. Hyvarinen, A.; Karhunen, J.; Oja, E. *Independent Component Analysis*; John Wiley & Sons, Inc.: Hoboken, NJ, USA, 2001; p. 505.
17. Abbassi, B.; Cheng, L.-Z. 3D Geophysical Post-Inversion Feature Extraction for Mineral Exploration through Fast-ICA. *Minerals* **2021**, *11*, 21. [CrossRef]
18. Abbassi, B.; Cheng, L.-Z. SFE2D: A Hybrid Tool for Spatial and Spectral Feature Extraction. In *Mining Technology*; Hammond, A., Donnelly, B., Ashwath, N., Eds.; IntechOpen: London, UK, 2021.
19. Honório, B.C.Z.; Sanchetta, A.C.; Leite, E.P.; Vidal, A.C. Independent component spectral analysis. *Interpretation* **2014**, *2*, SA21–SA29. [CrossRef]
20. Martínez-Vargas, A.; Domínguez-Guerrero, J.; Andrade, Á.G.; Sepúlveda, R.; Montiel-Ross, O. Application of NSGA-II algorithm to the spectrum assignment problem in spectrum sharing networks. *Appl. Soft Comput.* **2016**, *39*, 188–198. [CrossRef]
21. Remeseiro, B.; Bolon-Canedo, V. A review of feature selection methods in medical applications. *Comput. Biol. Med.* **2019**, *112*, 103375. [CrossRef]
22. Zhao, Z.A.; Liu, H. *Spectral Feature Selection for Data Mining*; Chapman and Hall/CRC: Boca Raton, FL, USA, 2018.
23. Guyon, I.; Nikravesh, M.; Gunn, S.; Zadeh, L.A. *Feature Extraction Foundations and Applications. Pattern Recognition*; Springer: Berlin, Germany, 2006; Volume 207.
24. Badar, A.Q.H. *Evolutionary Optimization Algorithms*; CRC Press: Boca Raton, FL, USA, 2022; p. 274.
25. Shapiro, A.F. The merging of neural networks, fuzzy logic, and genetic algorithms. *Insur. Math. Econ.* **2002**, *31*, 115–131. [CrossRef]
26. Van Coillie, F.; Verbeke, L.; Dewulf, R. Feature selection by genetic algorithms in object-based classification of IKONOS imagery for forest mapping in Flanders, Belgium. *Remote Sens. Environ.* **2007**, *110*, 476–487. [CrossRef]

27. Abbassi, B. Integrated Imaging through 3D Geophysical Inversion, Multivariate Feature Extraction, and Spectral Feature Selection. Ph.D. Thesis, Université du Québec à Montréal and Université du Québec en AbitibiTémiscamingue, Rouyn-Noranda, QC, Canada, 2018.
28. Liu, L.; Richards, J.P.; DuFrane, S.A.; Rebagliati, M. Geochemistry, geochronology, and fluid inclusion study of the Late Cretaceous Newton epithermal gold deposit, British Columbia. *Can. J. Earth Sci.* **2016**, *53*, 10–33. [CrossRef]
29. McClenaghan, L. *Geology and Genesis of the Newton Bulk-Tonnage Gold-Silver Deposit, Central British Columbia*; University of British Columbia, University of British Columbia Library: Vancouver, BC, Canada, 2013.
30. Pressacco, R. *Initial Resource Estimate for the Newton Project, Central British Columbia, Canada*; RPA Inc.: Williamsport, PA, USA, 2012; pp. NI43–NI101.
31. Abbassi, B.; Cheng, L.-Z.; Richards, J.P.; Hübert, J.; Legault, J.M.; Rebagliati, M.; Witherly, K. Geophysical properties of an epithermal Au-Ag deposit in British Columbia, Canada. *Interpretation* **2018**, *6*, T907–T918. [CrossRef]
32. Moreau, F.; Gibert, D.; Holschneider, M.; Saracco, G. Wavelet analysis of potential fields. *Inverse Probl.* **1997**, *13*, 165–178. [CrossRef]
33. Rosenblatt, F. *Principles of Neurodynamics: Perceptrons and the Theory of Brain Mechanisms*; Spartan Books: Washington, DC, USA, 1962.
34. Goodfellow, I.; Bengio, Y.; Courville, A. *Deep Learning*; The MIT Press: Cambridge, MA, USA, 2016.
35. Siddique, N.; Adeli, H. *Computational Intelligence: Synergies of Fuzzy Logic, Neural Networks and Evolutionary Computing*; John Wiley & Sons, Inc.: Hoboken, NJ, USA, 2013.
36. Zhang, G.P. Avoiding Pitfalls in Neural Network Research. *IEEE Trans. Syst. Man Cybern. Part C* **2007**, *37*, 3–16. [CrossRef]
37. Marquardt, D.W. An Algorithm for Least-Squares Estimation of Nonlinear Parameters. *J. Soc. Ind. Appl. Math.* **1963**, *11*, 431–441. [CrossRef]
38. Nocedal, J.; Wright, S. *Numerical Optimization*; Springer: Berlin, Germany, 2006.
39. Clerc, M. *Particle Swarm Optimization*; Wiley Online Library: Hoboken, NJ, USA, 2006.
40. Dorigo, M.; Stützle, T. *Ant Colony Optimization*; The MIT Press: Cambridge, MA, USA, 2004.
41. Haupt, R.L.; Haupt, S.E. *Practical Genetic Algorithms*; John Wiley: Hoboken, NJ, USA, 2003.
42. Tahmasebi, P.; Hezarkhani, A. A hybrid neural networks-fuzzy logic-genetic algorithm for grade estimation. *Comput. Geosci.* **2012**, *42*, 18–27. [CrossRef]
43. Srinivas, N.; Deb, K. Multiobjective optimization using nondominated sorting in genetic algorithms. *Evol. Comput.* **1995**, *2*, 221–248. [CrossRef]
44. Deb, K.; Pratap, A.; Agarwal, S.; Meyarivan, T. A Fast and Elitist Multiobjective Genetic Algorithm: NSGA-II. *IEEE Trans. Evol. Comput.* **2002**, *6*, 182–197. [CrossRef]
45. Yannibelli, V.; Pacini, E.; Monge, D.; Mateos, C.; Rodriguez, G. A Comparative Analysis of NSGA-II and NSGA-III for Autoscaling Parameter Sweep Experiments in the Cloud. *Sci. Program.* **2020**, *2020*, 17. [CrossRef]

MDPI
St. Alban-Anlage 66
4052 Basel
Switzerland
Tel. +41 61 683 77 34
Fax +41 61 302 89 18
www.mdpi.com

Minerals Editorial Office
E-mail: minerals@mdpi.com
www.mdpi.com/journal/minerals